de Gruyter Expositions in Mathematics 48

Stochastic Dynamics and Boltzmann Hierarchy

by

D. Ya. Petrina

Walter de Gruyter · Berlin · New York

Author

D. Ya Petrina (†)
formerly Institute of Mathematics
Ukrainian Academy of Sciences
Kiev, Ukraine

Translation and Typesetting

Dmitry V. Malyshev and Peter V. Malyshev
Institute of Mathematics
Ukrainian Academy of Sciences
Kiev, Ukraine

Mathematics Subject Classification 2000: 82-02, 76P05, 82B40, 82C40, 82D05.

Key words: Boltzmann equation, stochastic Boltzmann hierarchy, Itô–Liouville equation, boundary conditions, Hamilton dynamics, system of hard spheres.

⊚ Printed on acid-free paper which falls within the guidelines of the ANSI to ensure permanence and durability.

ISSN 0938-6572

ISBN 978-3-11-020804-7

Bibliographic information published by the Deutsche Nationalbibliothek

The Deutsche Nationalbibliothek lists this publication in the Deutsche Nationalbibliografie; detailed bibliographic data are available in the Internet at http://dnb.d-nb.de.

Printing and binding: Hubert & Co. GmbH & Co. KG, Göttingen.
Cover design: Thomas Bonnie, Hamburg.

Preface

In mathematical and statistical physics, it was generally accepted that the classical Boltzmann equation is based on the Hamilton equations. The fact of irreversibility of the Boltzmann equation and reversibility of the Hamilton equations leads to well-known paradoxes [Los, Poi1, Poi2, Zer].

At the same time, some arguments concerning the use of stochastic dynamics in deducing the Boltzmann equation were advanced by Boltzmann [Bol1] himself, and P. Ehrenfest and T. Ehrenfest [EE]. As early as 1935, Leontovich [Leo] proposed a stochastic dynamics of point particles in the phase space, postulated the Itô–Liouville equation for this dynamics, and introduced a hypothesis according to which the corresponding one-particle correlation function satisfies the Boltzmann equation in the thermodynamic limit.

For the spatially homogeneous Boltzmann equation in which the one-particle correlation function depends solely on the momentum and is independent of the position of the particle, Kac [Kac1, Kac2] proposed a stochastic dynamics and, in the mean-field approximation, deduced the Boltzmann equation in the thermodynamic limit. Skorokhod [Sko] proposed a stochastic dynamics in the phase space and deduced a nonlinear Boltzmann-type equation for the one-particle correlation function in the mean-field approximation in the thermodynamic limit. In all cases, the physical meaning of various types of stochastic dynamics and their relationship with the Hamiltonian dynamics was not clarified.

Bogolyubov [Bog1] also indicated that, in deducing the Boltzmann equation, the dynamics of particles "is interpreted as a random process..., and the efficient cross sections appearing in the equation of the random process are calculated by using the equations of classical mechanics." For the first time, he showed that the Boltzmann equation can be derived from the Hamiltonian dynamics as a result of a certain limit transition in a special solution of the hierarchy for correlation functions depending on time through the one-particle correlation function. Moreover, the cluster properties of the correlation functions and low densities are essentially used in this case. The problem of transformation (degeneration) of the Hamiltonian dynamics in the Bogolyubov limit was not studied.

The mathematical substantiation of the Bogolyubov method and the mechanism of appearance of stochastic dynamics from the Hamiltonian dynamics was absent for a fairly long period of time. For this reason, it was quite natural to try to solve this problem for a maximally simplified but still nontrivial model. To this end, Grad [Gra1, Gra2] studied a system of hard spheres and showed that, in the (thermodynamic) limit, as the diameter of the spheres tends to zero but the length of the free path of particles remains constant, all correlation functions turn into products of one-particle correlation functions, and the latter are solutions of the Boltzmann equation. This limiting procedure is called the Boltzmann–Grad limit.

At present, the mathematical procedure of deducing the Boltzmann equation in the Boltzmann–Grad limit can be regarded as, to a certain extent, completed due to the works by Lanford [Lan3], Cercignani, Illner, Pulvirenti [CIP, IllP1, IllP2], Spohn [Spo1, Spo2], Gerasimenko, and Petrina. The detailed rigorous proofs can be found in the works by Gerasimenko and Petrina [GeP1, GeP2, GeP3]. (See also the monograph by Cercignani, Gerasimenko, and Petrina [CGP]).

At the same time, the following question remained open: What dynamics does serve as a basis of the Boltzmann equation and the limiting BBGKY hierarchy (now called the Boltzmann hierarchy)? In the works by D. Ya. Petrina, K. D. Petrina [PeP1, PeP2, PeP3, Pet1], and M. Lampis [LaPe1, LaPe2, LaPe3, LaPe4], it is shown that the Hamiltonian dynamics of hard spheres in the Boltzmann–Grad limit degenerates into a certain stochastic dynamics of point particles. According to this stochastic dynamics, point particles move as free ones until they collide. Then they undergo elastic scattering, but the unit vector specifying the results of scattering is a random vector uniformly distributed over the unit sphere, etc.

However, in this case, we encounter the problem of determination of the corresponding correlation functions because the indicated stochastic dynamics differs from the free dynamics of noninteracting point particles on hypersurfaces of lower dimensionality neglected in traditional classical statistical mechanics. For this reason, it is necessary to introduce a new concept of correlation functions taking into account, in a certain way, the contributions of the hypersurfaces where the interaction of stochastic particles is specified. It can be shown that the solutions of the Boltzmann equation are also expressed via the contributions of these hypersurfaces. It is quite surprising that this fact was not discovered earlier.

For these correlation functions, the stochastic Boltzmann hierarchy is deduced with boundary conditions on the hypersurfaces where the positions of pairs of particles coincide. Note that, earlier, these boundary conditions were neglected in the ordinary Boltzmann hierarchy. The stochastic Boltzmann hierarchy is also obtained from the BBGKY hierarchy for a system of hard spheres in the Boltzmann–Grad limit if the boundary conditions are properly taken into account.

Thus, the stochastic Boltzmann hierarchy is deduced on the basis of the stochastic dynamics in exactly the same way as the BBGKY hierarchy is deduced on the basis of the Hamiltonian dynamics.

It is proved that the local (in time) solutions of the stochastic Boltzmann hierarchy exist for the initial data bounded in coordinates and exponentially decreasing in squared momenta. The global (in time) solutions exist for the initial data exponentially decreasing in the squared momenta and coordinates. If the initial data satisfy the condition of chaos, i.e., admit a representation in the form of products of one-particle correlation functions, then, outside the hypersurfaces of interaction of stochastic particles, the solutions of the stochastic hierarchy also satisfy the condition of chaos and the one-particle correlation function is a solution of the Boltzmann equation. The ordinary Boltzmann hierarchy without boundary conditions is solved in the entire phase space by the correlation functions represented in the form of the product of one-particle correlation functions satisfying the Boltzmann equation. Thus, the Boltzmann equation is deduced rigorously. It is shown that the Boltzmann equation is, in fact, based on the irreversible stochastic dynamics and, hence, there are no contradictions with the irreversibility of solutions of the Boltzmann equation.

The stochastic dynamics is very simple and possesses numerous properties of the Hamiltonian dynamics, namely, the trajectories with fixed random parameters, the operators of shift along the trajectories, and the hierarchy of equations for correlation functions with fixed random parameters in the boundary conditions. This enables us to use the results obtained for the BBGKY hierarchy for a system of hard spheres to prove the existence of solutions of the stochastic Boltzmann hierarchy and the properties of chaos.

The stochastic dynamics proposed by Kac in the momentum space is obtained from our stochastic dynamics in the phase space as a result of averaging over the coordinates. This clarifies its physical meaning. Note that the spatially homogeneous Boltzmann equation is derived from the stochastic Boltzmann hierarchy without using the mean-field approximation.

All results can be generalized to the case of Boltzmann equation with general differential cross section.

We now briefly describe the content of the monograph. It comprises the introduction and eight chapters.

In the first chapter, a critical survey of the results concerning the existence of solutions of the BBGKY hierarchy for a system of hard spheres and the justification of the Boltzmann–Grad limit is presented. Special attention is given to the boundary conditions for both the BBGKY hierarchy and the stochastic Boltzmann hierarchy.

In the second chapter, the stochastic dynamics is derived from the Hamiltonian dynamics of hard spheres in the Boltzmann–Grad limit. We deduce the Itô–Liouville equation and introduce the principle of duality according to which an ordinary function is associated with a generalized function concentrated on hypersurfaces of interaction of stochastic particles. These generalized functions are used to compute the contributions of the hypersurfaces to the correlation functions.

In the third chapter, the stochastic Boltzmann hierarchy with boundary conditions is derived from the stochastic dynamics of point particles.

In the fourth chapter, the existence of solutions of the stochastic Boltzmann hierarchy is proved and the property of chaos is established. These results are used to deduce the Boltzmann equation.

In the fifth chapter, the stochastic Kac dynamics in the momentum space is obtained from our stochastic dynamics in the phase space. It is shown that the spatially homogeneous Boltzmann equation can be derived from the stochastic Boltzmann hierarchy in the phase space without using the mean-field approximation.

In the sixth chapter, the results obtained for a system of hard spheres are generalized to systems of particles with arbitrary scattering cross section.

In the seventh chapter, we study a system of spheres with inelastic scattering used as a model of granular flows. A hierarchy of equations for correlation functions is deduced. This hierarchy contains the squared Jacobian of the phase trajectory unequal to one.

In the eighth chapter, we construct the solution of the Cauchy problem for the hierarchy in the space of sequences of summable functions. The group of evolution operators is obtained in the explicit form. The stochastic dynamics for granular flows corresponding to the Boltzmann equation is introduced.

The monograph contains two types of references: references of the first type are directly related to the problems analyzed in the book and are mentioned in it. References of the second type cover some other important problems of statistical mechanics.

Contents

Introduction

The entire nonequilibrium statistical mechanics is constructed according to the following scheme: the Hamilton equations for a system of particles $\to$ the initial data for the system are random and specified in the form of probability distribution functions in the phase space $\to$ the operator of shift along the trajectories acting upon the initial distribution function specifies the distribution function (state) at any time $\to$ the Liouville equation for the distribution function with initial data and boundary conditions is obtained $\to$ the BBGKY (Bogolyubov–Born–Green–Kirkwood–Yvon) equations for correlation functions are deduced $\to$ the thermodynamic limit transition is realized and the resulting sequence of correlation functions describes the states of infinite systems of particles.

This is the most general description of states of infinite systems of particles. A comprehensive survey of the results obtained in this direction for the last 50 years can be found in the series of our monographs (see [PGM3] and [CGP]).

Parallel with this general and abstract direction, another approach was also successfully developed. In this approach, the states of systems are described by the one-particle correlation function or, more generally, all correlation functions are represented as functionals of the one-particle correlation function. The one-particle correlation function is found as a solution of the nonlinear (second-order) Boltzmann equation or, in the general case, from the closed nonlinear equation with infinite nonlinearity.

For the first time, the Boltzmann equation was derived from the BBGKY hierarchy by Bogolyubov [Bog1] on the basis of the principle of weakening of correlations for low densities of particles. The mathematical justification of the Bogolyubov method is an extremely complicated problem whose solution is absent up to now.

In this connection, Grad [Gra1, Gra2] proposed to study a system of hard spheres. The dynamics of hard spheres is quite simple. Indeed, they move as free particles between collisions and, as a result of collisions, they undergo elastic scattering. Then they continue to move as free particles until the next collision, etc.

The BBGKY hierarchy for a system of hard spheres has the form [PGM3, CGP]

$$\begin{aligned}
&\frac{\partial F_s^a(t,x_1,\dots,x_s)}{\partial t}\\
&\quad= -\sum_{i=1}^{s} p_i\cdot\frac{\partial}{\partial q_i}F_s^a(t,x_1,\dots,x_s)\\
&\qquad+ a^2\sum_{i=1}^{s}\int dp_{s+1}\int_{S_+^2} d\eta_{is+1}\eta_{is+1}\cdot(p_i-p_{s+1})\\
&\qquad\times\big[F_{s+1}^a(t,x_1,\dots,q_i,p_i^*,\dots,x_s,q_i-a\eta_{is+1},p_{s+1}^*)\\
&\qquad\quad - F_{s+1}^a(t,x_1,\dots,q_i,p_i,\dots,x_s,q_i+a\eta_{is+1},p_{s+1})\big],\quad s\ge 1,\quad (1)
\end{aligned}$$

where a is the diameter of hard spheres, η_{ij} is a unit vector, $S_+^2 = S_+^2(\eta_{ij} \mid \eta_{ij}\cdot(p_i-p_j)>0)$, $q\in R^3$ is the coordinate of the center of a sphere, $p\in R^3$ is the momentum of the sphere, $x=(q,p)$ is the phase point of the sphere, and p_i^* and p_j^* are the momenta after the collision

$$p_i^* = p_i - \eta_{ij}\eta_{ij}\cdot(p_i-p_j),\quad p_j^* = p_j + \eta_{ij}\eta_{ij}\cdot(p_i-p_j), \tag{2}$$

where $\eta\cdot(p_i-p_j)$ is the scalar product of the vectors η and (p_i-p_j). The s-particle correlation function $F_s(t,x_1,\dots,x_s)$ is equal to zero on forbidden configurations W_s including all points for which the distance between the centers of at least one pair of spheres is less than a, i.e., $|q_i-q_j|<a$.

Hierarchy (1) should be equipped with the initial conditions

$$F_s^a(t,x_1,\dots,x_s)|_{t=0} = F_s^a(0,x_1,\dots,x_s),\quad s\ge 1,$$

and the following boundary conditions: if $q_i-q_j=a\eta_{ij}, \eta_{ij}\cdot(p_i-p_j)>0$, then the momenta p_i and p_j in the term $-\sum_{i=1}^{s} p_i\cdot\frac{\partial}{\partial q_i}F_s^a(t,x_1,\dots,x_s)$ should be replaced with the momenta p_i^* and p_j^* given by relations (2). These boundary conditions are very important and responsible for the collisions and elastic scattering of hard spheres. Without these boundary conditions, hierarchy (1) does not describe the system of hard spheres.

Unfortunately, in all works and monographs (except those written by D. Ya. Petrina, V. I. Gerasimenko and P. V. Malyshev [PGM3, CGP]), these boundary conditions are not taken into account explicitly, which sometimes leads to serious mistakes.

If we now pass to the renormalized correlations functions $a^{2s}F_s^a(t,x_1,\dots,x_s)$ and assume that the Boltzmann–Grad limit

$$\lim_{a\to 0} a^{2s}F_s^a(t,x_1,\dots,x_s) = F_s(t,x_1,\dots,x_s)$$

exists (in a certain sense, see [GPe1, GPe2, GPe3, CGP]) and the corresponding limit transition can be realized in hierarchy (1), then we get the hierarchy

$$\frac{\partial F_s(t,x_1,\dots,x_s)}{\partial t} = -\sum_{i=1}^{s} p_i\cdot\frac{\partial}{\partial q_i}F_s(t,x_1,\dots,x_s)$$

$$+\sum_{i=1}^{s}\int dp_{s+1}\int_{S_+^2} d\eta_{is+1}\eta_{is+1}\cdot(p_i-p_{s+1})$$
$$\times[F_{s+1}(t,x_1,\ldots,q_i,p_i^*,\ldots,x_s,q_i,p_{s+1}^*)$$
$$-F_{s+1}(t,x_1,\ldots,q_i,p_i,\ldots,x_s,q_i,p_{s+1})],\quad s\geq 1,\quad (3)$$

with the initial conditions

$$F_s(t,x_1,\ldots,x_s)|_{t=0}=F_s(0,x_1,\ldots,x_s)$$

and the following boundary conditions: for $q_i - q_j = 0$ and $\eta_{ij}\cdot(p_i-p_j)>0$, the momenta p_i and p_j in the term $-\sum_{i=1}^{s} p_i\cdot\frac{\partial}{\partial q_i}F_s(t,x_1,\ldots,x_s)$ should be replaced by the momenta p_i^* and p_j^* given by relation (2) [Pet5, LaPe3].

At first sight, these boundary conditions for hierarchy (3) are identical to those imposed for hierarchy (1). However, there exists a serious difference because, in the BBGKY hierarchy (1), the unit vector η is uniquely defined by the vectors q_i and q_j and directed from the center of the j-th sphere to the center of the i-th sphere. In hierarchy (3), the unit vector runs over the entire hemisphere S_+^2, it is random and uniformly distributed on S_+^2. The second difference between hierarchies (1) and (3) is connected with the fact that the set of forbidden configurations for $F_s(t,x_1,\ldots,x_s)$ turns into the empty set and the spheres turn into points. Unfortunately, these boundary conditions, both for hierarchy (3) and hierarchy (1), were not properly taken into account earlier excluding our works.

The following stochastic dynamics of point particles [PeP1, PeP2, PeP3, Pet1, LaPe1, LaPe2, LaPe3, LaPe4] plays the same role for hierarchy (3) as the Hamiltonian dynamics for the BBGKY hierarchy (1): Particles move as free until the positions of two particles coincide. In this case, the corresponding particles undergo elastic scattering according to relation (2). Then the particles again move as free ones until the next collision, etc. It is very important that, in this case, the vector η_{ij} in (2) is random and uniformly distributed on S_+^2. We say that this *dynamics is stochastic*. It differs from the dynamics of free particles only on the hyperplanes of lower dimensionality where the vectors $q_i - q_j$ are parallel to the vectors $p_i - p_j$:

$$q_i-q_j=\tau(p_i-p_j),\quad \tau\geq 0,\quad i,j\in\{1,\ldots,s\}.\quad (4)$$

This stochastic dynamics was introduced by the author, K. D. Petrina, and M. Lampis [PeP1, PeP2, PeP3, LaPe4, LaPe1, LaPe2, LaPe3], as a certain limit of the Hamiltonian dynamics of a system of hard spheres attained as the diameter of spheres approaches zero. Hierarchy (3) with initial and boundary conditions is called the stochastic Boltzmann hierarchy.

In this connection, we arrive at the fundamental problem of how to define the average values of observables over the states because, in traditional statistical mechanics, the average values are defined by Lebesgue integrals and the contribution of all sets of lower dimensionality is neglected. Hence, at first sight, the results of averaging over the states corresponding to this stochastic dynamics coincide with the results of averaging obtained for free systems.

However, this is not true if we introduce a new concept of averaging of observables taking into account, in a special way, the contributions of hypersurfaces (4) of interaction of point particles. We present the corresponding formula for a system of N particles and an infinitesimally small period of time Δt.

Let $S_N(-\Delta t)$ be an operator of shift along a certain stochastic trajectory acting upon an initial distribution function $f_N(x_1, \ldots, x_N)$. It is assumed that this function f_N is continuous and symmetric, i.e., invariant under permutations of its arguments. Then one has

$$\begin{aligned} S_N(-\Delta t) f_N(x_1, \ldots, x_N) &= f\big(X_1(-\Delta t, x_1, \ldots, x_N), \ldots, X_N(-\Delta t, x_1, \ldots, x_N)\big) \\ &= f_N(\Delta t, x_1, \ldots, x_N), \end{aligned}$$

where $\big(X_1(-\Delta t, x_1, \ldots, x_N), \ldots, X_N(-\Delta t, x_1, \ldots, x_N)\big)$ is the trajectory of N particles at time $-\Delta t, \Delta t > 0$, with fixed random parameters η_{ij} for the initial data $(x_1, \ldots, x_N)$ for $\Delta t = 0$. Let $\varphi_N(x_1, \ldots, x_N)$ be an observable in the form of a real function symmetric with respect to N phase variables. Then the average value of the observable $\varphi_N(x_1, \ldots, x_N)$ over the state $S_N(-\Delta t) f_N(x_1, \ldots, x_N)$ is given by the functional

$$\begin{aligned} &(S_N(-\Delta t) f_N, \varphi_N) \\ &\quad = \int f_N(q_1 - p_1 \Delta t, p_1, \ldots, q_N - p_N \Delta t, p_N) \\ &\qquad \times \varphi_N(x_1, \ldots, x_N) dx_1 \ldots dx_N \\ &\qquad + \sum_{i<j=1}^{N} \int_0^{\Delta t} d\tau \int_{S_+^2} d\eta_{ij} \eta_{ij} \cdot (p_i - p_j) \delta(q_i - p_i \tau - q_j + p_j \tau) \\ &\qquad \times \Big[f_N(q_1 - p_1 \Delta t, p_1, \ldots, q_i - p_i \tau - p_i^*(\Delta t - \tau), \\ &\qquad\qquad p_i^*, \ldots, q_j - p_j \tau - p_j^*(\Delta t - \tau), p_j^*, \ldots, q_N - p_N \Delta t, p_N) \\ &\qquad - f_N(q_1 - p_1 \Delta t, p_1, \ldots, q_i - p_i \Delta t, p_i \ldots, q_j - p_j \Delta t, \\ &\qquad\qquad p_j, \ldots, q_N - p_N \Delta t, p_N) \Big] \varphi_N(x_1, \ldots, x_N) dx_1 \ldots dx_N \\ &\quad = \int \bar{f}_N(\Delta t, x_1, \ldots, x_N) \varphi_N(t, x_1, \ldots, x_N) dx_1 \ldots dx_N, \end{aligned} \tag{5}$$

$$
\begin{aligned}
&\bar{f}_N(\Delta t, x_1, \dots, x_N) \\
&\quad = f_N(q_1 - p_1\Delta t, p_1, \dots, q_N - p_N\Delta t, p_N) \\
&\qquad + \sum_{i<j=1}^{N} \int_0^{\Delta t} d\tau \int_{S_+^2} d\eta_{ij}\eta_{ij} \cdot (p_i - p_j)\delta(q_i - p_i\tau - q_j + p_j\tau) \\
&\qquad\quad \times \Big[f_N(q_1 - p_1\Delta t, p_1, \dots, q_i - p_i\tau - p_i^*(\Delta t - \tau), \\
&\qquad\qquad\quad p_i^*, \dots, q_j - p_j\tau - p_j^*(\Delta t - \tau), p_j^*, \dots, q_N - p_N\Delta t, p_N) \\
&\qquad\qquad - f_N(q_1 - p_1\Delta t, p_1, \dots, q_i - p_i\Delta t, p_i, \dots, q_j - p_j\Delta t, \\
&\qquad\qquad\quad p_j, \dots, q_N - p_N\Delta t, p_N) \Big] \\
&\quad = \bar{S}_N(-\Delta t) f_N(x_1, \dots, x_N).
\end{aligned}
$$

The operator $\bar{S}_N(-\Delta t)$ is defined above. It is clear that the first term on the right-hand side of relation (5) is the contribution of the free system and the second term takes into account the contribution of hypersurfaces (4) of interaction of point particles. This fact is essential for the new concept of calculation of the average values of observables.

Note that, in deducing the functional representing the average value of the observable $\varphi_N(x_1, \dots, x_N)$, we have used the *principle of duality*, according to which the ordinary function

$$
\begin{aligned}
&S_N(-\Delta t) f_N(x_1, \dots, x_N) \\
&\qquad = f_N\big(X_1(-\Delta t, x_1, \dots, x_N), \dots, X_N(-\Delta t, x_1, \dots, x_N)\big) \\
&\qquad = F_N(\Delta t, x_1, \dots, x_N)
\end{aligned}
$$

with jumps on the hypersurfaces (4) of interaction of particles is associated with the generalized function $\bar{f}_N(\Delta t, x_1, \dots, x_N)$ with support on these hypersurfaces specified by the second term in relation (5) for $\bar{S}_N(-\Delta t) f_N(x_1, \dots, x_N)$.

It is quite surprising that the solutions of the Boltzmann equation are given by similar contributions of the same hypersurfaces.

The principle of duality can also be formulated for the Itô–Liouville equation and the stochastic hierarchy. Namely, the Itô–Liouville equation obtained in the sense of pointwise convergence for the distribution function of N particles $f_N(t, x_1, \dots, x_N) = S_N(-t) f_N(x_1, \dots, x_N)$ with fixed random vectors η_{ij} has the form

$$
\frac{\partial f_N(t, x_1, \dots, x_N)}{\partial t} = -\sum_{i=1}^{N} p_i \cdot \frac{\partial}{\partial q_i} f_N(t, x_1, \dots, x_N), \tag{6}
$$

$$f_N(t, x_1, \dots, x_N)|_{t=0} = f_N(x_1, \dots, x_N)$$

with boundary conditions according to which, for $q_i - q_j = 0$, $\eta_{ij} \cdot (p_i - p_j) \geq 0$, the momenta p_i and p_j on the right-hand side of (6) should be replaced with p_i^* and p_j^*.

In the weak sense, if the right-hand side of (6) is regarded as a generalized function, i.e., if (6) is integrated with a test function $\varphi_N(x_1, \dots, x_N)$ and averaged over the random vectors η_{ij}, then, according to the principle of duality, equation (6) takes the form

$$\begin{aligned}
&\frac{\partial \bar{f}_N(t, x_1, \dots, x_N)}{\partial t} \\
&\quad = -\sum_{i=1}^{N} p_i \cdot \frac{\partial}{\partial q_i} \bar{f}_N(t, x_1, \dots, x_N) \\
&\qquad + \sum_{i<j=1}^{N} \int_{S_+^2} d\eta_{ij}\, \eta_{ij} \cdot (p_i - p_j) \delta(q_i - q_j) \\
&\qquad\quad \times \Big[\bar{f}_N(t, x_1, \dots, q_i, p_i^*, \dots, q_j, p_j^*, \dots, x_N) \\
&\qquad\qquad - \bar{f}_N(t, x_1, \dots, q_i, p_i, \dots, q_j, p_j, \dots, x_N) \Big],
\end{aligned} \tag{6'}$$

$$\bar{f}_N(t, x_1, \dots, x_N)|_{\bar{t}=0} = f_N(x_1, \dots, x_N),$$

$$\bar{f}_N(t, x_1, \dots, x_N) = \lim_{n \to \infty} \prod_{i=1}^{n} \bar{S}_N(-\Delta t_i) f_N(x_1, \dots, x_N),$$

$$\sum_{i=1}^{n} \Delta t_i = t$$

without boundary conditions. By virtue of the principle of duality, $\bar{f}_N(t, x_1, \dots, x_N)$ denotes the generalized function specified by relation (5) as corresponding to the ordinary function $f_N(t, x_1, \dots, x_N)$.

Note that equation (6′) was postulated by Leontovich [Leo] but he did not derive the corresponding stochastic dynamics from the Hamiltonian dynamics of the system of hard spheres.

In the weak sense, according to the principle of duality, hierarchy (3) takes the following form:

$$\frac{\partial \bar{F}_s(t, x_1, \dots, x_s)}{\partial t} = -\sum_{i=1}^{s} p_i \cdot \frac{\partial}{\partial q_i} \bar{F}_s(t, x_1, \dots, x_s)$$

$$+\sum_{i<j=1}^{N}\int_{S_{+}^{2}} d\eta_{ij}\,\eta_{ij}\cdot(p_i-p_j)\delta(q_i-q_j)$$
$$\times\Big[\bar F_s(t,x_1,\dots,q_i,p_i^*,\dots,q_j,p_j^*,\dots,x_s)$$
$$-\bar F_s(t,x_1,\dots,q_i,p_i,\dots,q_j,p_j,\dots,x_s)\Big]$$
$$+\sum_{i=1}^{s}\int dp_{s+1}\int_{S_{+}^{2}} d\eta_{is+1}\eta_{is+1}\cdot(p_i-p_{s+1})$$
$$\times\Big[\bar F_{s+1}(t,x_1,\dots,q_i,p_i^*,\dots,x_s,q_i,p_{s+1}^*)$$
$$-\bar F_{s+1}(t,x_1,\dots,q_i,p_i,\dots,x_s,q_i,p_{s+1})\Big] \quad (3')$$

with the initial conditions $\bar F_s(t,x_1,\dots,x_s)|_{t=0} = F_s(0,x_1,\dots,x_s)$ but without boundary conditions. Thus $\bar F_s(t,x_1,\dots,x_s)$ is the generalized correlation function corresponding to the ordinary correlation function $F_s(t,x_1,\dots,x_s)$ according to the principle of duality. Note that the boundary conditions in (6′) and (3′) are taken into account by the terms with δ-functions. We emphasize that the stochastic Boltzmann hierarchy (3′) in the weak sense can be derived from the Itô–Liouville equation (6′) in the weak sense by simple integration over the $N-s$ phase variables in exactly the same way as when deducing the BBGKY hierarchy for the system of hard spheres. Functions $F_s(t,x_1,\dots,x_s)$ and $\bar F_s(t,x_1,\dots,x_s)$ coincide outside of the all hypersurfaces (4). $\bar F_1(t,x_1) = F_1(t,x_1)$ in the entire space.

The correlation functions $\bar F_s(t,x_1,\dots,x_s)$ can be defined as follows. According to (5), the s-particle correlation function is determined for an s-particle observable

$$\varphi_N(x_1,\dots,x_N) = \sum_{i_1<\dots<i_s} \varphi_s(x_1,\dots,x_s),$$

where summation is carried out over all numbers $i_1<\dots<i_s \subset \{1,\dots,N\}$, by the formula

$$(S_N(-\Delta t)f_N,\varphi_N) = (f_N(\Delta t),\varphi_N)$$
$$= \frac{1}{s!}\int \bar F_s(\Delta t,x_1,\dots,x_s)\varphi_s(x_1,\dots,x_s)dx_1\dots dx_s. \quad (7)$$

By using the group property of the operators $\bar S_N(-\Delta t)$ and $\bar S_N(-t)$, we can generalize relations (5), (7) for any time t and introduce hierarchy (3′) for the correlation functions $\bar F_s(t,x_1,\dots,x_s)$ without boundary conditions described above.

Hierarchy (3) can be derived from hierarchy (3′) by using the principle of duality or directly from equation (6) if we take into account the boundary conditions on the hypersurfaces $q_i - q_j = 0$ and compute, in a certain sense, the contributions of these hypersurfaces. In fact, the Itô–Liouville equation (6′) and the stochastic Boltzmann hierarchy in the weak sense (3′) enable us to compute automatically the contributions of hypersurfaces (4) of interaction of stochastic point particles by using a certain analogy with the BBGKY hierarchy. They are necessary only for this purpose. At the same time, the Itô–Liouville equation (6) and the stochastic Boltzmann hierarchy (3) should be regarded as basic.

It is now possible to present a scheme used to deduce hierarchy (3) with boundary conditions. Thus, we start from the stochastic dynamics → the initial data for the system are also random and given by a probability distribution function in the phase space → the operator of shift along the trajectories of the initial distribution function specifies the distribution function (state) at any time → the Itô–Liouville equation for the distribution function with initial and boundary conditions → the principle of duality according to which the ordinary distribution function is associated with a generalized function concentrated on the hypersurfaces of interaction of point stochastic particles → the new concept of averaging of observables taking into account the contributions of the hypersurfaces of interaction of particles → the hierarchy of equations for correlation functions with boundary and initial conditions.

Since hierarchy (3) is deduced on the basis of the stochastic dynamics, it is called the *true, or stochastic, Boltzmann hierarchy*. The Boltzmann equations are derived from this hierarchy as follows: If the initial data for the stochastic Boltzmann hierarchy satisfy the conditions of chaos, i.e., $F_s(0, x_1, \dots, x_s) = F_1(0, x_1) \dots F_1(0, x_s)$, $s \geq 1$, then, outside hyperplanes (4), the correlation functions $F_s(t, x_1, \dots, x_s)$ also satisfy the conditions of chaos, i.e.,

$$F_s(t, x_1, \dots, x_s) = F_1(t, x_1) \dots F_1(t, x_s),$$

and the one-particle correlation function satisfies the nonlinear Boltzmann equation

$$\frac{\partial F_1(t, x_1)}{\partial t} = -p_1 \frac{\partial}{\partial q_1} F_1(t, x_1) + \int dp_2 \int_{S_+^2} d\eta \eta \cdot (p_1 - p_2)$$
$$\times \Big[F_1(t, q_1, p_1^*) F_1(t, q_1, p_2^*) - F_1(t, q_1, p_1) F_1(t, q_1, p_2) \Big], \tag{8}$$
$$F_1(t, x)|_{t=0} = F_1(0, x_1).$$

Note that the Boltzmann equation does not contain boundary conditions.

Leontovich [Leo] indicated (without proof) that the Boltzmann equation (8) can be derived from equations (6′), (3′).

If the s-particle correlation functions are introduced as the products of solutions of the Boltzmann equation, i.e., $F_s(t, x_1, \dots, x_s) = F_1(t, x_1) \dots F_1(t, x_s)$, $s \geq 1$, then the resulting sequence of correlation functions satisfies hierarchy (3) with the initial conditions $F_s(0, x_1, \dots, x_s) = F_1(0, x_1) \dots F_1(0, x_s)$ but without boundary conditions. Hierarchy (3) without boundary conditions is called the *ordinary Boltzmann hierarchy*. Unlike the stochastic Boltzmann hierarchy with boundary conditions, it is impossible to indicate the dynamics of particles corresponding to this hierarchy and, hence, to realize a procedure similar to that carried out for the true stochastic Boltzmann hierarchy.

We now briefly formulate the scheme of the proof of the condition of chaos for $F_s(t, x_1, \ldots, x_s)$. The mild solutions of the stochastic Boltzmann hierarchy (3) are represented in the form of the following iterative series:

$$F_s(t,(x)_s) = \sum_{n=0}^{\infty} \int_0^t dt_1 \int_0^{t_1} dt_2 \ldots \int_0^{t_{n-1}} dt_n S_s(-t,(x)_s) S_s(t_1,(x)_s)$$

$$\times \sum_{i=1}^{s} \int dx_{s+1} \int_{S_+^2} d\eta_{is+1} \eta_{is+1} \cdot (p_i - p_{s+1}) \delta(q_i - q_{s+1})$$

$$\times \Big[S_{s+1}(-t_1,(x)_{s+1}^*) - S_{s+1}(-t_1,(x)_{s+1}) \Big] \ldots S_{s+n-1}(-t_{n-1},(x)_{s+n-1})$$

$$\times \sum_{i=1}^{s+n-1} \int dx_{s+n} \int_{S_+^2} d\eta_{is+n} \eta_{is+n} \cdot (p_i - p_{s+n}) \delta(q_i - q_{s+n})$$

$$\times \Big[S_{s+n}(-t_n,(x)_{s+n}^*) - S_{s+n}(-t_n,(x)_{s+n}) \Big] F_{s+n}(0,(x)_{s+n}), \tag{9}$$

$$(x)_s = (x_1, \ldots, x_s), \quad (x)_{s+n}^* = (x_1, \ldots, p_i^*, q_i, \ldots, x_s, q_{s+1}, p_{s+1}^*).$$

These series converge uniformly in $(x)_s$ on compact sets and in t in a finite time interval $0 < t < t_0$ for initial correlation functions bounded in coordinates and exponentially decreasing in squared momenta. Series (8) are globally convergent in time for the initial correlation functions exponentially decreasing in the squared coordinates and momenta. As the principal feature of the proof, one can mention the possibility of preserving the notion of operators of shift along the stochastic trajectories $S_{s+i}(\pm t, (x)_{s+i})$ with certain random vectors η_{ij}.

If the phase points $(x)_s$ lie outside hypersurfaces (4)

$$q_i - p_i \tau = q_j - p_j \tau, \quad i, j \in \{1, \ldots, s\}, \quad 0 \le \tau \le t,$$

then, the operators $S_{s+i}(\pm t, (x)_{s+i})$ in series (9) can be replaced by the operators $S_{s+i}^0(\pm t, (x)_{s+i})$ of free evolution of point particles and series (9) turns into a mild solution of the ordinary Boltzmann hierarchy satisfying the condition of chaos formulated above.

Note that the solutions (9) of the stochastic Boltzmann hierarchy (3) are, in fact, the Boltzmann–Grad limit of the solutions of the BBGKY hierarchy (1) in the entire phase space but not only outside hypersurfaces (4) $q_i - p_i \tau = q_j - p_j \tau, i, j \in \{1, \ldots, s\}$. At the same time, the solutions of the ordinary Boltzmann hierarchy are obtained as the

Boltzmann–Grad limit of the solutions of the BBGKY hierarchy (1) but only outside the indicated hypersurfaces.

To deduce a spatially-homogeneous Boltzmann equation

$$\frac{\partial F_1(t,p_1)}{\partial t} = \int dp_2 \int_{S_+^2} d\eta\,\eta\cdot(p_1-p_2) \times \Big[F_1(t,p_1^*)F_1(t,p_2^*) - F_1(t,p_1)F_1(t,p_2)\Big] \tag{10}$$

corresponding to the case where the one-particle correlation function is independent of coordinates, i.e., $F_1(t,q_1,p_1) = F_1(t,p_1)$, Kac [Kac1, Kac2] proposed a stochastic dynamics of N particles in the momentum space. For this dynamics, the Itô–Liouville equation in the mean-field approximation has the form

$$\frac{\partial f_N(t,p_1,\ldots,p_N)}{\partial t} = \frac{1}{N}\sum_{i<j=1}^{N}\int_{S_+^2} d\eta_{ij}\,\eta_{ij}\cdot(p_i-p_j) \times \Big[f_N(t,p_1,\ldots,p_i^*,\ldots,p_j^*,\ldots,p_N) - f_N(t,p_1,\ldots,p_i,\ldots,p_j,\ldots,p_N)\Big] \tag{11}$$

and the corresponding hierarchy of equations for the correlation functions can be written as follows:

$$\begin{aligned}\frac{\partial F_s^{(N)}(t,p_1,\ldots,p_N)}{\partial t} &= \frac{1}{N}\sum_{i<j=1}^{N}\int_{S_+^2} d\eta_{ij}\,\eta_{ij}\cdot(p_i-p_j)\\ &\quad\times\Big[F_s^{(N)}(t,p_1,\ldots,p_i^*,\ldots,p_j^*,\ldots,p_s) - F_s^{(N)}(t,p_1,\ldots,p_i,\ldots,p_j,\ldots,p_s)\Big]\\ &\quad+\frac{N-s}{N}\sum_{i=1}^{s}\int dp_{s+1}\int_{S_+^2} d\eta_{is+1}\,\eta_{is+1}\cdot(p_i-p_{s+1})\\ &\quad\times\Big[F_{s+1}^{(N)}(t,p_1,\ldots,p_i^*,\ldots,p_s,p_{s+1}^*) - F_{s+1}^{(N)}(t,p_1,\ldots,p_i,\ldots,p_s,p_{s+1})\Big],\end{aligned} \tag{12}$$

$$F_s^{(N)}(t,p_1,\ldots,p_s)|_{t=0} = F_s^{(N)}(0,p_1,\ldots,p_s), \quad 1\le s\le N.$$

If we now formally pass to the limit as $N\to\infty$ and assume that this limit transition can be realized in equation (12), then we get

$$\frac{\partial F_s(t,p_1,\ldots,p_s)}{\partial t} = \sum_{i=1}^{s} \int dp_{s+1} \int_{S_+^2} d\eta_{is+1}\eta_{is+1}\cdot(p_i - p_{s+1}) \times \Big[F_{s+1}(t,p_1,\ldots,p_i^*,\ldots,p_s,p_{s+1}^*) - F_{s+1}(t,p_1,\ldots,p_i,\ldots,p_s,p_{s+1})\Big], \qquad (13)$$

$$F_s(t,p_1,\ldots,p_s)|_{t=0} = F_s(0,p_1,\ldots,p_s), \quad s \geq 1,$$

$$F_s(t,p_1,\ldots,p_s) = \lim_{N\to\infty} F_s^{(N)}(t,p_1,\ldots,p_s).$$

Here, we have used the fact that the first term on the right-hand side of (12) tends to zero as $N \to \infty$ and

$$\lim_{N\to\infty} \frac{N-s}{N} = 1 \quad \text{for fixed } s.$$

Hierarchy (13) possesses the property of chaos, i.e., for

$$F_s(0,p_1,\ldots,p_s) = F_1(0,p_1)\ldots F_1(0,p_s),$$

it has the solutions

$$F_s(t,p_1,\ldots,p_s) = F_1(t,p_1)\ldots F_1(t,p_s),$$

and $F_1(t,p_1)$ is a solution of the spatially homogeneous Boltzmann equation (10).

Kac simply postulated equation (11). We now show that this equation follows from the stochastic dynamics in the phase space [LaPe2]. Substituting the spatially-homogeneous functions $\varphi_N(p_1,\ldots,p_N)$ and $f_N(p_1,\ldots,p_N)$ in the functional specifying the average value of an observable (5), we conclude that the first term diverges with the volume V of the domain Λ in the coordinate space as V^N and the second term diverges as V^{N-1}. Therefore, we introduce the following functional (for infinitesimally small Δt):

$$(\bar{S}_N(-\Delta t)f_N,\varphi_N) = \lim_{V\to\infty} \Big\{\frac{1}{V^N}\int_{\Lambda^N} f_N(p_1,\ldots,p_N) \times \varphi_N(p_1,\ldots,p_N)dq_1\ldots dq_N dp_1\ldots dp_N + \frac{1}{V^{N-1}}\frac{1}{N}\int_{\Lambda^N}\sum_{i<j=1}^{N}\int_0^{\Delta t} d\tau \int_{S_+^2} d\eta_{ij}\eta_{ij}\cdot(p_i - p_j)$$

$$\times \delta(q_i - p_i\tau - q_j + p_j\tau)\Big[f_N(p_1,\dots,p_i^*,\dots,p_j^*,\dots,p_N)$$

$$- f_N(p_1,\dots,p_i\dots,p_j,\dots,p_N)\Big]$$

$$\times \varphi_N(p_1,\dots,p_N)dq_1\dots dq_N dp_1\dots dp_N\Big\}$$

$$= \int \Big\{ f_N(p_1,\dots,p_N) + \Delta t \sum_{i<j=1}^{N} \int_{S_+^2} d\eta_{ij}\,\eta_{ij}\cdot(p_i - p_j)$$

$$\times \Big[f_N(p_1,\dots,p_i^*,\dots,p_j^*,\dots,p_N)$$

$$- f_N(p_1,\dots,p_i\dots,p_j,\dots,p_N)\Big]\Big\}\varphi_N(p_1,\dots,p_N)dp_1\dots dp_N, \quad (14)$$

where V is the volume of domain Λ. The function $f_N(t, p_1,\dots,p_N)$ defined by the formula

$$f_N(t,p_1,\dots,p_N) = \lim_{n\to\infty}\prod_{i=1}^{n}\bar{S}_N(-\Delta t_i)f_N(p_1,\dots,p_N),$$

$$\sum_{i=1}^{n}\Delta t_i = t,$$

where $\bar{S}_N(-\Delta t)$ is given by relation (14), satisfies equation (11) (without multiplier $1/N$), i.e., the Kac stochastic dynamics in the momentum space follows from our stochastic dynamics in the phase space.

Note that the spatially homogeneous Boltzmann equation directly follows from the stochastic Boltzmann hierarchy in the phase space (3) for spatially homogeneous chaotic initial data. Indeed, the correlation functions $F_s(t,(x)_s)$ (9) outside hypersurfaces (4) are also spatially homogeneous and satisfy the condition of chaos, and the one-particle correlation function satisfies the Boltzmann equation (8) which, in this case, reduces to equation (10). In this case, the mean-field approximation is not required.

Thus, the proposed method based on the stochastic dynamics in the phase space is more general than the Kac method based on the stochastic dynamics in the momentum space.

Our monograph is based on the works of the author, M. Lampis, and K. D. Petrina (see the list of references [PeP1, PeP2, PeP3, LaPe4, LaPe2, LaPe3]) in which the stochastic dynamics is introduced as the Boltzmann–Grad limit of the Hamiltonian dynamics of a system of hard spheres in the case where the diameter of hard spheres tends to zero. In these works, a new concept of averaging of observables is introduced and a sequence of correlation functions is constructed taking into account the contributions of the hypersurfaces of lower dimensionality where the interaction of point stochastic particles occurs. The stochastic Boltzmann hierarchy in the form of the ordinary Boltzmann hierarchy but with boundary conditions is deduced. For the stochastic Boltzmann hierarchy, the theorems on existence of solutions and the validity of the evolution of chaos are proved without using the mean-field approximation. On this basis, the Boltzmann equation is rigorously deduced both in the spatially inhomogeneous and spatially homogeneous cases. The Kac dynamics [Kac1, Kac2] in the momentum space is obtained as a result of a certain procedure averaging over the positions of particles.

As far as the spatially inhomogeneous Boltzmann equation (8) is concerned, we can emphasize that, prior to our works, it was deduced solely for a smoothed stochastic dynamics in which the δ-functions $\delta(q_i - q_j)$ in the Itô–Liouville equation ($6'$) are replaced with a smooth function $h(q_i, q_j)$, the collisions take place independently of the positions of particles, and the mean-field approximation and a smoothed kernel of collisions are used. We refer the reader to the monograph by Skorokhod [Sko] and the survey by Wagner [Wag] concerning the above mentioned results. All these restrictions are eliminated in a series of our works and the corresponding results are presented in the proposed monograph.

It should be stressed that, from our stochastic dynamics, we have derived the true standard not modified Boltzmann equation of kinetic theory. It seems that it is simpler to investigate directly the true stochastic Boltzmann hierarchy without modification.

Chapter 1

System of hard spheres

1.1 Introduction

In this chapter, we present a short survey of results about the Hamiltonian dynamics, the Liouville equation and the BBGKY (Bogolyubov–Born–Green–Kirkwood–Yvon) hierarchy of systems of hard spheres. We formulate problems and their solutions, theorems and ideas of their proofs. The main idea of the survey is to prepare readers to understand the problem of justification of the existence of the Boltzmann–Grad limit.

We present a critical survey of results obtained by numerous researchers in this field.

It seems to us that we are now able to formulate exactly rigorous results for systems of hard spheres because the period of pioneering attempts to solve some particular problems is exhausted and everything is clarified. The proof of results of this chapter can be found in our books [CGP, PGM3].

1.2 Hamiltonian dynamics of a system of hard spheres

1.2.1 Hamilton equations

Consider a system of N identical particles of unit mass $m = 1$ in the three-dimensional Euclidean space $\mathbb{R}^3$. Every particle is characterized by its position (coordinate) $q = (q^1, q^2, q^3)$ and momentum $p = (p^1, p^2, p^3)$. As usual, we combine q and p into a single point $x \equiv (q, p)$ of the phase space $\mathbb{R}^3 \times \mathbb{R}^3$. The state of system of N particles is characterized by N points in the phase space $(x_1, \dots, x_N)$.

We suppose that particles interact as hard spheres. This means that they interact via the pair potential $\phi(q)$ that is defined as follows:

$$\phi(q) = 0, \quad |q| > a, \quad \phi(q) = \infty, \quad |q| \le a, \tag{1.2.1}$$

where a is the diameter of spheres and q is the position of the center of a sphere.

The dynamics of N particles (hard spheres) is defined as follows: Particles move as the free ones until they collide when the distance between of their centers is equal to a. Then particles scatter elastically and continue to move as free ones until the next collision. The corresponding trajectories are defined as follows:

$$Q_i(t) = q_i + p_i t, \quad P_i(t) = p_i, \quad i = 1, \dots, N, \tag{1.2.2}$$

where q_i and p_i are the initial position and momentum of the particle with number i at time $t = 0$. $Q_i(t)$ and $P_i(t)$ are its position and momentum at time t. If $|Q_i(\tau) - Q_j(\tau)| = a$, then the i-th and j-th particles elastically collide, and after instantaneous collision their momenta are

$$\begin{aligned} P_i^*(\tau) &= P_i(\tau) - \eta_{ij}\,\eta_{ij}\cdot\big(P_i(\tau) - P_j(\tau)\big),\\ P_j^*(\tau) &= P_j(\tau) + \eta_{ij}\,\eta_{ij}\cdot\big(P_i(\tau) - P_j(\tau)\big), \end{aligned} \tag{1.2.3}$$

where the unit vector η_{ij} is defined as follows:

$$\eta_{ij} = \frac{Q_i(\tau) - Q_j(\tau)}{|Q_i(\tau) - Q_j(\tau)|}. \tag{1.2.4}$$

By τ we denote the time of collision of the i-th and j-th particles. It is the solution of the equation $|Q_i(\tau) - Q_j(\tau)| = a$. After the collision, the i-th and j-th particles move as free ones until the next collision, i.e.,

$$\begin{aligned} Q_i(t) &= Q_i(\tau) + P_i^*(\tau)(t-\tau), \quad P_i(t) = P_i^*(\tau),\\ Q_j(t) &= Q_j(\tau) + P_j^*(\tau)(t-\tau), \quad P_j(t) = P_j^*(\tau). \end{aligned} \tag{1.2.5}$$

It is easy to check that

$$\begin{aligned} P_i^*(\tau) + P_j^*(\tau) &= P_i(\tau) + P_j(\tau),\\ P_i^{*2}(\tau) + P_j^{*2}(\tau) &= P_i^2(\tau) + P_j^2(\tau). \end{aligned} \tag{1.2.6}$$

This means that, after the collision, the law of conservation of momentum and energy is true.

Later we will show that one can restrict oneself only to pair collisions; therefore, we will discuss in detail the corresponding Hamilton equation for two particles.

The above-defined trajectory for two particles is the solution of the Hamilton equations

$$\begin{aligned} \frac{dQ_1(t)}{dt} &= P_1(t),\\ \frac{dP_1(t)}{dt} &= \delta(t-\tau)\big(P_1^*(\tau) - P_1(\tau)\big),\\ \frac{dQ_2(t)}{dt} &= P_2(t),\\ \frac{dP_2(t)}{dt} &= \delta(t-\tau)\big(P_2^*(\tau) - P_2(\tau)\big), \end{aligned} \tag{1.2.7}$$

$$|Q_1(\tau) - Q_2(\tau)| = a$$

with initial data

$$Q_1(t)|_{t=0} = q_1, \quad P_1(t)|_{t=0} = p_1, \quad Q_2(t)|_{t=0} = q_2, \quad P_2(t)|_{t=0} = p_2.$$

It is easy to check that, for $t < \tau$, one has (1.2.2), and, for $t > \tau$, one has (1.2.5). One obtains the same trajectory from the Hamilton equations

$$\begin{aligned}
\frac{dQ_1(t)}{dt} &= P_1(t), \\
\frac{dP_1(t)}{dt} &= 0, \\
\frac{dQ_2(t)}{dt} &= P_2(t), \\
\frac{dP_2(t)}{dt} &= 0,
\end{aligned} \tag{1.2.8}$$

with the boundary condition according to which, at the time of collision τ $\big(|Q_1(\tau) - Q_2(\tau)| = a\big)$, the momenta $P_i(\tau),\ P_j(\tau)$ have jumps and become $P_i^*(\tau),\ P_j^*(\tau)$.

Thus, equations (1.2.7) with the δ-function $\delta(t-\tau)$ but without the boundary condition define the same trajectory (1.2.2)–(1.2.6) as equations (1.2.8) with the boundary condition. This means that equations (1.2.7) are equivalent to equations (1.2.8) with the boundary condition.

By using the identity

$$\delta(t-\tau) \equiv \delta(|Q_1(t) - Q_2(t)| - a) \times \left| \frac{Q_1(t) - Q_2(t)}{|Q_1(t) - Q_2(t)|} \cdot \big(P_1(t) - P_2(t)\big) \right|, \tag{1.2.9}$$

which can easily be proved, and the definition of δ-function and substituting the right-hand side of (1.2.9) for $\delta(t-\tau)$, one obtains equation (1.2.7) with the right-hand sides expressed through $Q_1(t),\ P_1(t),\ Q_2(t),\ P_2(t)$ as in the usual Hamilton equations.

For systems of N particles, one can also use equations with δ-functions but without boundary conditions or equivalent equations without δ-functions but with boundary conditions.

Denote by W_N the set of forbidden configurations in the phase space of N particles. W_N consists of the phase points such that $|q_i - q_j| < a$ for at least one pair $i, j \in \{1, \dots, N\}$ and arbitrary momenta $(p_1, \dots, p_N)$. This means that W_N consists of phase points such that the distance between the centers of at least two hard spheres is less than their diameter a.

1.2.2 Existence of trajectories

Denote by $X_i(t)$ the phase point of the i-th particle at time t, $X_i(t) = (Q_i(t), P_i(t))$, with initial phase point $x_i = (q_i, p_i)$ at time $t = 0$. Let $X(t) = (X_1(t), \dots, X_N(t))$;

to indicate the dependence of $X_i(t)$, $X(t)$ on the initial phase point $x = (x_1, \dots, x_N)$, we write

$$X_i(t) = X_i(t, x_1, \dots, x_N) = X_i(t, x),$$

$$X(t) = X(t, x_1, \dots, x_N) = X(t, x).$$

The solution $X(t, x)$ for $t \subset \mathbb{R}^1$ and fixed x defines a trajectory in the phase space.

Consider the problem of existence of trajectories $X(t, x)$ and its dependence on the initial phase point x. The trajectories do not exist for arbitrary phase points x. For example, for initial phase points such that a multiple (i.e., triple and of higher order) collisions occur. In this case one cannot define a trajectory after a collision. One cannot define a trajectory if, on a finite interval of time, there are infinitely many collisions.

It has been proved that all these initial data belong to certain hypersurfaces of lower dimension than the dimension of the phase space of N particles. At instant of collisions, the momenta of colliding pairs of particles have jumps according to (1.2.3) and the trajectory is discontinuous, but on the intervals between collisions the trajectory $X(t, x)$ is a continuous function of time t and initial phase points x. Obviously, the trajectory depends only on the interval of time, but not on initial and final times; therefore we consider $t = 0$ as the initial time the trajectory $X(t, x)$ has the group property

$$X(t_1 + t_2, x) = X\big(t_1, X(t_2, x)\big) = X\big(t_2, X(t_1, x)\big) \tag{1.2.10}$$

for arbitrary t_1, t_2.

We define a trajectory for the case where the collision of two particles, say, i-th and j-th, happens at the initial time $t = 0$, i.e., $q_i - q_j = a\eta$. The collision occurs instantaneously, and the phase points

$$x = x(q_1, p_1, \dots, q_i, p_i, \dots, q_i - a\eta, p_j, \dots, x_N)$$

and

$$x^* = x(q_1, p_1, \dots, q_i, p_i^*, \dots, q_i - a\eta, p_j^*, \dots, x_N)$$

will be considered as identical in the sense that they determine the same initial state for the system. We will choose this point as initial that particles after the collision should depart.

In this case, it is necessary to distinguish between positive or negative time. In the first case ($t > 0$), the i-th and j-th particles will depart with the same momenta if $\eta_{ij} \cdot (p_i - p_j) \geq 0$, i.e., one chooses x as the initial point. If $\eta_{ij} \cdot (p_i - p_j) \leq 0$, then one chooses x^* as the initial point. In the second case ($t < 0$), the i-th and j-th particles will depart if $\eta_{ij} \cdot (p_i - p_j) \geq 0$, i.e., one chooses x as the initial point. If $\eta_{ij} \cdot (p_i - p_j) \leq 0$, then one chooses x^* as the initial point. These boundary conditions of collisions at the initial time $t = 0$ play an important role in the Liouville equation.

It is clear that these boundary conditions of collisions, when momenta have jumps, are also satisfied at arbitrary time $t \neq 0$. Namely, if $Q_i(t) - Q_j(t) = a\eta$, $|t| \neq 0$, then $(P_i(t), P_j(t)) \to (P_i^*(t), P_j^*(t))$ for $\eta \cdot \big(P_i(t) - P_j(t)\big) \leq 0$ and $t > 0$, or for $\eta \cdot \big(P_i(t) - P_j(t)\big) \geq 0$ and $t < 0$.

We summarize the properties of trajectories described above in the following theorem:

Theorem 1.1. *For a system of N hard spheres, a unique trajectory $X(t,x)$ exists for almost all initial data outside certain hypersurfaces of lower dimension. The trajectory $X(t,x)$ is continuously differentiable on intervals of time between collisions with respect to time t and initial data x and has the group property (1.2.10). The trajectory satisfies the above-described boundary conditions of collisions.*

It is obvious that the trajectory $X(t,x)$ belongs to the admissible configuration, i.e., to the completion of the forbidden configuration.

1.2.3 Liouville theorem

We define a mapping T_t (a flow) of the phase space onto itself as a shift along the trajectory $X(t,x)$. Namely,

$$T_t x = X(t,x), \quad t \in \mathbb{R}^1, \quad x \in \mathbb{R}^{3N} \times \mathbb{R}^{3N},$$

for all x for which the trajectory $X(t,x)$ exists. This means that the mapping T_t is defined for almost all x, i.e., outside certain hypersurfaces. On time intervals between collisions, the Jacobian $\frac{\partial X(t,x)}{\partial x}$ of the mapping T_t is equal to one. It is easy to check that the transformation of momenta at the instant of collision (1.2.3) also has the Jacobian $\frac{\partial(P_i^*(t), P_j^*(t))}{\partial(P_i(t), P_j(t))}$ equal to one.

One can calculate the Jacobian $\frac{\partial X(t,x)}{\partial x}$ consequently on the intervals between collisions, where it is equal to one, and at the instants of collisions, where it is again equal to one. The obtained results are formulated as the Liouville theorem.

Theorem 1.2. *The Jacobian of the transformation T_t is equal to one.*

If D is a measurable domain and D_t is its image $D_t = T_t D$, then $VD_t = VD$, where V denotes the volume (measure). This is an equivalent formulation of the Liouville theorem.

1.3 Evolution operator for a system of hard spheres

1.3.1 Definition of the evolution operator

Denote by $f_N(x_1, \ldots, x_N) \equiv f_N(x)$ a measurable function defined on the phase space $\mathbb{R}^N \times (\mathbb{R}^N \setminus W_N)$ of the N-particle system which is invariant under the permutation of the arguments $x_1, \ldots, x_N$, i.e., $f_N(x_1, \ldots, x_N) = f_N(x_{i_1}, \ldots, x_{i_N})$, where $(i_1, \ldots, i_N)$ is an arbitrary permutation of indices $(1, \ldots, N)$.

The functions $f_N(x_1, \dots, x_N)$ are supposed to be equal to zero on the set W_N of forbidden configurations.

On this set of functions, we define the evolution operator $S_N(t)$, $t \in \mathbb{R}^1$, by the formula

$$(S_N(t) f_N)(x) = \begin{cases} f_N(X(t,x)) & \text{for } x \in \mathbb{R}^{3N} \times (\mathbb{R}^{3N} \setminus W_N), \\ 0 & \text{for } x \in W_N, \end{cases} \tag{1.3.1}$$

where $X(t,x)$ is the phase trajectory of our system. $S_N(t)$ is the operator of shift along the trajectory $X(t,x)$, and it is defined almost everywhere in the phase space outside certain hypersurfaces. Obviously, the operator $S_N(t)$ is linear and $S_N(0) = I$, where I is the unit operator.

Denote by $L^1(\mathbb{R}^{3N} \times (\mathbb{R}^{3N} \setminus W_N))$ the linear Banach space of summable functions which are symmetric (invariant under permutation) and equal to zero on the set W_N with the norm

$$\|f_N\| = \int dx_1 \dots dx_N |f_N(x_1, \dots, x_N)| = \int dx |f_N(x)|, \tag{1.3.2}$$

$$dx = dx_1 \dots dx_N.$$

Denote by L_0^1 the subset of the space L^1 consisting of continuously differentiable functions with compact support satisfying the condition

$$f_N(x_1, \dots, x_i, \dots, x_j, \dots, x_N) = f_N(x_1, \dots, x_i^*, \dots, x_j^*, \dots, x_N) \tag{1.3.3}$$

if $q_i - q_j = a\eta$, $E\ x_i^* = (q_i, p_i^*)$, $x_j^* = (q_j, p_j^*)$.

The subset L_0^1 contains functions equal to zero in some ε-neighborhood of the forbidden configurations (with above described properties).

Let us show that the function $(S_N(t) f_N)(x)$ is summable for arbitrary $f_N \in L_0^1$. It follows directly from the Liouville theorem that

$$\begin{aligned} \|S_N(t) f\| &= \int dx |f_N(X(t,x))| \\ &= \int dx \det\Big(\frac{\partial X(t,x)}{\partial x}\Big) |f_N(X(t,x))| \\ &= \int dx(t,x) |f_N(X(t,x))| = \int dx f_N(x). \end{aligned} \tag{1.3.4}$$

It follows from (1.3.4) that the operator $S_N(t)$ is isometric on L_0^1. The set L_0^1 is everywhere dense in L^1, and the operator $S_N(t)$ can be extended by continuity from L_0^1 on to the entire space L^1 as an isometric one, i.e., $\|S_N(t)\| = 1$.

From the group property of the trajectory $X(t,x)$ (1.2.10), one concludes that the family of operators $S_N(t)$ also has the group property

$$S_N(t_1+t_2) = S_N(t_1)S_N(t_2) = S_N(t_2)S_N(t_1) \tag{1.3.5}$$

for arbitrary t_1 and t_2.

We summarize the properties of the operator $S_N(t)$ in the space L^1 in the following theorem:

Theorem 1.3. *The family of operators $S_N(t)$, $t \in \mathbb{R}^1$, is a strongly continuous one-parameter group of isometric operators in L^1. On the set L^1_0, the infinitesimal generator of the group $S_N(t)$ is given by the Poisson bracket of a free system of particles with boundary conditions on $\partial(\mathbb{R}^{3N} \setminus W_N)$.*

The proof of this theorem can be found in books [CGP, PGM3]. We discuss the boundary conditions in detail.

First, consider positive time $t > 0$ (forward motion). On L^1_0, we have

$$\frac{dS_N(t)f_N(x)}{dt} = \sum_{i=1}^{N} p_i \cdot \frac{\partial}{\partial q_i} S_N(t)f_N(x) = \mathcal{H}_N S_N(t) f_N(x) \tag{1.3.6}$$

and the following boundary condition: for $q_i - q_j = a\eta$, $\eta \cdot (p_i - p_j) < 0$ momenta p_i and p_j, on the right-hand side of (1.3.6) should be replaced by p_i^*, p_j^*, given by (1.2.3). If $\eta \cdot (p_i - p_j) > 0$, then the momenta p_i, p_j do not change.

For negative time $t < 0$ (backward motion), one has the following boundary condition: for $q_i - q_j = a\eta$, $\eta \cdot (p_i - p_j) > 0$ momenta p_i, p_j should be replaced by p_i^*, p_j^* on the right-hand side of (1.3.6). If $\eta \cdot (p_i - p_j) < 0$, then the momenta p_i, p_j do not change. It follows from these boundary conditions that the group $S_N(t)$, in fact, consists of two semigroups $S_N^+(t)$, $t > 0$, and $S_N^-(t)$, $t < 0$, with different boundary conditions. But $S_N(\pm t)S_N(\mp t) = 1$, and one has the group $S_N(t)$, $-\infty < t < \infty$.

Note that the infinitesimal generator

$$\mathcal{H}_N f_N(x) = \sum_{i=1}^{N} p_i \cdot \frac{\partial}{\partial q_i} f_N(x) \tag{1.3.7}$$

with boundary conditions can be calculated on $f_N(x) \in L^1_0$ in the sense of pointwise convergence, or in the sense of strong convergence in L^1.

1.3.2 Liouville equation

We deduced formula (1.3.6) using the definition of the evolution operator $S_N(t)$ for given $f_N(x)$. Now consider this formula as the equation for the unknown function $f_N(t,x) = S_N(t)f_N(x)$ which depends on time t and phase point x and, at $t = 0$, is equal to a given function $f_N(x) = S_N(0) f_N(x)$. For the function $f_N(t, x)$, one obtains from (1.3.6) the Liouville equation

$$\frac{\partial}{\partial t} f_N(t, x) = \sum_{i=1}^{N} p_i \cdot \frac{\partial}{\partial q_i} f_N(t, x) = \mathcal{H}_N f_N(t, x), \qquad (1.3.8)$$

$$f_N(t, x)|_{t=0} = f_N(x) \in L_0^1,$$

with the same boundary conditions as for $S_N(t) f_N(x)$. We obtain the Liouville equation (1.3.8) from the fact that $S_N(t)$ is the operator of shift along the trajectory $S_N(t) f_N(x) = f_N(X(t, x)) = f_N(t, x)$.

Now consider the Liouville equation as a linear partial differential equation with initial data $f_N(t, x)|_{t=0} = f_N(x)$. It is well known that its solution is given by the formula $f(t, x) = f_N(X(t, x))$, where $X(t, x)$ is a solution of the corresponding equations of characteristics.

The equations of characteristics corresponding to (1.3.8) are the Hamilton equations

$$\frac{dP_i(t)}{dt} \equiv 0, \quad \frac{dQ_i(t)}{dt} = P_i(t), \quad i = 1, \dots, N \qquad (1.3.9)$$

with the initial conditions $P_i(t)|_{t=0} = p_i$, $Q_i(t)|_{t=0} = q_i$, and the boundary conditions according to which for $Q_i(t) - Q_j(t) = a\eta$ and for $t > 0$, momenta $(P_i(t), P_j(t))$ should be replaced by $(P_i^*(t), P_j^*(t))$ if $\eta \cdot \big(P_i(t) - P_j(t)\big) < 0$ and for $t < 0$ by $(P_i^*(t), P_j^*(t))$ if $\eta \cdot \big(P_i(t) - P_j(t)\big) > 0$.

Stress that these boundary conditions for the Hamiltonian equation (1.3.9) directly follow from the corresponding boundary condition for the Liouville equations or for the infinitesimal generator $\mathcal{H}_N$.

Unfortunately, these boundary conditions are absent in all books devoted to nonequilibrium classical statistical mechanics, excluding our monographs and papers (written together with V. Gerasimenko). In our opinion, this is a great mistake, which is a reason of the derivation of the incorrect Boltzmann hierarchy.

Thus, the Liouville equation only with boundary conditions is equivalent to the Hamilton equations of a system of hard spheres.

We differentiate the function $f_N(t, x) = S_N(t) f_N(x)$ with respect to time in the sense of pointwise convergence in the case of general differentiable functions $f_N(x) \in L^1$ which not necessarily belong to L_0^1. The function $f_N(t, x)$ for fixed x is a discontinuous function of t at the instant of collision τ that is the solution of the equation $Q_i(\tau, x) - Q_j(\tau, x) = a\eta$. At the instant of collision τ the function $f_N(t, x)$ has the jump $f_N(X^*(\tau, x)) - f_N(X(\tau, x))$, where

$$X^*(\tau, x) = \Big(Q_1(\tau, x), P_1(\tau, x), \dots, Q_i(\tau, x), P_i^*(\tau, x), \dots,$$
$$Q_j(\tau, x), P_j^*(\tau, x), \dots, Q_N(\tau, x), P_N(\tau, x)\Big).$$

According to the law of differentiation of generalized functions with jump at $t = \tau$, one obtains

$$\frac{\partial f_N(t,x)}{\partial t} = \sum_{i=1}^{N} p_i \cdot \frac{\partial}{\partial q_i} f_N(t,x) \\ + \delta(t-\tau)\Big[f_N(X^*(\tau,x)) - f_N(X(\tau,x))\Big]. \tag{1.3.10}$$

If one substitutes $p_i, q_i, i = 1,\ldots,N$ for $f_N(x)$, then one obtains from (1.3.10) the equations

$$\begin{aligned} \frac{dQ_i(t,x)}{dt} &= P_i(t,x), \\ \frac{dP_i(t,x)}{dt} &= \delta(t-\tau)\big[P_i^*(\tau,x) - P_i(\tau,x)\big], \\ \frac{dQ_j(t,x)}{dt} &= P_j(t,x), \\ \frac{dP_j(t,x)}{dt} &= \delta(t-\tau)\big[P_j^*(\tau,x) - P_j(\tau,x)\big], \\ \frac{dQ_l(t,x)}{dt} &= P_l(t,x), \\ \frac{dP_l(t,x)}{dt} &= 0, \quad l \neq i, \quad l \neq j, \quad 1 \leq l \leq N. \end{aligned} \tag{1.3.11}$$

and the initial conditions $X_i(t,x) = x_i, i = 1,\ldots,N$.

In Subsection 1.2.1 we showed that system (1.3.11) is equivalent to system (1.3.9) because their solutions coincide. In this sense the Liouville equation without δ-function (1.3.8) but with boundary conditions is equivalent to the Liouville equation with δ-function (1.3.10) but without boundary conditions.

It can be proved directly that solutions of the two kinds of Liouville equations (1.3.8) and (1.3.10) coincide. By the way, their equivalence follows directly from the well–known fact that the equations with one-dimensional δ-function are equivalent to the equations without δ-function but with the corresponding boundary conditions.

Indeed, let, for given x, τ be the first instant of collision. Then, for $t < \tau$, the solution of equation (1.3.10) with initial function $f_N(x)$ is $f_N(X(t,x))$, where $X(t,x)$ is the trajectory of free particles $Q_i(t,x) = q_i + p_i t$, $P_i(t,x) = p_i$, $i = 1,\ldots,N$. For $t \geq \tau$, equation (1.3.10) is again reduced to the Liouville equation of free particles

$$\frac{\partial f_N(t,x)}{\partial t} = \sum_{i=1}^{N} p_i \cdot \frac{\partial}{\partial q_i} f_N(t,x)$$

with initial data $f_N(t,x)|_{t=\tau} = f_N(X^*(t,x))$, and its solution is

$$f_N\big(Q_1(t,x), P_1(t,x), \dots, Q_i(\tau,x) + P_i^*(\tau,x)(t-\tau), P_i^*(\tau,x), \dots,$$
$$Q_j(\tau,x) + P_j^*(\tau,x)(t-\tau), P_j^*(\tau,x), \dots, Q_N(t,x), P_N(t,x)\big).$$

Thus, we obtain the same solution as for equation (1.3.8) without δ-function but with boundary conditions.

1.3.3 Evolution operator and Liouville equation for negative time

In the previous subsections, the Hamilton equation, the operator of evolution $S_N(t)$ and the Liouville equation were considered for arbitrary time $-\infty < t < \infty$. It was stressed that the boundary conditions for positive time (forward motion) and for negative time (backward motion) are different. In nonequilibrium statistical mechanics mainly negative time (backward motion) is used. Therefore, we especially discuss the Liouville equation and the corresponding boundary condition for negative time.

Consider the function

$$f_N(t,x) = S_N(-t) f_N(x) = f_N(X(-t,x)), \quad t \geq 0,$$
$$f_N(t,x)|_{t=0} = f_N(x). \tag{1.3.12}$$

As is common in nonequilibrium statistical mechanics, we denote the function $S_N(-t) f_N(x)$, $t > 0$, by $f_N(t,x)$.

It is obvious that the function $f_N(t,x)$ is a solution of the Liouville equation

$$\frac{\partial f_N(t,x)}{\partial t} = -\sum_{i=1}^{N} p_i \cdot \frac{\partial}{\partial q_i} f_N(t,x), \tag{1.3.13}$$
$$f_N(t,x)|_{t=0} = f_N(x)$$

with the following boundary conditions: If $q_i - q_j = a\eta$, $\eta \cdot (p_i - p_j) \geq 0$, then the momenta p_i, p_j should be replaced by p_i^*, p_j^* given by (1.2.3). If $q_i - q_j = a\eta$, $\eta \cdot (p_i - p_j) \leq 0$, then the momenta p_i, p_j do not change.

Unfortunately, in all available books and papers devoted to statistical mechanics of systems of hard spheres, excluding ours with V. I. Gerasimenko, the boundary condition are absent. This is a serious mistake which leads to another serious mistake in the BBGKY and Boltzmann hierarchy, as it will be shown in the next sections. Obviously, one can repeat, with some modifications results of the previous subsection and show that the Liouville equation (1.3.13) with boundary conditions is equivalent to the Liouville equation with δ-function but without boundary conditions.

1.4 BBGKY hierarchy for systems of hard spheres

1.4.1 Definition of correlation functions

The state of a system of N hard spheres in classical mechanics is described by the position of the centers of hard spheres and their momenta

$$\begin{aligned} X(t,x) &= \big(X_1(t,x),\ldots,X_N(t,x)\big) \\ &= \big(Q_1(t,x),P_1(t,x),\ldots,Q_N(t,x),P_N(t,x)\big) \end{aligned}$$

for arbitrary time $-\infty < t < \infty$. One supposes that the initial data of the system are known:

$$X(t,x)|_{t=0} = x = (x_1,\ldots,x_N) = (q_1,p_1,\ldots,q_N,p_N).$$

The state of the system of N hard spheres in classical statistical mechanics is described by a distribution function (density of probability) $D_N(t,x_1,\ldots,x_N) = D_N(t,x)$ defined in the phase space on admissible configurations and equal to zero on forbidden configurations W_N. One supposes that the initial distribution function $D_N(0,x_1,\ldots,x_N) = D_N(0,x)$ is known, normalized to unity, invariant with respect to the permutation of variables $x_1,\ldots,x_N$, i.e., symmetric, equal to zero on forbidden configurations, and such that $D_N(0,x) \subset L_0^1$.

We will consider $D_N(t,x)$ for positive time $t > 0$. As is known, it is defined through the initial function $D_N(0,x)$ as follows:

$$\begin{aligned} D_N(t,x_1,\ldots,x_N) = D_N(t,x) &= S_N(-t)D_N(0,x_1,\ldots,x_N) \\ &= D_N\big(0,X_1(-t,x),\ldots,X_N(-t,x)\big) \\ &= D_N\big(0,X_1(-t,x)\big). \end{aligned} \tag{1.4.1}$$

According to properties of trajectories and the operator $S_N(-t)$, the function $D_N(t,x_1,\ldots,x_N)$ is also symmetric and equal to zero on forbidden configurations. Indeed if x belongs to admissible configurations, then $X(-t,x)$ again belongs to it. If x belongs to forbidden configurations W_N, then $S_N(-t)D_N(0,x) = 0$. The function $D_N(t,x_1,\ldots,x_N)$ is symmetric with respect to the permutation of variables $x_1,\ldots,x_N$ because $D_N(0,x_1,\ldots,x_N)$ is symmetric and the trajectory $X(-t,x_1,\ldots,x_N)$ possesses the following property:

$$\begin{aligned} &\big(X_1(-t,x_{i_1},\ldots,x_{i_N}),\ldots,X_N(-t,x_{i_1},\ldots,x_{i_N})\big) \\ &\quad = \big(X_{i_1}(-t,x_1,\ldots,x_N),\ldots,X_{i_N}(-t,x_1,\ldots,x_N)\big). \end{aligned}$$

An arbitrary real symmetric test function $\varphi_N(x_1,\ldots,x_N) = \varphi_N(x)$ will be considered as observable. A function $\varphi_N(x_1,\ldots,x_N)$ is an s-particle observable of a system of N particles if it can be represented as follows:

$$\varphi_N(x_1,\ldots,x_N) = \sum_{i_1<\ldots<i_s} \varphi_s(x_{i_1},\ldots,x_{i_s}), \tag{1.4.2}$$

where $\varphi_s(x_1, \dots, x_s)$ is a real symmetric test function depending on s variables.

Consider the following functional average of the observable $\varphi_N(x_1, \dots, x_N)$ over the state $D_N(t, x)$:

$$(D_N(t), \varphi_N) = \int D_N(t, x_1, \dots, x_N)\varphi_N(x_1, \dots, x_N)dx_1 \dots dx_N$$
$$= \int D_N(t, x)\varphi_N(x)dx, \tag{1.4.3}$$

$$dx = dx_1 \dots dx_N, \qquad \int D_N(t, x)dx = 1.$$

It is obvious that functional (1.4.3) exists because $D_N(t) \in L^1$ and

$$|(D_N(t), \varphi_N)| \le \|D(0)\| \sup_x |\varphi_N(x)|,$$

according to the isometry of $S_N(-t)$. Let us calculate functional (1.4.3) for the s-particle observable (1.4.2).

Using the symmetry of $D_N(t, x_1, \dots, x_N)$ and $\varphi_N(x_1, \dots, x_N)$, one obtains

$$(D_N(t), \varphi_N) = \frac{N(N-1)\dots(N-s+1)}{s!} \int D_N(t, x_1, \dots, x_N)$$
$$\times \varphi_s(x_1, \dots, x_s)dx_1 \dots dx_N$$
$$= \frac{1}{s!} \int F_s^{(N)}(t, x_1, \dots, x_s)\varphi_s(x_1, \dots, x_s)dx_1 \dots dx_s, \tag{1.4.4}$$

where the s-particle correlation function $F_s^{(N)}(t, x_1, \dots, x_s)$ of the N-particle system is defined as follows:

$$F_s^{(N)}(t, x_1, \dots, x_s) = N(N-1)\dots(N-s+1)$$
$$\times \int D_N(t, x_1, \dots, x_s, x_{s+1}, \dots, x_N)dx_{s+1} \dots dx_N, \tag{1.4.5}$$

$$F_N^N(t, x_1, \dots, x_N) = N! D_N(t, x_1, \dots, x_N), \quad 1 \le s \le N.$$

This is the definition of an s-particle correlation function in the framework of canonical ensemble when the system consists of exactly N particles.

We obtained a new description of a state by the sequence of N correlation functions

$$F^{(N)}(t)$$
$$= \left(F_1^{(N)}(t, x_1), \dots, F_s^{(N)}(t, x_1, \dots, x_s), \dots, F_N^{(N)}(t, x_1, \dots, x_N)\right). \tag{1.4.6}$$

Obviously, the states $F^{(N)}(t)$ and $D_N(t) = D_N(t,x)$ are equivalent.

We introduce correlation functions in the framework of grand canonical ensemble. The state is defined by the infinite sequence of non-normalized distribution functions

$$D(t) = \Big(1, D_1(t,x_1), \ldots, D_n(t,x_1,\ldots,x_n), \ldots\Big),$$

where

$$D_n(t,x_1,\ldots,x_n) = S_n(-t)D_n(0,x_1,\ldots,x_n),$$

$$D_n(0,(x)_n) \in L^1.$$

This means that the system under consideration consists of an arbitrary number $n \geq 0$ of particles with some probability.

The functional average of an s-particle observable $\varphi_s(x_1,\ldots,x_s)$ (1.4.2), $n \geq s$, over the state $D(t)$ is defined as follows:

$$\frac{1}{\Xi}\sum_{n=s}^{\infty}\frac{1}{n!}\int D_n(t,x_1,\ldots,x_n)\varphi_n(x_1,\ldots,x_n)dx_1\ldots dx_n$$
$$= \frac{1}{s!}\int F_s(t,x_1,\ldots,x_s)\varphi_s(x_1,\ldots,x_s)dx_1\ldots dx_s, \quad (1.4.7)$$

where the s-particle correlation function in the framework of the grand canonical ensemble is defined as follows:

$$F_s(t,x_1,\ldots,x_s)$$
$$= \frac{1}{\Xi}\sum_{n\geq s}\frac{1}{(n-s)!}\int D_n(t,x_1,\ldots,x_s,x_{s+1},\ldots,x_n)dx_{s+1}\ldots dx_n, \quad (1.4.8)$$

$$\Xi = \sum_{n=0}^{\infty}\frac{1}{n!}\int D_n(t,(x)_n)d(x)_n = \sum_{n=0}^{\infty}\frac{1}{n!}\int D_n(0,(x)_n)d(x)_n,$$

$$F(t) = (F_1(t,x_1),\ldots,F_s(t,x_1,\ldots,x_s),\ldots);$$

Ξ is known as the grand partition function. The probability that one has exactly n particles in the system is equal to

$$\Xi^{-1}\int D_n(t,(x)_n)d(x)_n = \Xi^{-1}\int D_n(0,(x)_n)d(x)_n.$$

We suppose that series (1.4.8) are convergent.

1.4.2 Derivation of hierarchy of equations for correlation functions

First, we deduce equations for the correlations functions (1.4.5) in the framework of canonical ensemble. Differentiating the left and right-hand side of (1.4.5) with respect to time, using the Liouville equation (1.3.13) and the boundary condition and integrating over admissible configurations, one obtains

$$\begin{aligned}
&\frac{\partial F_s^{(N)}(t, x_1, \dots, x_s)}{\partial t} \\
&\quad = -\sum_{i=1}^{s} p_i \cdot \frac{\partial}{\partial q_i} F_s^{(N)}(t, x_1, \dots, x_s) \\
&\quad + a^2 \sum_{i=1}^{s} \int dp_{s+1} \int_{S_+^2} d\eta\eta \cdot (p_i - p_{s+1}) \\
&\qquad \times \Big[F_{s+1}^{(N)}(t, x_1, \dots, q_i, p_i^*, \dots, x_s, q_i - a\eta, p_{s+1}^*) \\
&\qquad - F_{s+1}^{(N)}(t, x_1, \dots, q_i, p_i, \dots, x_s, q_i + a\eta, p_{s+1}) \Big], \quad 1 \le s \le N, \quad (1.4.9)
\end{aligned}$$

with the boundary condition according to which, in the term

$$-\sum_{i=1}^{s} p_i \cdot \frac{\partial}{\partial q_i} F_s^{(N)}(t, x_1, \dots, x_s),$$

for $q_i - q_j = a\eta$ and $\eta \in S_+^2(\eta | \eta \cdot (p_i - p_j) \ge 0)$ momenta p_i, p_j should be replaced by p_i^*, p_j^*, and for $q_i - q_j = a\eta$, $\eta \in S_-^2(\eta | \eta \cdot (p_i - p_j) \le 0)$ momenta p_i, p_j do not change. We also have the initial condition

$$F_s^{(N)}(t, x_1, \dots, x_s)|_{t=0} = F_s^{(N)}(0, x_1, \dots, x_s), \quad 1 \le s \le N.$$

Note the these boundary conditions are of great importance; they follow from the boundary conditions of the Liouville equation (1.3.13) and correspond to the collision of hard spheres when spheres touch each other.

Note that the correlation functions $F_s^{(N)}(t, x_1, \dots, x_s)$ are equal to zero on the set of forbidden configurations $W_s((q)_s)$, i.e., $|(q_i - q_j)| < a$ for at least one pair $i, j \in \{1, \dots, s\}$.

We pass formally to the thermodynamic limit as $N \to \infty$ and suppose that, in some sense,

$$\lim_{N \to \infty} F_s^{(N)}(t, x_1, \dots, x_s) = F_s(t, x_1, \dots, x_s), \quad s \ge 1.$$

Then one obtains the limiting hierarchy of equations known as the BBGKY (Bogolyubov–Born–Green–Kirkwood–Yvon) hierarchy, namely,

$$\frac{\partial F_s(t, x_1, \ldots, x_s)}{\partial t}$$
$$= -\sum_{i=1}^{s} p_i \cdot \frac{\partial}{\partial q_i} F_s(t, x_1, \ldots, x_s)$$
$$+ a^2 \sum_{i=1}^{s} \int dp_{s+1} \int_{S_+^2} d\eta \eta \cdot (p_i - p_{s+1})$$
$$\times \Big[F_{s+1}(t, x_1, \ldots, q_i, p_i^*, \ldots, x_s, q_i - a\eta, p_{s+1}^*)$$
$$- F_{s+1}(t, x_1, \ldots, q_i, p_i, \ldots, x_s, q_i + a\eta, p_{s+1}) \Big], \quad s \geq 1 \qquad (1.4.10)$$

with the initial conditions $F_s(t, x_1, \ldots, x_s)|_{t=0} = F_s(0, x_1, \ldots, x_s)$ and the same boundary conditions as for (1.4.9).

It is easy to check that the correlation functions (1.4.8) defined in the framework of grand canonical ensemble satisfy the same hierarchy (1.4.10). Details of the derivation of the BBGKY hierarchy and justification of existence of the thermodynamic limit can be found in our, with V. Gerasimenko and P. Malyshev, books [PGM3].

1.4.3 Solution of the BBGKY hierarchy in the space of summable functions

We denote by $F(t)$ the sequence of correlation functions

$$F(t) = \big(F_1(t, x_1), \ldots, F_s(t, x_1, \ldots, x_s), \ldots\big)$$

and represent hierarchy (1.4.10) in the abstract form of an evolution equation for $F(t)$:

$$\frac{dF(t)}{dt} = \mathcal{H}F(t) + \mathcal{A}F(t), \qquad (1.4.11)$$
$$F(t)|_{t=0} = F(0),$$

where the operators $\mathcal{H}$ and A are defined by the first and the second term on the right-hand side of (1.4.10).

Consider the Cauchy problem for (1.4.10)–(1.4.11) in the Banach space L^1 which consists of sequences of symmetric summable functions defined on the phase space and equal to zero on forbidden configurations

$$f = \big(f_1(x_1), \ldots, f_s(x_1, \ldots, x_s), \ldots\big) \qquad (1.4.12)$$

and with obvious linear operations. The functions $f_s(x_1, \ldots, x_s)$ belong to the space $L^1(\mathbb{R}^{3s} \times \mathbb{R}^{3s} \setminus W_s) = L_s^1$ with the norm

$$\|f_s\| = \int |f_s((x)_s)| d(x)_s,$$

and sequences f (1.4.12) belong to the direct sum of spaces L_s^1:

$$L^1 = \sum_{s=1}^{\infty} \oplus L_s^1. \tag{1.4.13}$$

Denote by $S(-t)$ the direct sum of the operators $S_s(-t)$, i.e.,

$$S(-t) = \sum_{s=1}^{\infty} \oplus S_s(-t). \tag{1.4.14}$$

It is easy to check that the operators $S(-t), -\infty < t < \infty$, form a strongly continuous group isometric in L^1, $S(t_1 + t_2) = S(t_1)S(t_2) = S(t_2)S(t_1)$ for arbitrary t_1 and t_2. Its infinitesimal operator $\mathcal{H}$ is the direct sum of the operators $\mathcal{H}_s$:

$$\mathcal{H} = \sum_{s=1}^{\infty} \oplus \mathcal{H}_s. \tag{1.4.15}$$

We introduce an operator b which is an analog of the operator of annihilation in quantum field theory:

$$\left(\int dx f\right)_s (x_1, \dots, x_s) = \int f_{s+1}(x_1, \dots, x_s, x_{s+1}) dx_{s+1}. \tag{1.4.16}$$

The operator b is bounded in L^1 and

$$\left\| \int dx \right\| = 1.$$

We introduce the operator of evolution of hierarchy (1.4.10) as follows:

$$U(t) = e^{\int dx} S(-t) e^{-\int dx}, \quad -\infty < t < \infty. \tag{1.4.17}$$

It is a bounded, strongly continuous group in L^1,

$$U(t_1 + t_2) = U(t_1)U(t_2) = U(t_2)U(t_1) \tag{1.4.18}$$

for arbitrary t_1 and t_2, with infinitesimal generator that coincides with $\mathcal{H} + \mathcal{A}$ on an everywhere dense set L_0^1 consisting of finite sequences of functions $f_s(x_1, \dots, x_s) \in L_0^1(\mathbb{R}_s \times \mathbb{R}_s \setminus W_s)$.

According to well-known theorems from functional analysis, *the Cauchy problem for the BBGKY hierarchy (1.4.10)–(1.4.11) has a unique solution in* L^1, which can be represented in the form.

$$F(t) = U(t)F(0) = e^{\int dx} S(-t) e^{-\int dx} F(0). \tag{1.4.19}$$

This solution is strong for initial data $F(0) \subset L^1_0$ and is a generalized one for arbitrary $F(0) \subset L^1$.

For details of the proof, see the monographs [CGP] and [PGM3]. The solution of the BBGKY hierarchy (1.4.11) in L^1 describes only states of finite systems because the average number of particles in the state $F(t)$ is finite:

$$\bar{N} = \int F_1(t,x)dx = \int F_1(0,x)dx < \infty.$$

In order to describe states of infinite systems (with infinite $\bar{N}$), one should take initial data $F(0)$ from spaces that correspond to the perturbation of the equilibrium states with some temperature and density.

1.4.4 Solution of the BBGKY hierarchy

Denote by $E_{\xi,\beta}$ the Banach space consisting of sequences f of symmetric functions $f_s(x_1,\dots,x_s)$, $s \geq 1$, defined on the s-particle phase space and equal to zero on the set of forbidden configurations

$$f = \big(f_1(x_1),\dots, f_s(x_1,\dots,x_s),\dots\big)$$

with the norm

$$\|f\| = \sup_{s\geq 1} \frac{1}{\xi^s} \sup_{(x)_s} |f_s((x)_s)| e^{\beta \sum\limits_{i=1}^{s} \frac{p_i^2}{2}}, \tag{1.4.20}$$

where $\xi > 0$, $\beta > 0$, and β can be considered as inverse temperature of equilibrium states.

If one substitutes $F(0)$ from the space $E_{\xi,\beta}$ in (1.4.19), then, at first glance, $F(t)$ is meaningless. Indeed, the operator b is well defined in L^1, but $F(0)$ from $E_{\xi,\beta}$ does not belong to L^1 because $f_s((x)_s)$ are bounded with respect to $(q)_s$ and exponentially decreasing with respect to squared momenta:

$$\|f_s((x)_s)\| = \|f\|\xi^s e^{-\beta \sum\limits_{i=1}^{s} \frac{p_i^2}{2}}, \quad s \geq 1. \tag{1.4.21}$$

This assertion becomes obvious if one represents (1.4.19) componentwise, namely

$$F_s(t,(x)_s) = \sum_{n=0}^{\infty} \sum_{k=0}^{n} \frac{(-1)^k}{(n-k)!k!} \int S_{s+n-k}\big(-t,(x)_s, x_{s+1},\dots,x_{s+n-k}\big)$$
$$\times F_{s+n}\big(0,(x)_s, x_{s+1},\dots,x_{s+n}\big)dx_{s+1}\dots dx_{s+n}. \tag{1.4.22}$$

The initial functions $F_{s+n}\big(0,(x)_s, x_{s+1},\dots,x_{s+n}\big)$ satisfy inequality (1.4.21). The same inequality is satisfied by the functions

$$S_{s+n-k}\big(-t,(x)_s, x_{s+1},\dots,x_{s+n-k}\big) F_s\big(t,(x)_s, x_{s+1},\dots,x_{s+n-k}\big)$$

because $\sum_{i=1}^{s+n-k} \frac{p_i^2}{2}$ is invariant with respect to the dynamics of hard spheres. Therefore, at first glance, the integrals in (1.4.22) with respect to $q_{s+1}, \dots, q_{s+n}$ over admissible configurations are divergent.

In fact, using the fine cluster property of operators of evolution of hard spheres $S_{s+n-k}\big(-t,(x)_s, x_{s+1}, \dots, x_{s+n-k}\big)$, $0 \le k \le n$, we proved that, for fixed $(x)_s$ and fixed $p_{s+1}, \dots, p_{s+n}$, the expression

$$\sum_{k=0}^{n} \frac{(-1)^k}{(n-k)!k!} S_{s+n-k}\big(-t,(x)_s, x_{s+1}, \dots, x_{s+n-k}\big) \times F_{s+n}\big(0,(x)_s, x_{s+1}, \dots, x_{s+n}\big)$$

is different from zero in a bounded domain

$$V_{s+n}\big(t,(x)_s, p_{s+1}, \dots, p_{s+n}\big)$$

with respect the to variables $q_{s+1}, \dots, q_{s+n}$ (*domain of interaction*). Using this cluster property, one can prove that series (1.4.22) are uniformly convergent with respect to $(x)_s$ on compacta and for a finite interval of time $|t| < t_0$. For some special initial data $F(0) \subset E_{\xi,\beta}$, namely for $F(0)$ which are local perturbations of equilibrium states of a system of hard spheres, solutions (1.4.22) can be continued on to an arbitrary interval of time. Details of the proof can be found in books [CGP, PGM3].

In what follows, we will use the representation of solution (1.4.19) or (1.4.22) by series of iterations. Namely

$$F(t) = \sum_{n=0}^{\infty} \int_0^t dt_1 \int_0^{t_1} dt_2 \dots \int_0^{t_{n-1}} dt_n S(-t)S(t_1)\mathcal{A}S(-t_1)\dots S(t_n)\mathcal{A}S(-t_n)F(0),$$

or componentwise

$$\begin{aligned} F_s(t,(x)_s) = \sum_{n=0}^{\infty} a^{2n} \int_0^t dt_1 \int_0^{t_1} dt_2 \dots \int_0^{t_{n-1}} dt_n S_s(-t,(x)_s) S_s(t_1,(x)_s) \\ \times \sum_{i=1}^{s} \int dx_{s+1} \int_{S_+^2} d\eta_{is+1} \eta_{is+1} \cdot (p_i - p_{s+1}) \\ \times \Big[\delta(q_i - q_{s+1} - a\eta_{is+1}) S_{s+1}(-t_1,(x)_{s+1}^*) \\ - \delta(q_i - q_{s+1} + a\eta_{is+1}) S_{s+1}(-t_1,(x)_{s+1})\Big] \dots \\ \times S_{s+n-1}(t_{n-1},(x)_{s+n-1}) \dots \\ \times \sum_{i=1}^{s+n-1} \int dx_{s+n} \int_{S_+^2} d\eta_{is+n} \eta_{is+n} \cdot (p_i - p_{s+n}) \end{aligned}$$

$$\times \Big[\delta(q_i - q_{s+n} - a\eta_{is+n})S_{s+n}(-t_n,(x)^*_{s+n})$$

$$- \delta(q_i - q_{s+n} + a\eta_{is+n})S_{s+n}(-t_n,(x)_{s+n})\Big]$$

$$\times F_{s+n}(0,(x)_{s+n}), \tag{1.4.23}$$

where $(x)^*_{s+n} = (x_1,\dots,x^*_i,\dots,x_s,x^*_{s+1})$ in the i-th term and $S_s(\pm,(x)_s)$ is the operator of shift along the trajectory $X(\pm t,(x)_s)$ of s particles with initial data $(x)_s$ at $t=0$.

Note that, due to δ-functions, the integration in (1.4.23) is carried out over hypersurfaces of lower dimension in the phase space with respect to the variables $x_{s+1},\dots,x_{s+n}$. But the operators

$$S_{s+i}(\pm t,(x)_{s+i})$$

are well defined only outside certain hypersurfaces of lower dimension. Nevertheless, if one also takes into account the variables $t_1,\dots,t_n$, $\eta_{is+1},\dots,\eta_{is+n}$, then the integrand in (1.4.23) is well defined with respect to all variables of integration. The details can be found in books [CGP, PMG3].

It can be proved that the series of iterations is convergent uniformly with respect to $(x)_s$ on compacta and t on a finite interval $|t| < t_0$ for initial data $F(0) \in E_{\xi,\beta}$. For initial data that are local perturbations of equilibrium states for a system of hard spheres, solution (1.4.23) can be continued for arbitrary time $-\infty < t < \infty$. The series (1.4.23) is convergent for arbitrary time t for initial data from the space $\tilde{E}_{\xi,\beta}$, which consists of sequences of functions exponentially decreasing with respect to the squared momenta and coordinates. The proof of this statement can be found in books [CGP, PMG3, CIP], or in Chapter 4 of the present book.

Series (1.4.22) represent a *mild solution* of the BBGKY hierarchy (1.4.11). This means that it is a solution of the integral equation

$$F(t) = S(-t)F(0) + \int_0^t S(-t+\tau)F(\tau)d\tau, \tag{1.4.24}$$

which is equivalent to hierarchy (1.4.11) on a formal level.

1.4.5 BBGKY hierarchy with nonstandard normalization

Consider N particles (canonical ensemble) and let

$$D_N(t,x_1,\dots,x_N),$$

as usual, be the distribution function normalized to one. We introduce correlation functions by the formula

$$F^{(N)}_s(t,x_1,\dots,x_s) = \int D_N(t,x_1,\dots,x_s,x_{s+1},\dots,x_N)dx_{s+1}\dots dx_N. \tag{1.4.25}$$

Such a kind of correlation functions is used in many papers and books, for example, see [CIP].

The corresponding hierarchy can be derived by analogy with hierarchy (1.4.9):

$$\frac{\partial F_s^{(N)}(t,x_1,\dots,x_s)}{\partial t}$$
$$= -\sum_{i=1}^{s} p_i \cdot \frac{\partial}{\partial q_i} F_s^{(N)}(t,x_1,\dots,x_s)$$
$$+ (N-s)a^2 \sum_{i=1}^{s} \int dp_{s+1} \int_{S_+^2} d\eta\eta \cdot (p_i - p_{s+1})$$
$$\times \Big[F_{s+1}^{(N)}(t,x_1,\dots,q_i,p_i^*,\dots,x_s,q_i - a\eta, p_{s+1}^*)$$
$$- F_{s+1}^{(N)}(t,x_1,\dots,q_i,p_i,\dots,x_s,q_i + a\eta, p_{s+1})\Big], \quad 1 \le s \le N. \tag{1.4.26}$$

Again one has the known boundary conditions and initial data $F_s^{(N)}(t,(x)_s)|_{t=0} = F_s^{(N)}(0,(x)_s)$.

Note that the average of an s-particle observable $\varphi_N(x_1,\dots,x_N)$ over the state $D_N(t,x_1,\dots,x_N)$ (1.4.3) and the correlation function (1.4.25) are expressed by the formula

$$(D_N(t),\varphi_N) = \int D_N(t,x_1,\dots,x_N)\varphi_N(x_1,\dots,x_N)dx_1\dots dx_N$$
$$= \frac{N(N-1)-(N-s+1)}{s!} \int F_s^{(N)}(t,x_1,\dots,x_s)$$
$$\times \varphi_s(x_1,\dots,x_s)dx_1\dots dx_s. \tag{1.4.27}$$

It seems to us that the correlation functions (1.4.25) have the following disadvantage: The equilibrium correlation functions tend to zero in the thermodynamic limit. Suppose, as usual, that the N-particle system is contained in a finite domain Λ with volume V. The thermodynamic limit means that the number of particles N tends to infinity, the volume V tends to infinity (Λ in the limit becomes $\mathbb{R}^3$ without forbidden configurations), but $\frac{N}{V} = \frac{1}{v}$ is a constant density.

The normalized equilibrium distribution function is

$$D_{N,\Lambda}(x_1,\dots,x_N)$$
$$= Q(N,V)^{-1} \exp\Big(-\beta \sum_{i=1}^{N} \frac{p_i^2}{2m}\Big) \prod_{i<j=1}^{N} \Theta(|q_i - q_j| - a), \tag{1.4.28}$$

$$Q(N,V) = \int \exp\Big(-\beta \sum_{i=1}^{N} \frac{p_i^2}{2}\Big) \prod_{i<j=1}^{N} \Theta(|q_i - q_j| - a) \prod_{i=1}^{N} \chi_\Lambda(q_i) dx_1 \dots dx_N,$$

where $\chi_\Lambda(q)$ is the characteristic function of the domain Λ, $\Theta(t) = 1$ for $t > 0$, and $\Theta(t) = 0$ for $t < 0$, $t \in \mathbb{R}^1$. It is easy to see that $Q(N,V) \sim V^N$, and, therefore, $D_{N,\Lambda}(x_1,\dots,x_N)$ tends to zero in the thermodynamic limit as V^{-N} and $F_s^{(N)}(t,x_1,\dots,x_s)$, $s \geq 1$, tend to zero as V^{-N+s}.

Note that the factor V^{-N+s} cancels with the factor $N(N-1)\dots(N-s+1)$ in the correlation functions defined according to (1.4.5) because $\frac{N}{V} = \frac{1}{v}$ in the canonical ensemble.

1.5 Justification of the Boltzmann–Grad limit

1.5.1 Definition of the Boltzmann–Grad limit

Consider the BBGKY hierarchy (1.4.10), namely,

$$\begin{aligned}
\frac{\partial F_s^a(t,x_1,\dots,x_s)}{\partial t} &= -\sum_{i=1}^{s} p_i \cdot \frac{\partial}{\partial q_i} F_s^a(t,x_1,\dots,x_s) \\
&\quad + a^2 \sum_{i=1}^{s} \int dp_{s+1} \int_{S_+^2} d\eta\, \eta \cdot (p_i - p_{s+1}) \\
&\quad \times \big[F_{s+1}^a(t,x_1,\dots,q_i,p_i^*,\dots,x_s,q_i - a\eta, p_{s+1}^*) \\
&\quad - F_{s+1}^a(t,x_1,\dots,q_i,p_i,\dots,x_s,q_i + a\eta, p_{s+1})\big], \quad s \geq 1,
\end{aligned}$$

with the initial data $F_s^a(t,x_1,\dots,x_s)|_{t=0} = F_s^a(0,x_1,\dots,x_s)$, $s \geq 1$ and the boundary condition according to which, for $q_i - q_j = a\eta$, $\eta \cdot (p_i - p_j) \geq 0$, $i,j \in \{1,\dots,s\}$, the momenta p_i, p_j in $\sum_{i=1}^{s} p_i \cdot \frac{\partial}{\partial q_i} F_s^a(t,x_1,\dots,x_s)$ should be replaced by p_i^*, p_j^*. Analogous boundary conditions are already taken into account in the second term on the right-hand side of the hierarchy. The functions $F_s^a(t,x_1,\dots,x_s)$ are equal to zero on the forbidden configurations W_s.

(Instead of $F_s(t,(x)_s)$ we have used the denotation $F_s^a(t,(x)_s)$ in order to indicate that the correlation functions correspond to a system of hard spheres with diameter a.)

We introduce the sequence of renormalized correlation functions

$$\begin{aligned}
\tilde{F}^a(t) &= (a^2 F_1^a(t,x),\dots,a^{2s} F_s^a(t,x_1,\dots,x_s),\dots) \\
&= (\tilde{F}_1^a(t,x),\dots,\tilde{F}_s^a(t,x_1,\dots,x_s),\dots). \qquad (1.5.1)
\end{aligned}$$

For sequence (1.5.1), one obtains the following hierarchy:

$$\frac{\partial \tilde{F}_s^a(t,x_1,\dots,x_s)}{\partial t} = -\sum_{i=1}^{s} p_i \cdot \frac{\partial}{\partial q_i} \tilde{F}_s^a(t,x_1,\dots,x_s)$$

$$
\begin{aligned}
&+\sum_{i=1}^{s}\int dp_{s+1}\int_{S_+^2} d\eta\,\eta\cdot(p_i-p_{s+1})\\
&\times[\tilde{F}_{s+1}^a(t,x_1,\dots,q_i,p_i^*,\dots,x_s,q_i-a\eta,p_{s+1}^*)\\
&-\tilde{F}_{s+1}^a(t,x_1,\dots,q_i,p_i,\dots,x_s,q_i+a\eta,p_{s+1})] \qquad (1.5.2)
\end{aligned}
$$

with the initial data

$$\tilde{F}_s^a(t,x_1,\dots,x_s)|_{t=0}=a^{2s}F_s(0,x_1,\dots,x_s)=\tilde{F}_s(0,x_1,\dots,x_s)$$

and the same boundary conditions as for the correlation functions $F_s(t,x_1,\dots,x_s)$. Of course, $\tilde{F}_s(t,x_1,\dots,x_s)$ are equal to zero on W_s.

Now suppose that the whole sequence $\tilde{F}(t)$ of correlation functions has a limit as $a\to 0$ in some sense, namely,

$$\lim_{a\to 0}\tilde{F}_s^a(t,x_1,\dots,x_s)=F(t,x_1,\dots,x_s),\quad s\geq 1, \qquad (1.5.3)$$

and that this limiting correlation functions satisfy the hierarchy

$$
\begin{aligned}
\frac{\partial F_s(t,x_1,\dots,x_s)}{\partial t}&=-\sum_{i=1}^{s}p_i\cdot\frac{\partial}{\partial q_i}F_s(t,x_1,\dots,x_s)\\
&+\sum_{i=1}^{s}\int dp_{s+1}\int_{S_+^2} d\eta\,\eta\cdot(p_i-p_{s+1})\\
&\times[F_{s+1}(t,x_1,\dots,q_i,p_i^*,\dots,x_s,q_i,p_{s+1}^*)\\
&-F_{s+1}(t,x_1,\dots,q_i,p_i,\dots,x_s,q_i,p_{s+1})],\quad s\geq 1, \qquad (1.5.4)
\end{aligned}
$$

$$F_s(t,x_1,\dots,x_s)|_{t=0}=F_s(0,x_1,\dots,x_s),$$

which can be obtained from hierarchy (1.5.2) by putting $a=0$ in it. One has the following boundary conditions for hierarchy (1.5.4): *If* $q_i-q_j=0$ *and* $\eta\cdot(p_i-p_j)\geq 0$, *then the momenta* p_i,p_j *should be replaced by* p_i^*,p_j^*, $i,j\in\{1,\dots,s\}$. These boundary conditions directly follow from the boundary conditions for hierarchy (1.5.2) when $a\to 0$, and the point $q_i-a\eta$ tends to the point q_i for arbitrary $\eta\in S_+^2$.

Note that the unit vector η in the boundary conditions of the BBGKY hierarchy (1.5.2) is uniquely defined by the vectors q_i and q_j, namely $\eta=\frac{1}{a}(q_i-q_j)$. In hierarchy (1.5.4), the vector η is an arbitrary unit vector from $S_+^2(\eta|\eta\cdot(p_i-p_j)\geq 0)$. We will interpret it as a random vector uniformly distributed on S_+^2. This is the principal difference between hierarchies (1.5.2) and (1.5.4). Note that the set of forbidden configuration W_s is empty for $F_s(t,x_1,\dots,x_s)$.

The above-described limit as the diameter a of hard spheres tends to zero and the renormalized correlation functions (1.5.1) tend to the limit correlation functions (1.5.3) is known as the *Boltzmann–Grad limit*, and the limiting hierarchy (1.5.4) is known as the *proper Boltzmann hierarchy*. In connection with the random character of the vectors η, we will sometimes say that the obtained hierarchy (1.5.4) is the *stochastic Boltzmann hierarchy*, or the *proper stochastic Boltzmann hierarchy*.

Thus, we have derived a new kind of a stochastic hierarchy with boundary conditions on the hyperplanes of lower dimension $q_i = q_j$, $i, j \in \{1, \dots, s\}$, according to which, for $\eta_{ij} \cdot (p_i - p_j) \geq 0$, the momenta p_i, p_j should be replaced by p_i^*, p_j^* with random vectors $\eta_{ij} \subset S_+^2$.

Note that hierarchy (1.5.4) had usually been considered without boundary conditions at $q_i - q_j = 0$, which simply had been neglected. This had been connected with an erroneous version of the BBGKY hierarchy (1.4.10) for systems of hard spheres without boundary conditions. Unfortunately, in all books devoted to the BBGKY hierarchy of systems of hard spheres, excluding ours with Gerasimenko and Malyshev [PGM3] (see also [CGP]), these boundary conditions had been neglected and omitted. We will say that hierarchy (1.5.4) without boundary conditions is the *ordinary Boltzmann hierarchy*. In fact, the ordinary Boltzmann hierarchy is not the Boltzmann–Grad limit of the BBGKY hierarchy for systems of hard spheres (1.4.10) with boundary conditions. It is the hierarchy for the correlation functions $F_s(t, x_1, \dots, x_s)$ equal to the product of solutions of the Boltzmann equation. Namely, let $F_1(t, x_1)$ be the solution of the Boltzmann equation for hard spheres

$$\frac{\partial F_1(t,x_1)}{\partial t} = - p_1 \frac{\partial}{\partial q_1} F_1(t,x_1) + \int dp_2 \int_{S_+^2} d\eta \eta \cdot (p_1 - p_2) \times [F_1(t,x_1^*)F_1(t,x_2^*) - F_1(t,x_1)F_1(t,x_2)], \tag{1.5.5}$$

$$x_1 = (q_1, p_1), \quad x_2 = (q_1, p_2), \quad x_1^* = (q_1, p_1^*), \quad x_2^* = (q_1, p_2^*),$$

$$F_1(t,x_1)|_{t=0} = F_1(0,x_1).$$

Then the correlation functions

$$F_s(t, x_1, \dots, x_s) = F_1(t, x_1) \dots F_1(t, x_s), \quad s \geq 1, \tag{1.5.6}$$

are solutions of the ordinary Boltzmann hierarchy that coincides with hierarchy (1.5.4) but without the boundary condition and the following initial data:

$$F_s(t, x_1, \dots, x_s)|_{t=0} = F_1(0, x_1) \dots F_1(0, x_s). \tag{1.5.7}$$

1.5.2 Auxiliary lemmas

In this subsection, we describe the set on which the operators of evolution of systems of hard spheres $S_s^a(t)$ coincide with the operator of evolution of systems of free noninteracting particles $S_s^0(t)$. We will follow our (with Gerasimenko) papers [GeP1, GeP2, GeP3, PeG1, PeG2, PeG3, PeG4] and books [CGP, PGM3].

Let $\mathbb{R}^s_{p_o}$ be a ball of radius $p_o < \infty$ in the momentum subspace of a system of s hard spheres. Let $K^s_0 \in \mathbb{R}^{3s}\backslash W_s$ be an arbitrary compact set of the admissible configurations (positions) of a system of s hard spheres that consists of those configurations of hard spheres for which $|q_i - q_j| \geq a + a_0(a)$, $i, j \in \{1, \ldots, s\}$, $i \neq j$. Here $a_0(a)$ is a fixed function such that $a_0(a) \to 0$ and $a/a_0(a) \to 0$ as $a \to 0$.

Let us denote by $\mathfrak{M}_s \subset K^s_0 \times \mathbb{R}^s_{p_o}$ the set whose complement $(K^s_0 \times \mathbb{R}^s_{p_0})\backslash\mathfrak{M}_s$ consists of initial data of a system of s hard spheres such that the particles do not interact during the evolution for $t \in \mathbb{R}^1$, i.e., their trajectories coincide with those of free point particles:

$$(K^s_0 \times \mathbb{R}^s_{p_0})\backslash\mathfrak{M}_s = \Big\{(x_1, \ldots, x_s) \in K^s_0 \times \mathbb{R}^s_{p_0} | X_i(t, x_1, \ldots, x_s) \\ = X^0_i(t, x_1, \ldots, x_s) = X^0_i(t, x_i) = (q_i + p_i t, p_i), \\ i \in \{1, \ldots, s\}, \ \ t \in \mathbb{R}^1\Big\}. \tag{1.5.8}$$

In other words, s hard spheres can interact only if their initial data $(x_1, \ldots, x_s)$ belong to $\mathfrak{M}_s$.

We now describe the set $\mathfrak{M}_s$ and estimate its volume (measure).

Lemma 1.1 (on "sweeping up"). *The set $\mathfrak{M}_s$ consists of cones in the momentum space of i-th and j-th hard spheres with axis along the vector $q_i - q_j$ with base radius $4p_0\frac{a}{a_0(0)}$, height $2p_0$, and volume*

$$V_{ij} \leq \frac{32}{3}\pi p_0^3\Big(\frac{a}{a_0(0)}\Big)^2 \frac{4}{3}\pi p_0^3.$$

Of s hard spheres, we can form $\frac{1}{2}s(s-1)$ pairs and the corresponding cones with the common volume

$$V_s \leq \frac{1}{2}s(s-1)V_{ij} = \mathrm{const}(s, p_0)\Big(\frac{a}{a_0(a)}\Big)^2, \tag{1.5.9}$$

where $\mathrm{const}(s, p_0)$ depends on s and p_0.

The details of the proof of this lemma can be found in above mentioned our papers and books, but the idea is very simple. Two hard spheres can interact only if their relative momenta belong to the cone described above. The volume of the cone increases if the distance between the centers of hard spheres decreases and increases if the diameters of hard spheres increase. The estimate for V_s can be calculated using a very simple geometric picture.

From the definition of the set $\mathfrak{M}_s$ we obtain the following lemma:

Lemma 1.2. *The evolution operator $S^a_s(t, x_1, \ldots, x_s)$ of a system of s hard spheres coincides with the evolution operator $S^0_s(t, x_1, \ldots, x_s)$ of a non-interacting point-particle system on the set $K^s_0 \times \mathbb{R}^s_{p_0}\backslash\mathfrak{M}_s$, i.e., if initial data $(x_1, \ldots, x_s)$ belong to $K^s_0 \times \mathbb{R}^s_{p_0}\backslash\mathfrak{M}_s$.*

Thus, we know in what sense the group $S^a(t)$ of the evolution operators of systems of hard spheres coincides with the group of the operator $S^0(t)$ of a system of non-interacting point particles.

1.5.3 Convergence of solutions of the BBGKY hierarchy of a system of hard spheres to solutions of the ordinary Boltzmann hierarchy in the Boltzmann–Grad limit

Consider a solution of the BBGKY hierarchy for systems of hard spheres represented by the series of iterations (1.4.23), which converges on a finite interval of time $0 \le t \le t_0$ for initial data from the space $E_{\xi,\beta}$. Series (1.4.23) can easily be transformed into series representing a solution for the hierarchy of renormalized correlation functions (1.5.2).

Namely,

$$
\begin{aligned}
&\tilde{F}^a_s(t,(x)_s)\\
&= \sum_{n=0}^{\infty} \int_0^t dt_1 \int_0^{t_1} dt_2 \dots \int_0^{t_{n-1}} dt_n S^a_s(-t,(x)_s) S^a_s(t_1,(x)_s)\\
&\quad \times \sum_{i=1}^{s} \int dx_{s+1} \int_{S^2_+} d\eta_{is+1} \eta_{is+1} \cdot (p_i - p_{s+1})[\delta(q_i - q_{s+1} - a\eta_{is+1})\\
&\quad \times S^a_{s+1}(-t_1,(x)^*_{s+1}) - \delta(q_i - q_{s+1} + a\eta_{is+1}) S^a_{s+1}(-t_1,(x)_{s+1})]\\
&\quad \times \dots S^a_{s+n-1}(t_{n-1},(x)_{s+n-1}) \sum_{i=1}^{s+n-1} \int dx_{s+n} \int_{S^2_+} d\eta_{is+n}\\
&\quad \times \eta_{is+n} \cdot (p_i - p_{s+n})[\delta(q_i - q_{s+n} - a\eta_{is+n}) S^a_{s+n}(-t_n,(x)^*_{s+n})\\
&\quad - \delta(q_i - q_{s+n} + a\eta_{is+n}) S^a_{s+n}(-t_n,(x)_{s+n})] \tilde{F}^a_{s+n}(0,(x)_{s+n}),
\end{aligned}
\tag{1.5.10}
$$

$$
\tilde{F}^a_{s+n}(0,(x)_{s+n}) = a^{2(s+n)} F^a(0,(x)_{s+n}).
$$

(In (1.5.10), $S^a_s(\pm t,(x)_s)$ denote the operators of evolution of s hard spheres which were earlier denoted by $S_s(\pm t,(x)_s)$.)

Also consider a solution of the ordinary Boltzmann hierarchy (1.5.4) without boundary conditions. Then the operators $-\sum_{i=1}^{s} p_i \cdot \frac{\partial}{\partial q_i}$ without boundary conditions for $q_i = q_j$, $i,j \in \{1,\dots,s\}$ are the infinitesimal operators of the operator of evolution $S^0_s(-t,x_1,\dots,x_s)$ of systems of free point-particles. The solution of the ordinary Boltzmann hierarchy (1.5.4) can be represented by series of iterations analogous to (1.5.10), but with the operators $S^0_{s+i}(\pm t,x_1,\dots,x_{s+i})$ instead of the operators $S^a_{s+i}(\pm t,x_1,\dots,x_{s+i})$. Namely,

$$
\begin{aligned}
&F_s(t,(x)_s) \\
&\quad = \sum_{n=0}^{\infty} \int_0^t dt_1 \int_0^{t_1} dt_2 \dots \int_0^{t_{n-1}} dt_n S_s^0(-t,(x)_s) S_s^0(t_1,(x)_s) \\
&\qquad \times \sum_{i=1}^{s} \int dx_{s+1} \int_{S_+^2} d\eta_{is+1} \eta_{is+1} \cdot (p_i - p_{s+1}) \delta(q_i - q_{s+1}) \\
&\qquad \times [S_{s+1}^0(-t_1,(x)_{s+1}^*) - S_{s+1}^0(-t_1,(x)_{s+1})] \\
&\qquad \times \dots S_{s+n-1}^0(t_{n-1},(x)_{s+n-1}) \sum_{i=1}^{s+n-1} \int dx_{s+n} \int_{S_+^2} d\eta_{is+n} \\
&\qquad \times \eta_{is+n} \cdot (p_i - p_{s+n}) \delta(q_i - q_{s+n}) [S_{s+n}^0(-t_n,(x)_{s+n}^*) \\
&\qquad - S_{s+n}^0(-t_n,(x)_{s+n})] F_{s+n}(0,(x)_{s+n}). \qquad (1.5.11)
\end{aligned}
$$

Note that the operators $S_{s+n}^a(-t_n,(x)_{s+n}^*)$ and $S_{s+n}^0(-t_n,(x)_{s+n}^*)$, where $(x)_{s+n}^*$ in the i-th term means $(x_1,\dots,q_i,p_i^*,\dots,x_{s+n-1},q_{s+n},p_{s+n}^*)$, act on $\tilde{F}_{s+n}^a(0,(x)_{s+n}^*)$ and, correspondingly, on $F_{s+n}(0,(x)_{s+n}^*)$.

Suppose that

$$\lim_{a\to\infty} \tilde{F}_s^a(0,(x)_s) = F_s(0,(x)_s), \quad s \geq 1,$$

uniformly on $\mathbb{R}_{p_0}^{s+n} \times \mathbb{R}^{s(s+n)} \backslash W_{s+n}$. Our aim is to prove that

$$\lim_{a\to 0} \tilde{F}_s^a(t,(x)_s) = F_s(t,(x)_s)$$

uniformly on compacta from $K_0^s \times \mathbb{R}_{p_0}^s \backslash \mathfrak{M}_s$ on the time interval where series (1.5.10) and (1.5.11) converge.

The idea consists of the proof of the convergence of series (1.5.10) to series (1.5.11) term by term. We know that the operators $S_{s+i}^a(\pm t,(x)_{s+i})$ coincide with the operators $S_{s+i}^0(\pm t,(x)_{s+i})$ on $K_0^{s+i} \times \mathbb{R}_{p_0}^{s+i} \backslash \mathfrak{M}_{s+i}$. We can restrict ourselves to $K_0^s \times \mathbb{R}_{p_0}^s \backslash \mathfrak{M}_s$ with respect to the phase space of s particles with points $(x)_s$. But we cannot restrict ourselves to $K_0^{s+i} \times \mathbb{R}_{p_0}^{s+i} \backslash \mathfrak{M}_{s+i}$ with respect to the variables $(x_{s+1},\dots,x_{s+i})$, because $p_{s+i} \in \mathbb{R}^3$ and the positions q_{s+i} are shifted along the trajectories of systems of hard spheres and can take arbitrary values in $\mathbb{R}^3$ (on admissible configurations).

Nevertheless, this aim can be achieved by using Lemma 1.1 (on "sweeping up") and the fact that every term in (1.5.10) and (1.5.11) is represented by absolutely convergent integrals and that series (1.5.10) and (1.5.11) are uniformly convergent with respect to $(x)_s$ on compacta on admissible configurations and with respect to time on a finite interval.

By using the uniform convergence of both series, one can consider only a finite sum in (1.5.10) and (1.5.11). By using the absolute convergence of the integrals representing these finitely many terms, one can restrict the domain of integration with respect to momenta to $\mathbb{R}^{s+i}_{p_0}$. All q_{s+i} are replaced by $(q_1, \dots, q_s)$ according to δ-functions and are shifted by finite vectors because all p_i, $i \in (1, \dots, s+i)$, belong to $\mathbb{R}^{s+i}_{p_0}$.

The main obstacle to the application of Lemma 1.1 (on "sweeping up") is that some positions of particles with numbers $(s+1, \dots, s+i)$ do not satisfy the condition that the distance of pairs are greater than $a + a_0(a)$. It is necessary to estimate the measure (volume) of these dangerous sets of positions of particles.

By using very delicate estimates, we had been able to describe these dangerous sets, estimate their measure (volumes), and prove that their measures tend to zero as $a \to 0$. The detailed proof can be found in our papers [GeP1, GeP2, GeP3, PeG1, PeG2, PeG3, PeG4, PGM1, PGM2] and books [CGP, PGM3].

Thus, we can replace all operators of evolution of a system of hard spheres $S^a_{s+i}(\pm t, (x)_{s+i})$ in series (1.5.10) by the operators of evolution of free non-interacting point particles $S^0_{s+i}(\pm t, (x)_{s+i})$. Doing these procedures of replacements of $S^a_{s+i}(\pm t, (x)_{s+i})$ by $S^0_{s+i}(\pm t, (x)_{s+i})$, we make an error that tends to zero as $a \to 0$.

This means that

$$\lim_{a\to 0} \tilde{F}^a_s(t, (x)_s) = F_s(t, (x)_s) \tag{1.5.12}$$

uniformly with respect to $(x)_s$ and time on compacta from $K^s_0 \times \mathbb{R}^s_{p_0} \backslash \mathfrak{M}_s$ and on other finite time interval where series (1.5.10) and (1.5.11) converge.

Recall that these series converge on a finite interval of time for initial data $\tilde{F}^a(0, (x)_s)$, $F(0, (x)_s)$ from the space $E_{\xi,\beta}$ and for arbitrary time interval for initial data $\tilde{F}^a(0, (x)_s)$, $F(0, (x)_s)$ from the space $\tilde{E}_{\xi,\beta}$.

According to the definition, the set $\mathfrak{M}_s$ consists of cones with axis along the vectors $q_i - q_j$ that contain all vectors p_i, p_j for which hard spheres with numbers i and j collide $i, j \in \{1, \dots, s\}$. It is obvious that these cones degenerate into a vector parallel to the vectors $q_i - q_j$ as $a \to 0$. In this limit, hard spheres degenerate into points which can collide (interact) only if the vectors $p_i - p_j$ are parallel to the vectors $q_i - q_j$, i.e., $q_i - q_j = \tau(p_i - p_j)$, $\tau < t_0$.

The cones from the set $\mathfrak{M}_s$ contain all vectors $p_i - p_j$ such that, at the instants of collisions τ, the unit vectors

$$\frac{Q_i(\tau) - Q_j(\tau)}{|Q_i(\tau) - Q_j(\tau)|} = \eta_{ij}$$

take values from the entire semisphere $S^2_+(\eta_{ij} | \eta_{ij} \cdot (p_i - p_j) \geq 0)$. This means that, before the instant of collision τ, and as $a \to 0$, all trajectories degenerate into the unique trajectory of free point particles, such that $q_i - q_j = \tau(p_i - p_j)$, $0 \leq \tau \leq t$, and after collision we have the set of trajectories of free point particles with initial data

$(Q_i(\tau), p_i^*, Q_j(\tau), p_j^*)$, where

$$p_i^* = p_i - \eta_{ij}\, \eta_{ij} \cdot (p_i - p_j),$$
$$p_j^* = p_j + \eta_{ij}\, \eta_{ij} \cdot (p_i - p_j).$$

We can consider the vectors η_{ij} as random ones which are uniformly distributed on S_+^2. Thus, in the Boltzmann–Grad limit as $a \to 0$, the deterministic Hamiltonian dynamics reduces to the stochastic dynamics. In this stochastic dynamics, point particles move as free non-interacting ones until their positions coincide. After a collision, they again move as free point particles, but their momenta after a collision are defined by random unit vectors η_{ij}, and so on.

Note that the boundary conditions in the proper stochastic Boltzmann hierarchy contain these random vectors η_{ij}. Details about the stochastic dynamics will be presented in the next chapter.

1.5.4 Convergence of solutions of the BBGKY hierarchy of systems of hard spheres to solutions of the proper stochastic hierarchy in the Boltzmann–Grad limit

Now consider solutions of the proper stochastic hierarchy (1.5.4). It will be shown later that the operator of shift along the stochastic trajectory

$$S_N(-t) f_N(x) = f_N(X(-t,x)) = f_N(t,x) = f_N(t, x_1, \dots, x_N), \quad t > 0, \tag{1.5.13}$$

where $X(t,x)$ is the stochastic trajectory defined in the previous subsection, is well defined. The distribution function $f_N(t,x)$ satisfies the Itô–Liouville equation

$$\frac{\partial}{\partial t} f_N(t, x_1, \dots, x_N) = -\sum_{i=1}^{N} p_i \cdot \frac{\partial}{\partial q_i} f_N(t, x_1, \dots, x_N), \tag{1.5.14}$$

$$f_N(t,x)|_{t=0} = f_N(x)$$

with the following boundary condition: If $q_i = q_j$, $\eta \cdot (p_i - p_j) \geq 0$, then the momenta p_i, p_j on the right hand side of (1.5.14) should be replaced by p_i^*, p_j^*. This means that the operator

$$\mathcal{H}_N = -\sum_{i=1}^{N} p_i \cdot \frac{\partial}{\partial q_i}$$

with the above formulated boundary condition is the infinitesimal generator of the semigroup of operators $S_N(-t)$, $t > 0$.

It will be shown later that equation (1.5.14) can be represented in an equivalent form with δ-functions, but without boundary condition, as well as for hard spheres (1.3.10).

Solutions of the proper stochastic hierarchy can be represented by the following series of iterations:

$$
\begin{aligned}
F_s(t,(x)_s) = \sum_{n=0}^{\infty} \int_0^t dt_1 \int_0^{t_1} dt_2 \ldots \int_0^{t_{n-1}} & dt_n S_s(-t,(x)_s) S_s(t_1,(x)_s) \\
& \times \sum_{i=1}^{s} \int dx_{s+1} \int_{S_+^2} d\eta_{is+1} \eta_{is+1} \cdot (p_i - p_{s+1}) \delta(q_i - q_{s+1}) \\
& \times [S_{s+1}(-t_1,(x)_{s+1}^*) - S_{s+1}(-t_1,(x)_{s+1})] \\
& \times \ldots S_{s+n-1}(t_{n-1},(x)_{s+n-1}) \sum_{i=1}^{s+n-1} \int dx_{s+n} \int_{S_+^2} d\eta_{is+n} \\
& \times \eta_{is+n} \cdot (p_i - p_{s+n}) \delta(q_i - q_{s+n}) [S_{s+n}(-t_n,(x)_{s+n}^*) \\
& - S_{s+n}(-t_n,(x)_{s+n})] F_{s+n}(0,(x)_{s+n}). \qquad (1.5.15)
\end{aligned}
$$

It can be shown that series (1.5.15) is convergent as well as series (1.5.11).

Note that the operators $S_{s+i}(\mp t,(x)_{s+i})$ of shift along the stochastic trajectory differ from the operators $S_{s+i}^0(\mp t,(x)_{s+i})$ of shift along the trajectory of free noninteracting particles only on the hyperplanes $q_i - p_i\tau = q_j - p_j\tau$, $0 \le \tau \le t$, of lower dimension, $i, j \in \{1,\ldots,s+i\}$. Integration with respect to $q_{s+1},\ldots,q_{s+n}$ can be carried out by using δ-functions; therefore, all $q_{s+1},\ldots,q_{s+n}$ are expressed via $q_1,\ldots,q_s$. For fixed $q_1,\ldots,q_s$, the hyperplanes where stochastic particles interact are of lower dimension with respect to momenta $p_{s+1},\ldots,p_{s+n}$ and can be neglected in (1.5.15).

If the phase points $x_1,\ldots,x_s$ are also outside the hyperplanes $q_i - p_i\tau = q_j - p_j\tau$, $i, j \in \{1,\ldots,s\}$, then all operators $S_{s+i}(\mp t,(x)_{s+i})$, $i = 0,1,\ldots,n$, can be replaced by the operator $S_{s+i}^0(\mp t,(x)_{s+i})$. Thus, outside the hyperplanes $q_i - p_i\tau = q_j - p_j\tau$, $0 \le \tau \le t$, $i, j \in \{1,\ldots,s\}$, solutions (1.5.15) of the proper stochastic Boltzmann hierarchy coincide with solutions (1.5.11) of the ordinary Boltzmann hierarchy (1.5.11). Therefore, the stochastic Boltzmann hierarchy also has the chaos property, and the one-particle correlation function satisfies the Boltzmann equations (1.5.5).

In this respect, the stochastic Boltzmann hierarchy has the same chaos property as the ordinary Boltzmann hierarchy.

However, the proper stochastic Boltzmann hierarchy has two advantages.

The first is that the operators of the stochastic evolution $S_s(\pm t,(x)_s)$ are the Boltzmann–Grad limits of the operators of evolution of hard spheres $S_s^a(\mp t,(x)_s)$ in the entire phase space when the operators of evolution of free particles $S_s^0(\pm t,(x)_s)$ are the Boltzmann–Grad limits of $S^a(\mp t,(x)_s)$ only outside all hyperplanes $q_i - p_i\tau = q_j - p_j\tau$, $i, j \in \{1,\ldots,s\}$, $0 \le \tau \le t$.

The second is that the proper stochastic Boltzmann hierarchy can be derived from the stochastic dynamics by the same method as the BBGKY hierarchy for hard spheres can be derived from the Hamiltonian dynamics. It will also be shown that the well-known stochastic dynamics and hierarchy in the momentum space can be obtained by averaging over the space of position (coordinate) from our stochastic dynamics in the phase space.

Note that, for the ordinary Boltzmann hierarchy, the corresponding dynamics does not exist because the operator $-\sum_{i=1}^{s} p_i \cdot \frac{\partial}{\partial q_i}$ without boundary conditions corresponds to the dynamics of free noninteracting particles, and the integral operator on the right–hand side of the hierarchy is the same as for the stochastic hierarchy.

Chapter 2

Stochastic dynamics as the limit of the Hamiltonian dynamics of hard spheres

2.1 Introduction

The stochastic dynamics that corresponds to the Boltzmann equation and hierarchy has recently been proposed by D. Ya. Petrina, K. D. Petrina, and M. Lampis in [PeP1, PeP2, PeP3, LaPe1, LaPe2, LaPe3, LaPe4].

In the present chapter, we prove that the stochastic dynamics is a certain limit of averages over the sphere of the Hamiltonian dynamics of a system of hard spheres as their diameter tends to zero (the Boltzmann–Grad limit). We define the set (set of hypersurfaces) of interaction on which the stochastic dynamics differs from the Hamiltonian dynamics of free particles.

By using the concept of the set of interaction, we define the operator of evolution for the stochastic dynamics and its infinitesimal operator in the weak sense. We prove that the operator of evolution of stochastic particles and its infinitesimal operator are the limits of the averages of the operator of evolution of a system of hard spheres and its infinitesimal operator, respectively, over the sphere as its diameter tends to zero.

In this chapter, we present the rigorous derivation of the new concept of the stochastic dynamics of a system of point particles as the limit of the average over the sphere of the Hamiltonian dynamics of a system of hard spheres as their diameter tends to zero. We also define the action of the operator of evolution of stochastic particles on continuously differentiable functions as usual functions, i.e., numerically. The infinitesimal operator of the operator of evolution of stochastic particles is also defined in the sense of pointwise convergence. It is equal to the infinitesimal operator of the operator of evolution of free particles with a certain boundary condition or an equivalent one-dimensional δ-function. As a generalized function, its action coincides with the action of the infinitesimal operator defined in the weak sense.

Thus, two equivalent representations of the evolution operator and the corresponding infinitesimal operator are defined.

The crucial feature of this new concept of the stochastic dynamics and the associated notion of the average of an observable over states of a system of point particles governed by the stochastic dynamics is the fact that they take into account hypersur-

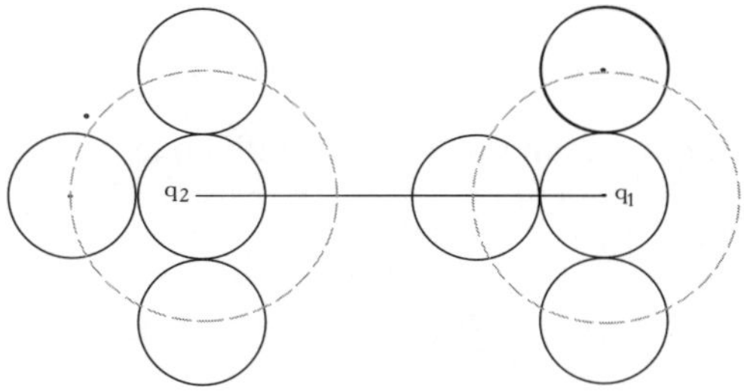

Fig. 1

faces of lower dimension than the phase space. In the traditional statistical mechanics, hypersurfaces of lower dimension are neglected. In subsequent chapters, we show that, in the solutions of the Boltzmann equation and the Boltzmann hierarchy, the same hypersurfaces of lower dimension are taken into account. Namely, the terms of the series of iterations for the Boltzmann equation, as well as for the Boltzmann hierarchy, can be represented via the integrals over the hypersurfaces of lower dimension on which the stochastic particles interact.

Thus, the new concept of the stochastic dynamics corresponds to the Boltzmann equation and the Boltzmann hierarchy.

2.2 Stochastic trajectories as the limit of the Hamiltonian trajectories of hard spheres

2.2.1 Hamiltonian trajectories of hard spheres

First, we consider two hard spheres with diameter a and mass 1. Denote by $(q_1, p_1) = x_1$ and $(q_2, p_2) = x_2$ the positions of their centers and momenta; x_1 and x_2 are their phase points.

We fix the initial momenta p_1, p_2 and consider the position q_1^0, q_2^0 such that the vector $q_1^0 - q_2^0$ is parallel to the vector $p_1 - p_2$ and $(p_1 - p_2) \cdot (q_1^0 - q_2^0) < 0$. Then, for given q_2^0, we consider the semisphere $q_2^0 - a\eta$, where $|\eta| = 1$, $\eta \subset S_-^2$ $(\eta \mid \eta \cdot (p_1 - p_2) < 0)$. As the initial position of the first sphere, we take the point q_1^0 and, as the initial positions of the second sphere, we take the points $q_2^0 - a\eta$, $\eta \subset S_-^2$ (Fig. 1).

We consider positive (increasing) time $t \geq 0$, and $t = 0$ is the initial time. It is obvious that, at the time

$$\tau = \frac{|q_1^0 - q_2^0|}{|p_1 - p_2|}, \tag{2.2.1}$$

the particles collide and touch each other at the point $q_1^0 - \frac{a}{2}\eta + p_1\tau$. After the elastic collision, their momenta are

$$p_1^* = p_1 - \eta\eta\cdot(p_1 - p_2), \quad p_2^* = p_2 + \eta\eta\cdot(p_1 - p_2), \quad \eta \subset S_-^2. \tag{2.2.2}$$

The corresponding Hamiltonian trajectory is defined as

$$\begin{aligned}
&Q_1(t) = q_1 + p_1 t, \quad P_1(t) = p_1,\\
&Q_2(t) = q_2 + p_2 t, \quad P_2(t) = p_2, \quad t \le \tau,\\
&Q_1(t) = q_1 + p_1\tau + p_1^*(t-\tau), \quad P_1(t) = p_1^*,\\
&Q_2(t) = q_2 + p_2\tau + p_2^*(t-\tau), \quad P_2(t) = p_2^*, \quad t > \tau,
\end{aligned} \tag{2.2.3}$$

for all $q_2 = q_2^0 - a\eta$, $\eta \subset S_-^2$, and fixed $q_1 = q_1^0$. Denote by

$$X^a(t) = \big(Q_1(t), P_1(t), Q_2(t), P_2(t)\big)$$

the trajectories (2.2.3) of two hard spheres.

2.2.2 Stochastic trajectories

We have defined by (2.2.3) the bunch of trajectories characterized by vectors $\eta \subset S_-^2$. Now let the diameter a tend to zero. Then, for $t \le \tau$, the entire bunch of trajectories (2.2.3) is shrunk to the single trajectory

$$\begin{aligned}
&Q_1(t) = q_1^0 + p_1 t, \quad P_1(t) = p_1,\\
&Q_2(t) = q_2^0 + p_2 t, \quad P_2(t) = p_2, \quad t \le \tau.
\end{aligned} \tag{2.2.4}$$

For $t > \tau$, the arbitrary trajectory (2.2.3) with fixed $\eta \subset S_-^2$ converges to the trajectory

$$\begin{aligned}
&Q_1(t) = q_1^0 + p_1\tau + p_1^*(t-\tau), \quad P_1(t) = p_1^*,\\
&Q_2(t) = q_2^0 + p_2\tau + p_2^*(t-\tau), \quad P_2(t) = p_2^*, \quad t \ge \tau,
\end{aligned} \tag{2.2.5}$$

with the same $\eta \subset S_-^2$.

We now define the stochastic trajectories for two point particles. We assume that particles move as free ones until their positions coincide:

$$\begin{aligned}
&Q_1(t) = q_1 + p_1 t, \quad P_1(t) = p_1,\\
&Q_2(t) = q_2 + p_2 t, \quad P_2(t) = p_2.
\end{aligned} \tag{2.2.6}$$

If their positions coincide at time τ, i.e.,

$$Q_1(\tau) = Q_2(\tau),$$

then they collide instantaneously and their momenta change jumpwise, namely,

$$P_1(\tau+0) = P_1(\tau) + \eta\,\eta\cdot(P_1(\tau) - P_2(\tau)) = p_1^*,$$
$$P_2(\tau+0) = P_2(\tau) + \eta\,\eta\cdot(P_1(\tau) - P_2(\tau)) = p_2^*, \tag{2.2.7}$$

if $\eta\cdot(p_1 - p_2) \le 0$, where $|\eta| = 1$, $\eta \subset S^2_-$. If the vector $\eta \subset S^2_+$, $\eta\cdot(p_1 - p_2) \ge 0$, then, after collisions, the momenta p_1, p_2 do not change. We assume that $\eta \subset S^2 = S^2_- \cup S^2_+$ is a random vector with the constant density of probability $\mathcal{X}(\eta) = 1/4\pi$, and the particles after collision may have vectors p_1^*, p_2^* with $\eta \subset S^2_-$ with the same constant probability. Thus, the stochasticity consists of the fact that, with the same probability, the particles after collision have the momenta p_1^*, p_2^* defined by (2.2.2) with a random vector $\eta \subset S^2_-$. For $t > \tau$, the particles again move as free ones:

$$Q_1(t) = q_1 + p_1\tau + p_1^*(t-\tau), \quad P_1(t) = p_1^*,$$
$$Q_2(t) = q_2 + p_2\tau + p_2^*(t-\tau), \quad P_2(t) = p_2^*, \quad t > \tau, \quad \eta \subset S^2_-,$$
$$Q_1(t) = q_1 + p_1 t, \quad P_1(t) = p_1,$$
$$Q_2(t) = q_2 + p_2 t, \quad P_2(t) = p_2, \quad t > \tau, \quad \eta \subset S^2_+. \tag{2.2.8}$$

It is obvious that the limit trajectories of hard spheres (as $a \to 0$) (2.2.4), (2.2.5) coincide with the stochastic trajectories (2.2.6)–(2.2.8) for the same $\eta \subset S^2_-$.

The stochastic dynamics is defined by the stochastic trajectories, whereas the Hamiltonian dynamics is defined by the Hamiltonian trajectories.

If $q_1 = q_2$ at initial time $t = 0$, then the stochastic trajectories are defined as follows:

$$Q_1(t) = q_1 + p_1^* t, \quad P_1(t) = p_1^*,$$
$$Q_2(t) = q_1 + p_2^* t, \quad P_2(t) = p_2^*, \quad t > 0, \tag{2.2.9}$$

where p_1^* and p_2^* are defined by (2.2.2) with $\eta \subset S^2_-$. If $\eta \subset S^2_+$, then

$$Q_1(t) = q_1 + p_1 t, \quad P_1(t) = p_1,$$
$$Q_2(t) = q_2 + p_2 t, \quad P_2(t) = p_2, \quad t > 0. \tag{2.2.10}$$

Let us stress that, for the stochastic dynamics, the state of particles at the instant of collision is defined not only by their positions and momenta, but also by a random vector $\eta \subset S^2 = S^2_- \cup S^2_+$ with constant density of probability on S^2. Note that the union of points (2.2.9) of the stochastic trajectories with respect to random $\eta \subset S^2$, $t > 0$, and q_1, p_1, p_2 forms a set of the same dimension as the phase space.

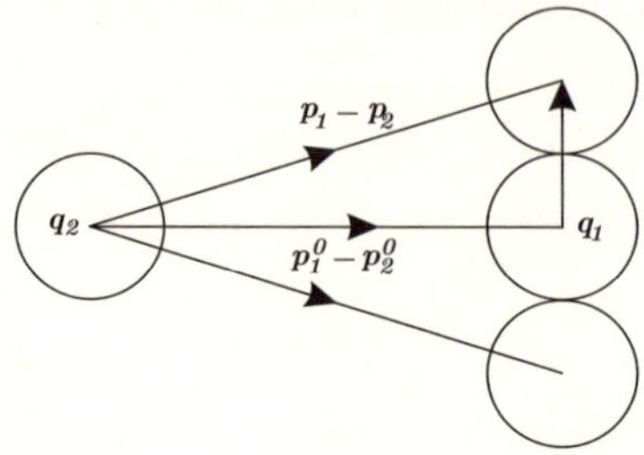

Fig. 2

If hard spheres touch each other at the initial time $t = 0$, i.e., $q_2 = q_1 - a\eta$, then, for $\eta \subset S_-^2$, we have

$$Q_1(t) = q_1 + p_1^* t, \quad P_1(t) = p_1^*,$$
$$Q_2(t) = q_1 - a\eta + p_2^* t, \quad P_2(t) = p_2^*, \quad t > 0. \tag{2.2.11}$$

If $\eta \subset S_+^2$, then

$$Q_1(t) = q_1 + p_1 t, \quad P_1(t) = p_1,$$
$$Q_2(t) = q_1 - a\eta + p_2 t, \quad P_2(t) = p_2, \quad t > 0. \tag{2.2.12}$$

As $a \to 0$ the trajectories (2.2.11) and (2.2.12) converge to the trajectories (2.2.9) and (2.2.10), respectively.

Note that there exists another possibility to obtain stochastic trajectories. Namely, we fix the positions q_1^0 and q_2^0 of two hard spheres and fix the difference of their momenta $p_1^0 - p_2^0$ that is parallel to the vector $q_1^0 - q_2^0$.

Let

$$\tau = \frac{|q_1^0 - q_2^0|}{|p_1^0 - p_2^0|} > 0$$

be the instant of collision.

Consider all initial momenta (p_1, p_2) such that spheres collide at time τ and the unit vectors η in the direction of their centers at the instant of collision belong to the semisphere S_-^2 (Fig. 2). The initial momenta (p_1, p_2) depend on η, $(p_1, p_2) = (p_1(\eta), p_2(\eta))$, and there is a one-to-one correspondence between η and $(p_1(\eta), p_2(\eta))$. After collisions, the hard spheres have momenta (2.2.2) with $\eta \subset S_-^2$. The Hamiltonian trajectories with the above-described initial data are again represented by formula (2.2.3) with $q_1 = q_1^0$ and $q_2 = q_2^0$. We say that the collection of these trajectories is the cone of trajectories.

It is obvious that the momenta $(p_1(\eta), p_2(\eta))$ are continuous functions of a and tend to (p_1^0, p_2^0) as $a \to 0$. The limit of the cone of trajectories of hard spheres (2.2.3) as $a \to 0$ coincides with the stochastic trajectories (2.2.6)–(2.2.8) for the same $\eta \subset S_-^2$, $q_1 = q_1^0$, $q_2 = q_2^0$, $p_1 = p_1^0$, and $p_2 = p_2^0$.

2.2.3 Convergence of Hamiltonian trajectories to stochastic trajectories

We now consider the problem of the convergence of the Hamiltonian trajectories of two hard spheres (2.2.3) to the stochastic trajectories (2.2.4), (2.2.5) as the diameter a tends to zero. For this purpose, we associate the Hamiltonian trajectory with the stochastic trajectory with the same vector η.

Lemma 2.1. *The Hamiltonian trajectory of two hard spheres (2.2.3) converges to the stochastic trajectory (2.2.4), (2.2.5) [or (2.2.6)–(2.2.8)] with the same η uniformly in time, and the following estimate holds:*

$$\|X^a(t) - X(t)\| \leq a, \quad t \geq 0, \tag{2.2.13}$$

where $\|\cdot\|$ is the Euclidean norm in the phase space.

Proof. Estimate (2.2.13) follows directly from the definitions (2.2.3)–(2.2.8) of the Hamiltonian and stochastic trajectories. Lemma 2.1 holds for the bunch and the cone of trajectories.

Remark 2.1. It follows from the Hamiltonian dynamics of hard spheres that particles can collide only if $(q_1 - q_2) \cdot (p_1 - p_2) \leq 0$ or $\eta \cdot (p_1 - p_2) \leq 0$ (for initial data $q_2 = q_2^0 - a\eta$, $q_1 = q_1^0$, $\eta \subset S_-^2$). This means that, in the stochastic dynamics, particles also collide only if $\eta \cdot (p_1 - p_2) \leq 0$. But after collision, we have $\eta \cdot (p_1^* - p_2^*) \leq -\eta \cdot (p_1 - p_2) \geq 0$ and, thus, $\eta \subset S_+^2$ with respect to the momenta p_1^*, p_2^*. In order to avoid repeated collisions at the same instant of time, we require that, for $\eta \subset S_+^2$, momenta do not change.

Let us show that the convergence of Hamiltonian trajectories to stochastic ones implies the following weak convergence: For continuous functions $f_2(x_1, x_2) = f_2(x)$, consider the functional

$$\begin{gathered}\frac{1}{2\pi a^2} \int_0^t dt' \int_{S_-^2} d\eta a^2 f_2(X^a(t', x^a)), \\ x^a = (q_1^0, p_1, q_2^0 - a\eta, p_2), \quad \eta \subset S_-^2,\end{gathered} \tag{2.2.14}$$

for the Hamiltonian dynamics. Here, $d\eta$ is an element of the unit sphere $|\eta| = 1$ and $X^a(t, x^a)$ denote trajectories (2.2.3) with initial data x^a. Thus, functional (2.2.14) is the average over the semisphere $a\eta$, $\eta \subset S_-^2$, of the bunch of trajectories of hard spheres with initial data $x^a = (q_1^0, p_1, q_2^0 - a\eta, p_2)$, $\eta \subset S_-^2$, for fixed (q_1^0, p_1, q_2^0, p_2), and the vector $p_1 - p_2$ is parallel to the vector $q_1^0 - q_2^0$.

Also consider the functional

$$\begin{gathered}\frac{1}{2\pi} \int_0^t dt' \int_{S_-^2} d\eta f_2(X(t', x)), \\ x = (q_1^0, p_1, q_2^0, p_2), \quad \eta \subset S_-^2,\end{gathered} \tag{2.2.15}$$

for the stochastic dynamics. Here, $X(t,x)$ are the stochastic trajectories (2.2.6)–(2.2.12) with the initial data $x = (q_1^0, p_1, q_2^0, p_2)$, $\eta \subset S_-^2$. We now show that there exists the limit of functional (2.2.14) as $a \to 0$, and that it coincides with functional (2.2.15).

Lemma 2.2. *The following formula holds uniformly with respect to t:*

$$\lim_{a\to 0} \frac{1}{2\pi a^2} \int_0^t dt' \int_{S_-^2} d\eta a^2 f_2(X^a(t', x^a)) = \frac{1}{2\pi} \int_0^t dt' \int_{S_-^2} d\eta f_2(X(t', x)).$$

Proof. We have

$$\lim_{a\to 0} \frac{1}{2\pi a^2} \Big| \int_0^t dt' \int_{S_-^2} d\eta a^2 [f_2(X^a(t', x^a)) - f_2(X(t', x))] \Big|$$
$$\leq \lim_{a\to 0} \frac{1}{2\pi} \int_0^t dt' \int_{S_-^2} d\eta | f_2(X^a(t', x^a)) - f_2(X(t', x))| = 0$$

because, as follows from Lemma 2.1,

$$\lim_{a\to 0} \big[f_2(X^a(t', x^a)) - f_2(X(t', x)) \big] = 0$$

uniformly with respect to $0 \leq t' \leq t$ and $\eta \subset S_-^2$. Thus, according to Lemmas 2.1 and 2.2, the stochastic trajectories are the uniform and weak limits of the Hamiltonian trajectories in the above-described sense.

Note that, in the case of the cone of trajectories, the integration in (2.2.14) is carried out over the semisphere $\eta \subset S_-^2$ and the initial momenta $(p_1(\eta), p_2(\eta))$ depend on η as described above. The limit of the Hamiltonian dynamics of hard spheres is the stochastic dynamics, but not the free one.

It is easy to extend all results obtained above to negative (decreasing) time, $t \leq 0$. Namely, it is sufficient to replace the semispheres S_-^2 and S_+^2 by the semispheres S_+^2 and S_-^2, respectively, in all constructions for positive (increasing) time, $t \geq 0$. For example, if $q_1 - q_2 = 0$ at the initial time $t = 0$, then the stochastic trajectories are defined as follows:

$$\begin{aligned} Q_1(t) &= q_1 + p_1^* t, \quad P_1(t) = p_1^*, \\ Q_2(t) &= q_2 + p_2^* t, \quad P_2(t) = p_2^*, \quad t < 0, \end{aligned} \tag{2.2.16}$$

where p_1^*, p_2^*, are defined as in (2.2.2) for $\eta \subset S_+^2$. If $\eta \subset S_-^2$, then

$$\begin{aligned} Q_1(t) &= q_1 + p_1 t, \quad P_1(t) = p_1, \\ Q_2(t) &= q_2 + p_2 t, \quad P_2(t) = p_2, \quad t < 0. \end{aligned} \tag{2.2.17}$$

The union of the points $X(t)$ (2.2.17) with respect to the random vectors $\eta \subset S^2_+$, $t < 0$ and q_1, p_1, p_2 forms a set of the same dimension as the phase space.

If hard spheres touch each other at the initial time $t = 0$, i.e., $q_2 = q_1 - a\eta$, then, for $\eta \subset S^2_+$, we have

$$\begin{aligned} &Q_1(t) = q_1 + p_1^* t, \quad P_1(t) = p_1^*, \\ &Q_2(t) = q_1 - a\eta + p_2^* t, \quad P_2(t) = p_2^*, \quad t < 0. \end{aligned} \tag{2.2.18}$$

If $\eta \subset S^2_-$, then

$$\begin{aligned} &Q_1(t) = q_1 + p_1 t, \quad P_1(t) = p_1, \\ &Q_2(t) = q_1 - a\eta + p_2 t, \quad P_2(t) = p_2, \quad t < 0. \end{aligned} \tag{2.2.19}$$

It is obvious that the Hamiltonian trajectories of hard spheres (2.2.18)–(2.2.19) converge as $a \to 0$ to the stochastic trajectories (2.2.16)–(2.2.17) with the same η.

2.3 New representation of Hamiltonian and stochastic trajectories

2.3.1 Representation of Hamiltonian trajectories

In this section, we introduce a new representation of the trajectories of interacting hard spheres and interacting stochastic point particles. In both cases, there also exist free trajectories for noninteracting hard spheres and stochastic point particles.

In order to interact, the positions of centers of hard spheres q_1, q_2 must belong at a certain time $\tau \geq 0$ to the hypersurface

$$q_1 = q_2 + a\eta, \quad \eta \subset S^2, \tag{2.3.1}$$

with arbitrary p_1, p_2. For given fixed p_1, p_2, q_1, hypersurface (2.3.1) is parameterized by the vector $\eta \subset S^2$, $\eta = \frac{q_1 - q_2}{|q_1 - q_2|}$.

In order to interact, the positions of the stochastic particles must belong at a certain time $\tau \geq 0$ to the hypersurface

$$q_1 = q_2, \quad \eta \subset S^2, \tag{2.3.2}$$

with arbitrary p_1, p_2. In the case of stochastic dynamics, at every point of hypersurface (2.3.2), there exist random vectors $\eta \subset S^2$ with constant probability density $\chi(\eta) = 1/4\pi$. In this case, hypersurface (2.3.2) is characterized by points $q_1 = q_2, p_1, p_2$, and the random vectors η.

Consider the initial positions such that hard spheres interact at a certain time τ on the interval $[0, t]$. In order to obtain them it is necessary to shift the points $q_1 = q_2 + a\eta$

"backward" in time on the time interval $[-\tau, 0]$. According to the dynamics of hard spheres, we get

$$(q_1 - p_1\tau, p_1, q_1 - a\eta - p_2\tau, p_2), \quad \eta \subset S^2_-, \ (p_1 - p_2)\cdot\eta \le 0,$$
$$(q_1 - p_1^*\tau, p_1^*, q_1 - a\eta - p_2^*\tau, p_2^*), \quad \eta \subset S^2_+, \ (p_1 - p_2)\cdot\eta \ge 0. \tag{2.3.3}$$

Denote the collection of points (2.3.3) with arbitrary q_1, p_1, p_2, $\eta \subset S^2_-$, $\eta \subset S^2_+$, $0 \le \tau \le t$ by D^a_{-t}. Obviously, the domain D^a_{-t} is of full Lebesgue measure. We define the infinitesimal and entire volume of D^a_{-t} in the next section. Hard spheres with initial data (2.3.3) from the domain D^a_{-t} interact (touch one another) at time $t = \tau$ during the "forward" evolution on the time interval $[0, t]$. For $t = \tau - 0$, points (2.3.3) are shifted along the Hamiltonian trajectory to the points

$$(q_1, p_1, q_1 - a\eta, p_2), \quad \eta \subset S^2_-,$$
$$(q_1, p_1^*, q_1 - a\eta, p_2^*), \quad \eta \subset S^2_+. \tag{2.3.4}$$

For $t = \tau + 0$, they turn into the points

$$(q_1, p_1^*, q_1 - a\eta, p_2^*), \quad \eta \subset S^2_-,$$
$$(q_1, p_1, q_1 - a\eta, p_2), \quad \eta \subset S^2_+. \tag{2.3.5}$$

Finally, for $t > \tau$, points (2.3.3) turn into the points

$$\big(q_1 + p_1^*(t-\tau), p_1^*, q_1 - a\eta + p_2^*(t-\tau), p_2^*\big), \quad \eta \subset S^2_-,$$
$$\big(q_1 + p_1(t-\tau), p_1, q_1 - a\eta + p_2(t-\tau), p_2\big), \quad \eta \subset S^2_+. \tag{2.3.6}$$

Note that points (2.3.4) are the states of hard spheres before a collision because $\eta\cdot(p_1 - p_2) \le 0$ for $\eta \subset S^2_-$, and $\eta\cdot(p_1^* - p_2^*) \le -\eta\cdot(p_1 - p_2) \le 0$ for $\eta \subset S^2_+$, i.e., the vectors $\eta \subset S^2_+$ belong to the semisphere S^2_- with respect to the momenta p_1^*, p_2^*. Analogously, points (2.3.5) are the states after a collision because $\eta\cdot(p_1 - p_2) \ge 0$ for $\eta \subset S^2_+$, and $\eta\cdot(p_1^* - p_2^*) \le -\eta\cdot(p_1 - p_2) \ge 0$ for $\eta \subset S^2_-$, i.e., the vectors $\eta \subset S^2_-$ belong to the semisphere S^2_+ with respect to the momenta p_1^*, p_2^*.

Note that the collection of points (2.3.6) with respect to $\eta \subset S^2_-$, $\eta \subset S^2_+$, $0 \le \tau \le t$, q_1, p_1, p_2 is of the same dimension as the phase space. We call the domain D^a_{-t} the *domain of interaction of two hard spheres*.

2.3.2 Representation of stochastic trajectories

Now consider the stochastic particles and again shift the points $q_1 = q_2$, $\eta \subset S^2$, p_1, p_2 of hypersurface (2.3.2) "backward" in time on the interval $[-\tau, 0]$. According to the dynamics of stochastic particles, we obtain

$$\begin{aligned} &(q_1 - p_1\tau, p_1, q_1 - p_2\tau, p_2), \quad \eta \subset S_-^2, \\ &(q_1 - p_1^*\tau, p_1^*, q_1 - p_2^*\tau, p_2^*), \quad \eta \subset S_+^2. \end{aligned} \tag{2.3.7}$$

Denote the collection of points (2.3.7) with arbitrary $q_1, p_1, p_2, \eta \subset S_-^2, \eta \subset S_+^2$, $0 \le \tau \le t$ by D_{-t}. We call the set D_{-t} the *set of interaction of two stochastic particles*.

Stochastic particles with initial data from the set D_{-t} interact at time τ during the forward evolution on the time interval $[0, t]$ because, for $t = \tau$, particles touch each other. For $t = \tau - 0$, points (2.3.7) are shifted along the stochastic trajectories to the points

$$\begin{aligned} &(q_1, p_1, q_1, p_2), \quad \eta \subset S_-^2, \\ &(q_1, p_1^*, q_1, p_2^*), \quad \eta \subset S_+^2. \end{aligned} \tag{2.3.8}$$

For $t = \tau + 0$, they turn into the points

$$\begin{aligned} &(q_1, p_1^*, q_1, p_2^*), \quad \eta \subset S_-^2, \\ &(q_1, p_1, q_1, p_2), \quad \eta \subset S_+^2. \end{aligned} \tag{2.3.9}$$

For $t > \tau$, points (2.3.7) of the set D_{-t} turn into the points

$$\begin{aligned} &\big(q_1 + p_1^*(t-\tau), p_1^*, q_1 + p_2^*(t-\tau), p_2^*\big), \quad \eta \subset S_-^2, \\ &\big(q_1 + p_1(t-\tau), p_1, q_1 + p_2(t-\tau), p_2\big), \quad \eta \subset S_+^2. \end{aligned} \tag{2.3.10}$$

It is obvious that the collection of points (2.3.10) with respect to $\eta \subset S_-^2, \eta \subset S_+^2$, $0 \le \tau \le t, q_1, p_1, p_2$ is of the same dimension as the phase space.

Thus, we have obtained a new representation of all Hamiltonian trajectories of hard spheres (2.3.6) and all stochastic trajectories of stochastic particles (2.3.10) that interact at a certain time τ on the interval $[0, t]$. We have also obtained a complete description of the initial phase points (2.3.3) and (2.3.7) for the trajectories of interacting hard spheres and interacting stochastic particles, respectively. We have shown that the points of the domain of interaction D_{-t}^a and the set of interaction D_{-t} are the states before collisions for the "forward" motion of hard spheres and stochastic particles, respectively.

All results obtained above can easily be reformulated for negative (decreasing) time. The domain D_t^a is the collection of points

$$\begin{aligned} &(q_1 + p_1^*\tau, p_1^*, q_1 - a\eta + p_2^*\tau, p_2^*), \quad \eta \subset S_-^2, \quad \eta \cdot (p_1 - p_2) \le 0, \\ &(q_1 + p_1\tau, p_1, q_1 - a\eta + p_2\tau, p_2), \quad \eta \subset S_+^2, \quad \eta \cdot (p_1 - p_2) \ge 0. \end{aligned} \tag{2.3.11}$$

For $-t < 0$, points (2.3.11) turn into the points

$$\begin{gathered}\big(q_1 + p_1(-t+\tau), p_1, q_1 - a\eta + p_2(-t+\tau), p_2\big), \quad \eta \subset S^2_-, \\ \big(q_1 + p_1^*(-t+\tau), p_1^*, q_1 - a\eta + p_2^*(-t+\tau), p_2^*\big), \quad \eta \subset S^2_+, \quad 0 \le \tau \le t,\end{gathered} \tag{2.3.12}$$

during the "backward" evolution on the interval $[-t, 0]$.

The set D_t is the collection of points

$$\begin{gathered}(q_1 + p_1^*\tau, p_1^*, q_1 + p_2^*\tau, p_2^*), \quad \eta \subset S^2_-, \\ (q_1 + p_1\tau, p_1, q_1 + p_2\tau, p_2), \quad \eta \subset S^2_+, \quad 0 \le \tau \le t.\end{gathered} \tag{2.3.13}$$

For $-t < 0$, points (2.3.13) turn into the points

$$\begin{gathered}\big(q_1 + p_1(-t+\tau), p_1, q_1 + p_2(-t+\tau), p_2\big), \quad \eta \subset S^2_-, \\ \big(q_1 + p_1^*(-t+\tau), p_1^*, q_1 + p_2^*(-t+\tau), p_2^*\big), \quad \eta \subset S^2_+, \quad 0 \le \tau \le t,\end{gathered} \tag{2.3.14}$$

during the "backward" evolution on the interval $[-t, 0]$.

Note that the sets D_{-t} (2.3.7) and D_t (2.3.13) have the following characteristic property: The vectors of differences of positions are parallel to the vectors of differences of momenta. Indeed, for D_{-t}, we have

$$\begin{gathered}q_1 - p_1\tau - q_1 + p_2\tau = \tau(p_2 - p_1), \quad \eta \subset S^2_-, \\ q_1 - p_1^*\tau - q_1 + p_2^*\tau = \tau(p_1^* - p_2^*), \quad \eta \subset S^2_+.\end{gathered} \tag{2.3.15}$$

For D_t, we have

$$\begin{gathered}q_1 + p_1^*\tau - q_1 - p_2^*\tau = \tau(p_1^* - p_2^*), \quad \eta \subset S^2_-, \\ q_1 + p_1\tau - q_1 - p_2\tau = \tau(p_2 - p_1), \quad \eta \subset S^2_+.\end{gathered} \tag{2.3.16}$$

In the previous section, it has been indicated that, according to the stochastic dynamics, particles may interact only if the vector of difference of their initial positions is parallel to the vector of difference of their initial momenta. We have shown above that all points of the set D_{-t} taken as initial data for the "forward" evolution satisfy this condition. Similarly, all points of the domain D_t taken as initial data for the "backward" evolution also satisfy this condition.

Remark 2.2. It should be stressed that the collection of points (x_1, x_2) (2.3.7) or (2.3.13) of the set D_{-t} or D_t, $0 \le \tau \le t$, has lower dimension in the phase space because it consists of points such that the differences of their positions are parallel to the differences of their momenta. It is obvious that points (2.3.7) with $\eta \subset S^2_-$ do not change in the phase space as η varies on S^2_-.

It is quite easy to check that the Jacobian of the transformation $(x_1, x_2) \to (q_1 - p_1\tau, p_1, q_1 - p_2\tau, p_2)$, $\eta \subset S^2_-$, $\tau \in [o,t)$, $q_1 \in \mathbb{R}^3$, $p_1 \in \mathbb{R}^3$, $p_2 \in \mathbb{R}^3$, is equal to zero. One can also check that the Jacobian of the transformation $(x_1, x_2) \to (q_1 - p_1^*\tau, p_1^*, q_1 - p_2^*\tau, p_2^*)$, $\eta \subset S^2_+$, $\tau \in [o,t)$, $q_1 \in \mathbb{R}^3$, $p_1 \in \mathbb{R}^3$, $p_2 \in \mathbb{R}^3$, is also equal to zero. In the case of hard spheres, it was established that the corresponding Jacobian of the transformation $(x_1, x_2) \to (q_1 - p_1\tau, p_1, q_1 - a\eta - p_2\tau, p_2)$, $\eta \subset S^2_-$, $0 \le \tau \le t$, or $(x_1, x_2) \to (q_1 - p_1^*\tau, p_1^*, q_1 - a\eta - p_2^*\tau, p_2^*)$, $\eta \subset S^2_+$, $0 \le \tau \le t$, is different from zero and equal to $a^2|\eta \cdot (p_1 - p_2)|$ [CIP]. This means that the domain D^a_{-t} has full Lebesgue measure in the phase space (with respect to x_1, x_2). It is also true for the domain D^a_t.

2.4 Functional for a system of two hard spheres

2.4.1 Domain of interaction and functional

Consider the following functional:

$$(S_2^a(t)f_2 - S_2^0(t)f_2, \varphi_2)$$
$$= \frac{1}{2}\int \big[S_2^a(t)f_2(x_1,x_2) - S_2^0(t)f_2(x_1,x_2)\big]\varphi_2(x_1,x_2)dx_1dx_2, \quad t \ge 0, \tag{2.4.1}$$

where f_2 is a continuously differentiable function, φ_2 is a test function, f_2 and φ_2 are equal to zero on forbidden configurations, $S_2^a(t)$ is the operator of evolution of two hard spheres, $S_2^a(t)f_2(x_1,x_2) = f_2(X^a(t,x_1,x_2))$, and $S_2^0(t)$ is the operator of evolution of free point particles: $S_2^0(t)f_2(x_1,x_2) = f_2(X^0(t,x_1,x_2))$. By $X^a(t,x_1,x_2)$ and $X^0(t,x_1,x_2)$ we denote the trajectories of two hard spheres and two free particles, respectively. The properties of the operators $S_2^a(t)$ and $S_2^0(t)$ in the space L_1 are described in detail in [PGM, CGP] and summarized in Chapter 1. For what follows, it is convenient to define the operator $S_2^a(t)$ on forbidden configurations $|q_1 - q_2| < a$ as the operator of free evolution.

The function

$$S_2^a(t)f_2(x_1,x_2) - S_2^0(t)f_2(x_1,x_2) \tag{2.4.2}$$

is different from zero in the domain of interaction D^a_{-t} because, for initial data from D^a_{-t}, hard spheres interact at time τ on the interval $[0,t]$ and, according to (2.3.6), we have

$$\begin{aligned}
&S_2^a(t)f_2(x_1,x_2) - S_2^0(t)f_2(x_1,x_2)\\
&\quad = f_2\big(q_1 + p_1^*(t-\tau), p_1^*, q_1 - a\eta + p_2^*(t-\tau), p_2^*\big)\\
&\qquad - f_2\big(q_1 + p_1(t-\tau), p_1, q_1 - a\eta + p_2(t-\tau), p_2\big), \quad \eta \subset S_-^2,\\
&S_2^a(t)f_2(x_1,x_2) - S_2^0(t)f_2(x_1,x_2)\\
&\quad = f_2\big(q_1 + p_1(t-\tau), p_1, q_1 - a\eta + p_2(t-\tau), p_2\big)\\
&\qquad - f_2\big(q_1 + p_1^*(t-\tau), p_1^*, q_1 - a\eta + p_2^*(t-\tau), p_2^*\big), \quad \eta \subset S_+^2.
\end{aligned} \tag{2.4.3}$$

For initial data outside the domain D_{-t}^a, function (2.4.2) is equal to zero because hard spheres do not interact and their trajectories coincide with free ones.

Thus, functional (2.4.1) can be represented in the following form:

$$\begin{aligned}
&(S_2^a(t)f_2 - S_2^0(t)f_2, \varphi_2)\\
&\quad = \frac{1}{2}\int_{D_{-t}^a} \big[S_2^a(t)f_2(x_1,x_2) - S_2^0(t)f_2(x_1,x_2)\big]\varphi_2(x_1,x_2)dx_1dx_2\\
&\quad = \frac{1}{2}\int_0^t d\tau \int_{S_-^2} d\eta \int dq_1dp_1dp_2a^2|\eta\cdot(p_1-p_2)|\\
&\qquad \times \big[f_2(q_1 + p_1^*(t-\tau), p_1^*, q_1 - a\eta + p_2^*(t-\tau), p_2^*)\\
&\qquad\qquad - f_2(q_1 + p_1(t-\tau), p_1, q_1 - a\eta + p_2(t-\tau), p_2)\big]\\
&\qquad \times \varphi_2(q_1 - p_1\tau, p_1, q_1 - a\eta - p_2\tau, p_2)\\
&\qquad + \frac{1}{2}\int_0^t d\tau \int_{S_+^2} d\eta \int dq_1dp_1dp_2a^2|\eta\cdot(p_1-p_2)|\\
&\qquad \times \big[f_2(q_1 + p_1(t-\tau), p_1, q_1 - a\eta + p_2(t-\tau), p_2)\\
&\qquad\qquad - f_2(q_1 + p_1^*(t-\tau), p_1^*, q_1 - a\eta + p_2^*(t-\tau), p_2^*)\big]\\
&\qquad \times \varphi_2(q_1 - p_1^*\tau, p_1^*, q_1 - a\eta - p_2^*\tau, p_2^*)\\
&\quad = \int_0^t d\tau \int_{S_-^2} d\eta \int dq_1dp_1dp_2a^2|\eta\cdot(p_1-p_2)|\\
&\qquad \times \big[f_2(q_1 + p_1^*(t-\tau), p_1^*, q_1 - a\eta + p_2^*(t-\tau), p_2^*)\\
&\qquad\qquad - f_2(q_1 + p_1(t-\tau), p_1, q_1 - a\eta + p_2(t-\tau), p_2)\big]\\
&\qquad \times \varphi_2(q_1 - p_1\tau, p_1, q_1 - a\eta - p_2\tau, p_2)
\end{aligned}$$

$$= \int_0^t d\tau \int_{S_-^2} d\eta \int dq_1 dp_1 dq_2 dp_2 a^2 |\eta \cdot (p_1 - p_2)| \delta(q_1 - q_2 - a\eta)$$
$$\times \big[f_2(q_1 + p_1^*(t-\tau), p_1^*, q_2 + p_2^*(t-\tau), p_2^*)$$
$$- f_2(q_1 + p_1(t-\tau), p_1, q_2 + p_2(t-\tau), p_2) \big]$$
$$\times \varphi_2(q_1 - p_1\tau, p_1, q_2 - p_2\tau, p_2)$$
$$= \int_0^t d\tau \int_{S_-^2} d\eta \int dq_1 dp_1 dq_2 dp_2 a^2 |\eta \cdot (p_1 - p_2)|$$
$$\times \delta(q_1 + p_1\tau - q_2 - p_2\tau - a\eta)$$
$$\times \big[f_2(q_1 + p_1\tau + p_1^*(t-\tau), p_1^*, q_2 + p_2\tau + p_2^*(t-\tau), p_2^*)$$
$$- f_2(q_1 + p_1 t, p_1, q_2 + p_2 t, p_2) \big] \varphi_2(q_1, p_1, q_2, p_2). \tag{2.4.4}$$

In (2.4.4), we have used the variables $\tau, \eta, q_1, p_1, p_2$ in the domain D_{-t}^a and the corresponding Jacobian [CIP]; in the second term with S_+^2, we have used the variables p_1^*, p_2^*, instead of p_1, p_2, taking into account that the corresponding Jacobian is equal to one. The infinitesimal volume of the domain D_{-t}^a in the variables $\tau, \eta, q_1, p_1, p_2$ is equal to $dx_1 dx_2 = d\tau d\eta dq_1 dp_1 dp_2 a^2 |\eta \cdot (p_1 - p_2)|$ [CIP].

One can see that the term with S_+^2 coincides with the term with S_-^2 because the points from D_{-t}^a with S_+^2 represent the states before a collision in the "forward" evolution as well as the points from D_{-t}^a with S_-^2. Indeed, in the second term in (2.4.4), we have

$$\eta \cdot (p_1^* - p_2^*) = -\eta \cdot (p_1 - p_2) \le 0, \quad \eta \subset S_+^2, \tag{2.4.5}$$

and, thus, $\eta \subset S_-^2$ with respect to the momenta p_1^*, p_2^*. In the first term, $\eta \cdot (p_1 - p_2) \le 0$, $\eta \subset S_-^2$, and we also have points from D_{-t}^a in the states before a collision. This explains the factor $\frac{1}{2}$ in functional (2.4.1), (2.4.4).

In what follows, we use the domain D_{-t}^a as the set $q_1 + p_1\tau - q_2 - p_2\tau - a\eta = 0$, $q_1 \in \mathbb{R}^3$, $p_1 \in \mathbb{R}^3$, $p_2 \in \mathbb{R}^3$, $0 \le \tau \le t$, $\eta \subset S_-^2$. Formula (2.4.4) can be identically represented in the following form:

$$(S_2^a(t) f_2, \varphi_2) = \int S_2^a(t) f_2(x_1, x_2) \varphi_2(x_1, x_2) dx_1 dx_2$$
$$= \int S_2^0(t) f_2(x_1, x_2) \varphi_2(x_1, x_2) dx_1 dx_2$$
$$+ \int \big[S_2^a(t) f_2(x_1, x_2) - S_2^0(t) f_2(x_1, x_2) \big] \varphi_2(x_1, x_2) dx_1 dx_2$$
$$= (S_2^0(t) f_2, \varphi_2) + \Big(\big[S_2^a(t) f_2 - S_2^0(t) f_2 \big], \varphi_2 \Big), \tag{2.4.4$'$}$$

$$S_2^a(t) f_2(x_1, x_2) = S_2^0(t) f_2(x_1, x_2) + \big[S_2^a(t) f_2(x_1, x_2) - S_2^0(t) f_2(x_1, x_2) \big].$$

Note that the function $\left[S_2^a(t)f_2(x_1,x_2) - S_2^0(t)f_2(x_1,x_2)\right]$ is different from zero in the domain D^a_{-t}.

Formula (2.4.4′) defines a functional that is the average of the observable $\varphi_2(x_1,x_2)$ over the state $S_2^a(t)f_2(x_1,x_2)$.

2.4.2 Derivative of functional

Let us define the derivative of functional (2.4.1), (2.4.4) with respect to time at $t = 0$. From (2.4.4), we have

$$\frac{d}{dt}(S_2^a(t)f_2 - S_2^0(t)f_2,\varphi_2)|_{t=0}$$

$$= \int_{S^2_-} d\eta \int dq_1dq_2dp_1dp_2a^2|\eta\cdot(p_1-p_2)|[f_2(q_1,p_1^*,q_2,p_2^*) - f_2(q_1,p_1,q_2,p_2)]$$

$$\times\,\delta(q_1-q_2-a\eta)\varphi_2(q_1,p_1,q_2,p_2) \tag{2.4.6}$$

for arbitrary test functions φ_2. This means that

$$\frac{d}{dt}(S_2^a(t)f_2(x_1,x_2) - S_2^0(t)f_2(x_1,x_2)|_{t=0}$$

$$= a^2\int_{S^2} d\eta\delta(q_1-q_2-a\eta)|\eta\cdot(p_1-p_2)|\Theta(-\eta\cdot(p_1-p_2))$$

$$\times\left[f_2(q_1,p_1^*,q_2,p_2^*) - f_2(q_1,p_1,q_2,p_2)\right]$$

$$= \delta(|q_1-q_2|-a)\left|\frac{(q_1-q_2)}{|q_1-q_2|}\cdot(p_1-p_2)\right|\Theta\Big(-\frac{(q_1-q_2)}{|q_1-q_2|}\cdot(p_1-p_2)\Big)$$

$$\times\left[f_2(q_1,p_1^*,q_2,p_2^*) - f_2(q_1,p_1,q_2,p_2)\right] \tag{2.4.7}$$

in the sense of generalized functions. From (2.4.6), we get

$$\frac{d}{dt}S_2^a(t)f_2(x_1,x_2)|_{t=0}$$

$$= \sum_{i=1}^{2} p_i\cdot\frac{\partial}{\partial q_i}f_2(x_1,x_2) + a^2\int_{S^2} d\eta\delta(q_1-q_2-a\eta)|\eta\cdot(p_1-p_2)|$$

$$\times\,\Theta(-\eta\cdot(p_1-p_2))\big[f_2(q_1,p_1^*,q_2,p_2^*) - f_2(q_1,p_1,q_2,p_2)\big]. \tag{2.4.8}$$

We have obtained a well-known expression [BoB] with a boundary condition according to which, in the first term on the right-hand side of (2.4.8), one must replace p_1, p_2 by p_1^*, p_2^* for $q_1-q_2-a\eta = 0$, $\eta\subset S^2_-$.

Consider the derivative of functional (2.4.1), (2.4.4) with respect to time at $t \neq 0$. We have

$$\begin{aligned}
&\frac{d}{dt}(S_2^a(t)f_2 - S_2^0(t)f_2, \varphi_2) \\
&\quad = \int_0^t d\tau \int_{S_-^2} d\eta \int dq_1 dp_1 dq_2 dp_2 a^2 |\eta\cdot(p_1 - p_2)| \delta(q_1 + p_1\tau - q_2 - p_1\tau - a\eta) \\
&\qquad \times \Big[\Big(p_1^* \frac{\partial}{\partial q_1} + p_2^* \frac{\partial}{\partial q_2}\Big) \\
&\qquad \times f_2(q_1 + p_1\tau + p_1^*(t-\tau), p_1^*, q_2 + p_2\tau + p_2^*(t-\tau), p_2^*) \\
&\qquad - \Big(p_1 \frac{\partial}{\partial q_1} + p_2 \frac{\partial}{\partial q_2}\Big) f_2(q_1 + p_1 t, p_1, q_2 + p_2 t, p_2)\Big] \varphi_2(q_1, p_1, q_2, p_2) \\
&\qquad + \int_{S_-^2} d\eta \int dq_1 dp_1 dq_2 dp_2 a^2 |\eta\cdot(p_1 - p_2)| \delta(q_1 + p_1 t - q_2 - p_2 t - a\eta) \\
&\qquad \times \big[f_2(q_1 + p_1 t, p_1^*, q_2 + p_2 t, p_2^*) - f_2(q_1 + p_1 t, p_1, q_2 + p_2 t, p_2)\big] \\
&\qquad \times \varphi_2(q_1, p_1, q_2, p_2).
\end{aligned} \tag{2.4.9}$$

This formula means that

$$\frac{d}{dt}(S_2^a(t)f_2(x_1, x_2)) = \Big(p_1 \frac{\partial}{\partial q_1} + p_2 \frac{\partial}{\partial q_2}\Big)(S_2^0(t)f_2(x_1, x_2)) \tag{2.4.10}$$

if $q_1 - q_2 \neq (p_2 - p_1)\tau + a\eta$ for all $0 \le \tau \le t$, i.e., for $(x_1, x_2) \notin D_{-t}^a$.

If $q_1 - q_2 = (p_2 - p_1)\tau + a\eta$ for some $0 \le \tau \le t$, i.e., $(x_1, x_2) \in D_{-t}^a$, then (2.4.9) yields

$$\begin{aligned}
&\frac{d}{dt} S_2^a(t) f_2(x_1, x_2) \\
&\quad = \Big(p_1^* \frac{\partial}{\partial q_1} + p_2^* \frac{\partial}{\partial q_2}\Big) \\
&\qquad \times f_2(q_1 + p_1\tau + p_1^*(t-\tau), p_1^*, q_2 + p_2\tau + p_1^*(t-\tau), p_2^*) \\
&\qquad + \int_{S_-^2} d\eta a^2 |\eta\cdot(p_1 - p_2)| \delta(q_1 + p_1 t - q_2 - p_2 t - a\eta) \\
&\qquad \times \big[f_2(q_1 + p_1 t, p_1^*, q_2 + p_2 t, p_2^*) - f_2(q_1 + p_1 t, p_1, q_2 + p_2 t, p_2)\big].
\end{aligned}$$

2.5 Functional for a system of two stochastic particles

2.5.1 Functional of stochastic particles as the limit of the functional of hard spheres

Consider the functional

$$\frac{1}{4\pi a^2}\Big(S_2^a(t)f_2 - S_2^0(t)f_2, \varphi_2\Big), \tag{2.5.1}$$

which is the average of functional (2.4.1) over the sphere $q_1 - q_2 - a\eta = 0$. It is easy to see from (2.4.4) that, for continuous functions f_2 and φ_2, the following limit exists:

$$\begin{aligned}
&\lim_{a\to 0}\frac{1}{4\pi a^2}\Big(S_2^a(t)f_2 - S_2^0(t)f_2, \varphi_2\Big)\\
&= \frac{1}{4\pi}\frac{1}{2}\int_0^t d\tau \int_{S_-^2} d\eta \int dq_1 dp_1 dp_2 |\eta\cdot(p_1-p_2)|\\
&\quad\times\big[f_2(q_1 + p_1^*(t-\tau), p_1^*, q_1 + p_2^*(t-\tau), p_2^*)\\
&\qquad - f_2(q_1 + p_1(t-\tau), p_1, q_1 + p_2(t-\tau), p_2)\big]\\
&\quad\times\varphi_2(q_1 - p_1\tau, p_1, q_1 - p_2\tau, p_2)\\
&\quad+\frac{1}{4\pi}\frac{1}{2}\int_0^t d\tau \int_{S_+^2} d\eta \int dq_1 dp_1 dp_2 |\eta\cdot(p_1-p_2)|\\
&\quad\times\big[f_2(q_1 + p_1(t-\tau), p_1, q_1 + p_2(t-\tau), p_2)\\
&\qquad - f_2(q_1 + p_1^*(t-\tau), p_1^*, q_1 + p_2^*(t-\tau), p_2^*)\big]\\
&\quad\times\varphi_2(q_1 - p_1^*\tau, p_1^*, q_1 - p_2^*\tau, p_2^*)\\
&= \frac{1}{4\pi}\int_0^t d\tau \int_{S_-^2} d\eta \int dq_1 dp_1 dp_2 |\eta\cdot(p_1-p_2)|\\
&\quad\times\big[f_2(q_1 + p_1^*(t-\tau), p_1^*, q_1 + p_2^*(t-\tau), p_2^*)\\
&\qquad - f_2(q_1 + p_1(t-\tau), p_1, q_1 + p_2(t-\tau), p_2)\big]\\
&\quad\times\varphi_2(q_1 - p_1\tau, p_1, q_1 - p_2\tau, p_2)\\
&= \frac{1}{4\pi}\int_0^t d\tau \int_{S_-^2} d\eta \int dq_1 dp_1 dq_2 dp_2 |\eta\cdot(p_1-p_2)|\delta(q_1-q_2)\\
&\quad\times\big[f_2(q_1 + p_1^*(t-\tau), p_1^*, q_2 + p_2^*(t-\tau), p_2^*)\\
&\qquad - f_2(q_1 + p_1(t-\tau), p_1, q_2 + p_2(t-\tau), p_2)\big]
\end{aligned}$$

$$\times \varphi_2(q_1 - p_1\tau, p_1, q_2 - p_2\tau, p_2)$$

$$= \frac{1}{4\pi}\int_0^t d\tau \int_{S^2_-} d\eta \int dq_1 dp_1 dq_2 dp_2 |\eta \cdot (p_1 - p_2)|$$
$$\times \delta(q_1 + p_1\tau - q_2 - p_2\tau)$$
$$\times \big[f_2(q_1 + p_1\tau + p_1^*(t-\tau), p_1^*, q_2 + p_2\tau + p_2^*(t-\tau), p_2^*)$$
$$- f_2(q_1 + p_1 t, p_1, q_2 + p_2 t, p_2)\big]\varphi_2(q_1, p_1, q_2, p_2)$$

$$= \frac{1}{4\pi}\int_0^t d\tau \int_{S^2_-} d\eta \int dq_1 dp_1 dq_2 dp_2 |\eta \cdot (p_1 - p_2)|$$
$$\times \delta(q_1 + p_1\tau - q_2 - p_2\tau)$$
$$\times \big[S_2(t) f_2(x_1, x_2) - S_2^0(t) f_2(x_1, x_2)\big]\varphi_2(x_1, x_2)$$
$$= (S_2(t) f_2 - S_2^0(t) f_2, \varphi_2), \tag{2.5.2}$$

where $S_2(t)$ is the evolution operator of the stochastic dynamics, $S_2(t) f_2(x_1, x_2) = f_2(X(t, x_1, x_2))$, and $X(t, x_1, x_2)$ is the stochastic trajectory. The last equalities in (2.5.2) define the functional $(S_2(t) f_2 - S_2^0(t) f_2, \varphi_2)$ for two stochastic particles.

It follows from the definition of the stochastic dynamics (2.3.10) on the set D_{-t} (2.3.7) that

$$\lim_{a\to 0} \frac{1}{4\pi a^2}(S_2^a(t) f_2 - S_2^0(t) f_2, \varphi_2) = (S_2(t) f_2 - S_2^0(t) f_2, \varphi_2)$$
$$= \frac{1}{4\pi}\int_0^t d\tau \int_{S^2_-} d\eta \int dq_1 dp_1 dq_2 dp_2 |\eta \cdot (p_1 - p_2)|$$
$$\times \delta(q_1 + p_1\tau - q_2 - p_2\tau)$$
$$\times \big[f_2(q_1 + p_1\tau + p_1^*(t-\tau), p_1^*, q_2 + p_2\tau + p_2^*(t-\tau), p_2^*)$$
$$- f_2(q_1 + p_1 t, p_1, q_2 + p_2 t, p_2)\big]$$
$$= \frac{1}{4\pi}\int_{D_{-t}} \big[S_2(t) f_2(x_1, x_2) - S_2^0(t) f_2(x_1, x_2)\big]\varphi_2(x_1, x_2) dx_1 dx_2. \tag{2.5.3}$$

In what follows, we use the set D_{-t}, $q_1 + p_1\tau - q_2 - p_2\tau = 0$, $q_1 \in \mathbb{R}^3$, $q_2 \in \mathbb{R}^3$, $p_1 \in \mathbb{R}^3$, $p_2 \in \mathbb{R}^3$, $0 \le \tau \le t$, $\eta \in S^2_-$.

It is obvious that the function

$$S_2(t) f_2(x_1, x_2) - S_2^0(t) f_2(x_1, x_2)$$
$$= f_2(q_1 + p_1^*(t-\tau), p_1^*, q_2 + p_2^*(t-\tau), p_2^*)$$

$$- f_2(q_1 + p_1(t-\tau), p_1, q_2 + p_2(t-\tau), p_2), \quad \eta \in S^2_-, \quad (x_1, x_2) \in D_{-t},$$

is different from zero for $(x_1, x_2) \in D_{-t}$ because, for initial data $(x_1, x_2) \in D_{-t}$, stochastic particles interact at $t = \tau$ and the stochastic trajectories differ from free ones.

In the set D_{-t}, we have used the variables $\tau, \eta, q_1, p_1, p_2$ and the infinitesimal volume $|\eta \cdot (p_1 - p_2)| d\tau d\eta dq_1 dp_1 dp_2$. This volume has been directly defined in formula (2.5.2) and coincides with the infinitesimal volume $a^2 |\eta \cdot (p_1 - p_2)| d\tau d\eta dq_1 dp_1 dp_2$ used in (2.4.4) for $a^2 = 1$.

We now introduce a functional $(\bar{S}_2(t) f_2, \varphi_2)$ associated with the stochastic dynamics of two particles. Using formulas (2.5.2) and (2.5.3), we define the functional $(\bar{S}_2(t) f_2, \varphi_2)$ with arbitrary test functions $\varphi_2(x_1, x_2)$ as follows:

$$\begin{aligned}
&(\bar{S}_2(t) f_2, \varphi_2) \\
&\quad = (S_2^0(t) f_2, \varphi_2) + (S_2(t) f_2 - S_2^0(t) f_2, \varphi_2) \\
&\quad = \int dq_1 dp_1 dq_2 dp_2 \, f_2(q_1 + p_1 t, p_1, q_2 + p_2 t, p_2) \varphi_2(q_1, p_1, q_2, p_2) \\
&\qquad + \frac{1}{4\pi} \int_0^t d\tau \int_{S^2_-} d\eta \int dq_1 dp_1 dq_2 dp_2 |\eta \cdot (p_1 - p_2)| \delta(q_1 + p_1 \tau - q_2 - p_2 \tau) \\
&\qquad\quad \times \big[f_2(q_1 + p_1 \tau + p_1^*(t-\tau), p_1^*, q_2 + p_2 \tau + p_2^*(t-\tau), p_2^*) \\
&\qquad\qquad - f_2(q_1 + p_1 t, p_1, q_2 + p_2 t, p_2) \big] \varphi_2(q_1, p_1, q_2, p_2).
\end{aligned} \tag{2.5.4}$$

According to (2.5.4), the functional average $(\bar{S}_2(t) f_2, \varphi_2)$ of the observable $\varphi_2(x_1, x_2)$ over the state $S_2(t) f_2(x_1, x_2)$ consists of two terms. The first one coincides with the corresponding functional of free dynamics. The second one takes into account the hypersurfaces $q_1 + p_1 \tau = q_2 + p_2 \tau$, $0 \le \tau \le t$, on which the stochastic particles interact and where the stochastic dynamics differs from free dynamics. On these hypersurfaces, we use the measure $\delta(q_1 + p_1 \tau - q_2 - p_2 \tau) \times |\eta \cdot (p_1 - p_2)| dq_1 dp_1 dq_2 dp_2$ and the averaging procedure with respect to the random vector η. The second functional was obtained as the limit when $a \to 0$ for the average of the corresponding functional (2.4.4) for hard spheres over a sphere. We stress that this is a crucial point in the definition (2.5.4) of the functional $(\bar{S}_2(t) f_2, \varphi_2)$ because, in traditional statistical mechanics, sets of lower dimension than the phase space are neglected.

It follows from (2.5.4) that

$$\begin{aligned}
\bar{S}_2(t) f(x_1, x_2) &= f_2(q_1 + p_1 t, p_1, q_2 + p_2 t, p_2) \\
&\quad + \frac{1}{4\pi} \int_0^t d\tau \int_{S^2_-} d\eta \, |\eta \cdot (p_1 - p_2)| \delta(q_1 + p_1 \tau - q_2 - p_2 \tau) \\
&\quad \times \big[f_2(q_1 + p_1 \tau + p_1^*(t-\tau), p_1^*, q_2 + p_2 \tau + p_2^*(t-\tau), p_2^*)
\end{aligned}$$

$$- f_2(q_1 + p_1 t, p_1, q_2 + p_2 t, p_2)\big]. \tag{2.5.5}$$

We now stress the principal difference between the operator $S_2(t)$ and $\bar{S}_2(t)$. According to the definition of the stochastic dynamics, the following statements are true:

(i) if $q_1 + p_1\tau \neq q_2 + p_2\tau$ for all $0 \leq \tau \leq t$, then

$$S_2(t) f_2(x_1, x_2) = f(q_1 + p_1 t, p_1, q_2 + p_2 t, p_2),$$

(ii) if $q_1 + p_1\tau = q_2 + p_2\tau$ for some $0 \leq \tau \leq t$ and $\eta \in S^2_-$, then

$$S_2(t) f_2(x_1, x_2) = f(q_1 + p_1\tau + p_1^*(t-\tau), p_1^*, q_2 + p_2\tau + p_2^*(t-\tau), p_2^*),$$

and if $\eta \in S^2_+$, then

$$S_2(t) f(x_1, x_2) = f(q_1 + p_1 t, p_1, q_2 + p_2 t, p_2).$$

Thus, as a usual function $S_2(t) f_2(x_1, x_2)$ is defined by above written formulas (i) and (ii), formula (2.5.5) defines the generalized function $\bar{S}_2(t) f_2(x_1, x_2)$ corresponding to $S_2(t) f_2(x_1, x_2)$; it defines how to integrate the usual function $S_2(t) f_2(x_1, x_2)$ with a test function $\varphi_2(x_1, x_2)$, or how to calculate the functional $(\bar{S}_2(t) f_2, \varphi_2)$ that is the average of the observable $\varphi_2(x_1, x_2)$ over the state $S_2(t) f_2(x_1, x_2)$.

Remark 2.3. Note that the integrand in functional (2.5.2), regarded as a function of phase points (x_1, x_2), is defined on the set D_{-t}, which consists of points with the following characteristic property: the vectors of differences of positions are parallel to the vectors of differences of momenta. This means that the dimension of the set D_{-t} is lower than the dimension of the domain D^a_{-t} with $a = 1$ or the dimension of the entire phase space. But the integrand, regarded as a function of τ, η, q_1, p_1, p_2, is different from zero on the domain $0 \leq \tau \leq t$, $\eta \subset S^2$, $q_1 \in \mathbb{R}^3$, $p_1 \in \mathbb{R}^3$, $p_2 \in \mathbb{R}^3$, and functional (2.5.2) exists for test functions $\varphi_2(x_1, x_2)$ with compact support and for continuous functions f_2.

The main difference between functionals (2.4.4) and (2.5.2) is that the integrand of functional (2.4.4)

$$\begin{aligned} &\big[f_2(q_1 + p_1^*(t-\tau), p_1^*, q_1 - a\eta + p_2^*(t-\tau), p_2^*) \\ &\quad - f_2(q_1 + p_1(t-\tau), p_1, q_1 - a\eta + p_2(t-\tau), p_2)\big] \\ &\quad \times \varphi_2(q_1 - p_1\tau, p_1, q_1 - a\eta - p_2\tau, p_2) \end{aligned}$$

depends on the points

$$(x_1, x_2) = \big(q_1 + p_1^*(t-\tau), p_1^*, q_1 - a\eta + p_2^*(t-\tau), p_2^*\big)$$

in the first term or on the points

$$(x_1, x_2) = \big(q_1 + p_1(t-\tau), p_1, q_1 - a\eta + p_2(t-\tau), p_2\big)$$

in the second term, and the corresponding Jacobian of the transformation $(x_1, x_2) \to (\tau, \eta, q_1, p_1, p_2)$ is different from zero and equal to $a^2|\eta \cdot (p_1 - p_2)|$ in both cases.

The integrand of functional (2.5.2)

$$\begin{aligned}\big[f_2(q_1 + p_1^*(t-\tau), p_1^*, q_1 + p_2^*(t-\tau), p_2^*)&\\ - f_2(q_1 + p_1(t-\tau), p_1, q_1 + p_2(t-\tau), p_2)\big]&\\ \times \varphi_2(q_1 - p_1\tau, p_1, q_1 - p_2\tau, p_2)&\end{aligned} \tag{2.5.6}$$

depends on the points

$$(x_1, x_2) = \big(q_1 + p_1^*(t-\tau), p_1^*, q_1 + p_2^*(t-\tau), p_2^*\big)$$

in the first term or on the points

$$(x_1, x_2) = \big(q_1 + p_1(t-\tau), p_1, q_1 + p_2(t-\tau), p_2\big)$$

in the second term with parallel differences of positions and momenta, and, thus, this set of points has lower dimension than the dimension of $D^a_{-t}|_{a=1} = D^1_{-t}$ and the Jacobian of the transformation $(x_1, x_2) \to (\tau, \eta, q_1, p_1, p_2)$ on this set is equal to zero. The integrand, regarded as a function of $0 \le \tau \le t$, $\eta \subset S^2$, $q_1 \in \mathbb{R}^3$, $p_1 \in \mathbb{R}^3$, $p_2 \in \mathbb{R}^3$, is a continuous function with compact support, and, thus, functional (2.5.2) exists.

2.5.2 Derivative of functional with respect to time

We now define the derivative of functional (2.5.4) with respect to time at $t = 0$. From (2.5.4), we obtain

$$\begin{aligned}&\frac{d}{dt}(\bar{S}_2(t)f_2, \varphi_2)|_{t=0}\\ &\quad = \frac{d}{dt}(S_2^0(t)f_2, \varphi_2) + \frac{d}{dt}\Big(S_2(t)f_2 - S_2^0(t)f_2, \varphi_2\Big)|_{t=0}\\ &\quad = \int dq_1dp_1dq_2dp_2\Big(\sum_{i=1}^{2} p_i \cdot \frac{\partial}{\partial q_i} f_2(x_1, x_2)\Big)\varphi_2(x_1, x_2)\\ &\qquad + \frac{1}{4\pi}\int_{S^2_-} d\eta \int dq_1dq_2dp_1dp_2|\eta \cdot (p_1 - p_2)|\delta(q_1 - q_2)\\ &\qquad\qquad \times \Big[f_2(q_1, p_1^*, q_2, p_2^*) - f_2(q_1, p_1, q_2, p_2)\Big]\varphi_2(q_1, p_1, q_2, p_2)\end{aligned}$$

$$= \int dq_1 dp_1 dq_2 dp_2 \Big(\sum_{i=1}^{2} p_i \cdot \frac{\partial}{\partial q_i} f_2(x_1, x_2) \Big) \varphi_2(x_1, x_2)$$

$$+ \int_{S^2} d\eta \int dq_1 dq_2 dp_1 dp_2 \chi(\eta) \Theta(-\eta \cdot (p_1 - p_2)) |\eta \cdot (p_1 - p_2)| \delta(q_1 - q_2)$$

$$\times \Big[f_2(q_1, p_1^*, q_2, p_2^*) - f_2(q_1, p_1, q_2, p_2) \Big] \varphi_2(q_1, p_1, q_2, p_2) \tag{2.5.7}$$

for arbitrary test functions φ_2. It follows from (2.5.7) that

$$\frac{d}{dt} \bar{S}_2(t) f_2(x_1, x_2)|_{t=0}$$

$$= \sum_{i=1}^{2} p_i \cdot \frac{\partial}{\partial q_i} f_2(x_1, x_2) + \frac{1}{4\pi} \int_{S^2} d\eta \Theta(-\eta \cdot (p_1 - p_2)) |\eta \cdot (p_1 - p_2)| \delta(q_1 - q_2)$$

$$\times \big[f_2(q_1, p_1^*, q_2, p_2^*) - f_2(q_1, p_1, q_2, p_2) \big]. \tag{2.5.8}$$

We have the boundary condition according to which, in the first term on the right-hand side of (2.5.7)–(2.5.8), one must replace p_1, p_2 by p_1^*, p_2^* for $q_1 = q_2$, $\eta \subset S^2_-$. Comparing (2.4.7) with (2.5.7), we obtain

$$\lim_{a \to 0} \frac{1}{4\pi a^2} \frac{d}{dt} (S_2^a(t) f_2 - S_2^0(t) f_2, \varphi_2)|_{t=0}$$

$$= \frac{d}{dt} (S_2(t) f_2 - S_2^0(t) f_2, \varphi_2)|_{t=0}. \tag{2.5.9}$$

Consider the derivative of functional (2.5.4) with respect to time at $t \neq 0$. We have

$$\frac{d}{dt} (\bar{S}_2(t) f_2, \varphi_2)$$

$$= \frac{d}{dt} (S_2^0(t) f_2, \varphi_2) + \frac{d}{dt} (S_2(t) f_2 - S_2^0(t) f_2, \varphi_2)$$

$$= \int dq_1 dp_1 dq_2 dp_2 \Big(\sum_{i=1}^{2} p_i \cdot \frac{\partial}{\partial q_i} f_2(q_1 + p_1 t, p_1, q_2 + p_2 t, p_2) \Big) \times \varphi_2(x_1, x_2)$$

$$+ \frac{1}{4\pi} \int_0^t d\tau \int_{S^2_-} d\eta \int dq dp_1 dq_2 dp_2 |\eta \cdot (p_1 - p_2)|$$

$$\times \delta(q_1 + p_1 \tau - q_2 - p_2 \tau) \Big[\Big(p_1^* \frac{\partial}{\partial q_1} + p_2^* \frac{\partial}{\partial q_2} \Big)$$

$$\times f_2(q_1 + p_1 \tau + p_1^*(t - \tau), p_1^*, q_2 + p_2 \tau + p_2^*(t - \tau), p_2^*)$$

$$- \Big(p_1 \frac{\partial}{\partial q_1} + p_2 \frac{\partial}{\partial q_2}\Big) f_2(q_1 + p_1 t, p_1, q_2 + p_2 t, p_2)\Big] \varphi_2(q_1, p_1, q_2, p_2)$$

$$+ \frac{1}{4\pi} \int_{S^2_-} d\eta \int dq_1 dp_1 dq_2 dp_2 |\eta \cdot (p_1 - p_2)| \delta(q_1 + p_1 t - q_2 - p_2 t)$$

$$\times \big[f_2(q_1 + p_1 t, p_1^*, q_2 + p_2 t, p_2^*) - f_2(q_1 + p_1 t, p_1, q_2 + p_2 t, p_2)\big]$$

$$\times \varphi_2(q_1, p_1, q_2, p_2). \tag{2.5.10}$$

This formula means that

$$\frac{d}{dt} \bar{S}_2(t) f_2(x_1, x_2) = \Big(p_1 \frac{\partial}{\partial q_1} + p_2 \frac{\partial}{\partial q_2}\Big) S_2^0(t) f_2(x_1, x_2)$$

if $q_1 - q_2 \neq (p_2 - p_1)\tau$ for all $0 \leq \tau \leq t$, i.e., for $(x_1, x_2) \notin D_{-t}$.

If $q_1 - q_2 = (p_2 - p_1)\tau$ for some $0 \leq \tau \leq t$, i.e., for $(x_1, x_2) \in D_{-t}$, then (2.5.10) yields

$$\frac{d}{dt} \bar{S}_2(t) f_2(x_1, x_2)$$

$$= \Big(p_1^* \frac{\partial}{\partial q_1} + p_2^* \frac{\partial}{\partial q_2}\Big) \times f_2(q_1 + p_1 \tau + p_1^*(t - \tau), p_1^*, q_2 + p_2 \tau + p_2^*(t - \tau), p_2^*)$$

$$+ \frac{1}{4\pi} \int_{S^2_-} d\eta \, |\eta \cdot (p_1 - p_2)| \big[f_2(q_1 + p_1 t, p_1^*, q_2 + p_2 t, p_2^*)$$

$$- f_2(q_1 + p_1 t, p_1, q_2 + p_2 t, p_2)\big] \delta(q_1 + p_1 t - q_2 - p_2 t).$$

The results obtained above can be regarded as a proof of the following theorem:

Theorem 2.1. *The average of the evolution operator of the system of two hard spheres over the sphere $q_1 - q_2 - a\eta = 0$ (2.5.1) converges as $a \to 0$ to the evolution operator of the stochastic dynamics of two point particles (2.5.2), (2.5.3). The average of the infinitesimal operator of the system of two hard spheres over the sphere $q_1 - q_2 - a\eta = 0$ converges as $a \to 0$ to the infinitesimal operator of the stochastic dynamics of two point particles (2.5.8)–(2.5.9). In both cases, the convergence is understood in the weak sense (in the sense of generalized functions).*

All results can be extended to the operators $S_2^a(-t)$, $S_2(-t)$, $S_2^0(-t)$, $t \leq 0$. It is sufficient to replace the domains D_{-t}^a, D_{-t} by D_t^a, D_t and the operators $S_2^a(t)$, $S_2(t)$, $S_2^0(t)$ by $S_2^a(-t)$, $S_2(-t)$, $S_2^0(-t)$. We obtain

$$\frac{d}{dt} S_2^a(-t) f_2(x_1, x_2)|_{t=0}$$

$$= -\sum_{i=1}^{2} p_i \cdot \frac{\partial}{\partial q_i} f_2(x_1, x_2)$$

$$+ a^2 \int_{S^2} d\eta \delta(q_1 - q_2 - a\eta)|\eta \cdot (p_1 - p_2)|\Theta(\eta \cdot (p_1 - p_2))$$
$$\times \left[f_2(q_1, p_1^*, q_2, p_2^*) - f_2(q_1, p_1, q_2, p_2) \right], \tag{2.5.11}$$

$$\frac{d}{dt} \bar{S}_2(-t) f_2(x_1, x_2)|_{t=0}$$
$$= -\sum_{i=1}^{2} p_i \cdot \frac{\partial}{\partial q_i} f_2(x_1, x_2)$$
$$+ \frac{1}{4\pi} \int_{S^2} d\eta \delta(q_1 - q_2)|\eta \cdot (p_1 - p_2)|\Theta(\eta \cdot (p_1 - p_2))$$
$$\times \left[f_2(q_1, p_1^*, q_2, p_2^*) - f_2(q_1, p_1, q_2, p_2) \right].$$

In the operators $-\sum_{i=1}^{2} p_i \cdot \frac{\partial}{\partial q_i}$, we should replace the momenta p_1, p_2 by p_1^*, p_2^* if $q_1 - q_2 - a\eta = 0$, $\eta \subset S_+^2$, for hard spheres or $q_1 = q_2$, $\eta \subset S_+^2$, for stochastic particles.

2.6 General case of many-particle system

2.6.1 Functional for many hard spheres

The Hamiltonian dynamics of N hard spheres is defined as follows: hard spheres move as free particles until two of them collide; after a collision, their momenta change according to (2.2.2). Denote by $X^a(t,x)$ the trajectory of N hard spheres with initial data $x = (x_1, \dots, x_N)$, $X^a(t,x) = (X_1^a(t,x), \dots, X_N^a(t,x))$, $X_i^a(t,x) = (Q_i^a(t,x), P_i^a(t,x))$, $X^a(t,x)|_{t=0} = x$. Recall that $Q_i(t)$ is the position of the center of i-th sphere at instant of time t.

The stochastic dynamics of N point particles is defined as follows: particles move as free ones until two of them collide; after a collision, their momenta change according to (2.2.2), but with a random vector η. Only pair collisions are considered. Denote by $X(t,x)$ the trajectory of N stochastic particles:

$$X(t,x) = (X_1(t,x), \dots, X_N(t,x)),$$
$$X_i(t,x) = (Q_i(t,x), P_i(t,x)),$$
$$X(t,x)|_{t=0} = x.$$

All arguments presented in the previous sections can be applied to N-particle systems of hard spheres and stochastic particles or, more exactly, to every pair of colliding particles. We are interested in relations between the operators of evolution of N-particle systems of hard spheres and stochastic particles.

Consider the following functional for the case of N hard spheres:

$$(S_N^a(t)f_N - S_N^0(t)f_N, \varphi_N)$$
$$= \frac{1}{2}\int \big[S_N^a(t)f_N(x_1,\dots,x_N)$$
$$- S_N^0(t)f_N(x_1,\dots,x_N)\big]\varphi_N(x_1,\dots,x_N)dx_1\dots dx_N, \qquad (2.6.1)$$

where $S_N^a(t)$ is the operator of evolution (shift along the trajectory) of N hard spheres and $S_N^0(t)$ is the operator of evolution of N free particles. We assume that the test function φ_N is equal to zero on the forbidden configurations and on some neighborhood of the intersection of two or more hypersurfaces $|q_i - q_j| = a, i, j \in \{1,\dots,N\}$, and the function f_N is continuous.

Denote by $D_{-t}^a = D_{-t}^{a,N}$ the domain of initial data in the phase space such that spheres interact on the time interval $[0,t]$, $t > 0$. It is obvious that, for initial data from D_{-t}^a, hard spheres interact in "forward" evolution on $[0,t]$ and the function $\big[S_N^a(t) - S_N^0(t)\big]f_N(x_1,\dots,x_N)$ is not equal to zero for $x \in D_{-t}^a$ and is equal to zero for $x \notin D_{-t}^a$ because hard spheres do not interact for such initial data and $X^a(t,x) = X^0(t,x)$. Thus, we have

$$\big(S_N^a(t)f_N - S_N^0(t)f_N, \varphi_N\big)$$
$$= \frac{1}{2}\int_{D_{-t}^a} \big[S_N^a(t)f_N(x_1,\dots,x_N)$$
$$- S_N^0(t)f_N(x_1,\dots,x_N)\big] \times \varphi_N(x_1,\dots,x_N)dx_1\dots dx_N. \qquad (2.6.2)$$

(For forbidden initial data, we set $X^a(t,x) = X^0(t,x)$).

In order to define the infinitesimal operator of $S_N^a(t)$, we restrict ourselves to an infinitesimal $t = \Delta t$. Then functional (2.6.1), (2.6.2) can be represented in the following form:

$$(S_N^a(\Delta t)f_N - S_N^0(\Delta t)f_N, \varphi_N)$$
$$= \sum_{i<j=1}^{N} \int_0^t d\tau \int_{S_-^2} d\eta_{ij} \int dq_i dp_i dp_j dx_1 \dots \overset{i}{\vee} \dots \overset{j}{\vee} \dots dx_N$$
$$\times a^2|\eta_{ij}\cdot(p_i - p_j)|\big[f_N(q_1 + p_1\Delta t, p_1, \dots, q_i + p_i^*(\Delta t - \tau), p_i^*, \dots,$$
$$q_i - a\eta_{ij} + p_j^*(\Delta t - \tau), p_j^*, \dots, q_N + p_N t, p_N)$$
$$- f_N(q_1 + p_1\Delta t, p_1, \dots, q_i + p_i(\Delta t - \tau), p_i, \dots,$$
$$q_i - a\eta_{ij} + p_j(\Delta t - \tau), p_j, \dots, q_N + p_N\Delta t, p_N)\big]$$

$$\times \varphi_N(q_1, p_1, \ldots, q_i - p_i\tau, p_i, \ldots, q_i - a\eta_{ij} - p_j\tau, p_j, \ldots, q_N, p_N), \tag{2.6.3}$$

$$|\eta_{ij}| = 1, \quad i, j \in \{1, \ldots, N\}.$$

Here, the sign $\overset{i}{\vee} \ldots \overset{j}{\vee}$ means that dx_i and dx_j are omitted. Formula (2.6.3) has the following equivalent representation:

$$(S_N^a(\Delta t) f_N, \varphi_N) = (S_N^0(\Delta t) f_N, \varphi_N) + (S_N^a(\Delta t) f_N - S_N^0(\Delta t) f_N, \varphi_N). \tag{2.6.3$'$}$$

The functional $(S_N^a(\Delta t) f_N, \varphi_N)$ defines the average of the observable $\varphi_N(x_1, \ldots, x_N)$ over the state $S_N^a(\Delta t) f_N(x_1, \ldots, x_N)$.

2.6.2 Derivative of functional with respect to time

It is obvious that

$$\begin{aligned}
&\frac{d}{dt}(S_N^a(t) f_N - S_N^0(t) f_N, \varphi_N)|_{t=0} \\
&= \sum_{i<j=1}^{N} \int_{S_-^2} d\eta_{ij} \int dq_i dp_i dp_j dx_1 \ldots \overset{i}{\vee} \ldots \overset{j}{\vee} \ldots dx_N a^2 |\eta_{ij} \cdot (p_i - p_j)| \\
&\quad \times \big[f_N(q_1, p_1, \ldots, q_i, p_i^*, \ldots, q_i - a\eta_{ij}, p_j^*, \ldots, q_N, p_N) \\
&\quad\quad - f_N(q_1, p_1, \ldots, q_i, p_i, \ldots, q_i - a\eta_{ij}, p_j, \ldots, q_N, p_N) \big] \\
&\quad \times \varphi_N(q_1, p_1, \ldots, q_i, p_i, \ldots, q_i - a\eta_{ij}, p_j, \ldots, q_N, p_N).
\end{aligned} \tag{2.6.4}$$

Using (2.6.4), we obtain

$$\begin{aligned}
&\frac{d}{dt} S_N^a(t) f_N(x_1, \ldots, x_N)|_{t=0} \\
&= \sum_{i=1}^{N} p_i \cdot \frac{\partial}{\partial q_i} f_N(x_1, \ldots, x_N) \\
&\quad + \sum_{i<j=1}^{N} \int_{S_-^2} d\eta_{ij} a^2 \delta(q_i - q_j - a\eta_{ij}) |\eta_{ij} \cdot (p_1 - p_2)| \\
&\quad \times \big[f_N(x_1, \ldots, x_i^*, \ldots, x_j^*, \ldots, x_N) \\
&\quad\quad - f_N(x_1, \ldots, x_i, \ldots, x_j, \ldots, x_N) \big],
\end{aligned} \tag{2.6.5}$$

$$x_i^* = (q_i, p_i^*), \quad x_j^* = (q_j, p_j^*).$$

In the free Poisson bracket, one should replace (p_i, p_j) by (p_i^*, p_j^*) if $q_i - q_j - a\eta_{ij} = 0$, $\eta \in S^2_-$ (boundary condition). Formula (2.6.5) can be represented in the following identical form (see [BoB]):

$$\begin{aligned}
&\frac{d}{dt} S_N^a(t) f_N(x_1, \dots, x_N)|_{t=0} \\
&\quad = \sum_{i=1}^{N} p_i \cdot \frac{\partial}{\partial q_i} f_N(x_1, \dots, x_N) \\
&\qquad + \sum_{i<j=1}^{N} |\eta_{ij} \cdot (p_1 - p_2)| \Theta(-\eta_{ij} \cdot (p_1 - p_2)) \delta(|q_i - q_j| - a) \\
&\qquad \times \big[f_N(x_1, \dots, x_i^*, \dots, x_j^*, \dots, x_N) - f_N(x_1, \dots, x_i, \dots, x_j, \dots, x_N) \big],
\end{aligned} \tag{2.6.6}$$

$$\eta_{ij} = \frac{q_i - q_j}{|q_i - q_j|}.$$

2.6.3 Limit of the average of the functional for hard spheres and the functional of stochastic particles

Now consider the average of functional (2.6.3) over the sphere $|q_i - q_j| = a$:

$$\frac{1}{4\pi a^2} (S_N^a(\Delta t) f_N - S_N^0(\Delta t) f_N, \varphi_N). \tag{2.6.7}$$

It is obvious that, for continuous f_N and φ_N, there exists the limit of functional (2.6.7) as $a \to 0$:

$$\begin{aligned}
&\lim_{a \to 0} \frac{1}{4\pi a^2} (S_N^a(\Delta t) f_N - S_N^0(\Delta t) f_N, \varphi_N) \\
&\quad = \sum_{i<j=1}^{N} \frac{1}{4\pi} \int_0^t d\tau \int_{S^2_-} d\eta_{ij} \int dq_i dp_i dp_j dx_1 \dots \overset{i}{\vee} \dots \overset{j}{\vee} \dots dx_N \\
&\qquad \times |\eta_{ij} \cdot (p_1 - p_2)| \big[f_N(q_1 + p_1 \Delta t, p_1, \dots, q_i + p_i^*(\Delta t - \tau), p_i^*, \dots, \\
&\qquad\qquad q_i + p_j^*(\Delta t - \tau), p_j^*, \dots, q_N + p_N t, p_N) \\
&\qquad\quad - f_N(q_1 + p_1 \Delta t, p_1, \dots, q_i + p_i(\Delta t - \tau), p_i, \dots, \\
&\qquad\qquad q_i + p_j(\Delta t - \tau), p_j, \dots, q_N + p_N t, p_N) \big] \\
&\qquad \times \varphi_N(q_1, p_1, \dots, q_i - p_i \tau, p_i, \dots, q_i - p_j \tau, p_j, \dots, q_N, p_N).
\end{aligned} \tag{2.6.8}$$

Note that the stochastic dynamics of N point particles is defined in full analogy with the stochastic dynamics of two stochastic point particles. Namely, particles move freely until the positions of two particles coincide. Then they elastically collide, after the collision their momenta change according to (2.2.2) with the random vector $\eta \subset S^2_-$, and all particles continue to move freely until the next collision, and so on. For more details see the next section.

The last equality in (2.6.8) is the definition of the functional $(S_N(t)f_N - S^0_N(t)f_N, \varphi_N)$ for N stochastic particles. As follows from (2.6.8) the contributions of the hypersurfaces where the stochastic point particles interact are taken into account. The operator of evolution of N stochastic point particles for infinitesimal time Δt is defined as follows:

$$(S_N(\Delta t)f_N(x_1,\dots,x_N) = f_N(q_1 + p_1\Delta t, p_1,\dots,q_N + p_N\Delta t, p_N)$$

if $q_i + p_i\tau \neq q_j + p_j\tau$ for all pairs $i, j \in \{1,\dots,N\}$ and $0 \leq \tau \leq \Delta t$,

$$\begin{aligned} &S_N(\Delta t)f_N(x_1,\dots,x_N) \\ &\quad = f_N(q_1 + p_1\Delta t, p_1,\dots,q_i + p_i\tau + p_i^*(\Delta t - \tau), p_i^*,\dots, \\ &\qquad q_j + p_j\tau + p_j^*(\Delta t - \tau), p_j^*,\dots,q_N + p_N\Delta t, p_N) \end{aligned} \tag{2.6.9}$$

if $q_i + p_i\tau = q_j + p_j\tau$ for some $0 \leq \tau \leq \Delta t, i, j \in \{1,\dots,N\}, \eta_{ij}\cdot(p_i - p_j) \leq 0$, and

$$\begin{aligned} S_N(\Delta t)f_N(x_1,\dots,x_N) = f_N(&q_1 + p_1\Delta t, p_1,\dots,q_i + p_i\Delta t,\dots, \\ &p_i,\dots,q_j + p_j\Delta t, p_j,\dots,q_N + p_N\Delta t, p_N) \end{aligned}$$

if $q_i + p_i\tau = q_j + p_j\tau$ for some $0 \leq \tau \leq \Delta t, i, j \in \{1,\dots,N\}, \eta_{ij}\cdot(p_i - p_j) \geq 0$. The vectors p_i^*, p_j^* is defined by formula (2.2.2).

We neglect cases where two and more pairs of particles collide simultaneously on the time interval $0 \leq \tau \leq \Delta t$.

According to the definition of stochastic dynamics and by analogy with the two-particle system, functional (2.6.8) is equal to the following functional for the stochastic dynamics:

$$\begin{aligned} &\lim_{a\to 0}\frac{1}{4\pi a^2}(S^a_N(\Delta t)f_N - S^0_N(\Delta t)f_N, \varphi_N) \\ &\quad = (S_N(\Delta t)f_N - S^0_N(\Delta t)f_N, \varphi_N), \end{aligned} \tag{2.6.10}$$

where $S_N(\Delta t)$ is the evolution operator of N stochastic particles.

Using (2.6.8), we define the following functional for N stochastic particles with infinitesimal Δt:

$$
\begin{aligned}
&(\bar{S}_N(\Delta t) f_N, \varphi_N) \\
&\quad = \int dq_1 dp_1 dq_N dp_N\, f_N(q_1 + p_1\Delta t, p_1, \dots, q_N + p_N\Delta t, p_N) \\
&\qquad \times \varphi_N(q_1, p_1, \dots, q_N, p_N) \\
&\qquad + \sum_{i<j=1}^{N} \frac{1}{4\pi} \int_0^{\Delta t} d\tau \int_{S^2_-} d\eta_{ij} \int dq_1 dp_1 \dots dq_N dp_N |\eta_{ij} \cdot (p_1 - p_2)| \\
&\qquad \times \delta(q_i + p_i\tau - q_j - p_j\tau) \\
&\qquad \times \big[f_N(q_1 + p_1\Delta t, p_1, \dots, q_i + p_i\tau + p_i^*(\Delta t - \tau), p_i^*, \dots, \\
&\qquad\qquad q_j + p_j\tau + p_j^*(\Delta t - \tau), p_j^*, \dots, q_N + p_N\Delta t, p_N) \\
&\qquad - f_N(q_1 + p_1\Delta t, p_1, \dots, q_i + p_i\Delta t, p_i, \dots, q_j + p_j\Delta t, p_j, \dots, \\
&\qquad\qquad q_N + p_N\Delta t, p_N)\big] \varphi_N(q_1, p_1, \dots, q_i, p_i, \dots, q_j, p_j, \dots, q_N, p_N) \\
&\quad = (S_N^0(\Delta t) f_N, \varphi_N) + (S_N(\Delta t) f_N - S_N^0(\Delta t) f_N, \varphi_N).
\end{aligned} \tag{2.6.11}
$$

The functional $(\bar{S}_N(\Delta t) f_N, \varphi_N)$ (2.6.11) is the average of the observable φ_N over the state $S_N(t) f_N$. It consists of two terms: the first one is the average for the free dynamics and the second one takes into account the hypersurfaces $q_i + p_i\tau = q_j + p_j\tau$, $0 \le \tau \le t$, $i, j \in \{1, \dots, N\}$, where the stochastic particles interact. Here lies the principal difference between the traditional and stochastic statistical mechanics: in the traditional statistical mechanics, sets of lower dimension than the phase space are neglected.

Formula (2.6.11) holds for arbitrary test functions, and therefore, it defines the generalized function $\bar{S}_N(\Delta t) f_N(x_1, \dots, x_N)$ as follows:

$$
\begin{aligned}
&\bar{S}_N(\Delta t) f_N(x_1, \dots, x_N) \\
&\quad = f_N(q_1 + p_1\Delta t, p_1, \dots, q_N + p_N\Delta t, p_N) \\
&\qquad + \frac{1}{4\pi} \sum_{i<j=1}^{N} \int_0^{\Delta t} d\tau \int_{S^2_-} d\eta_{ij} |\eta_{ij} \cdot (p_i - p_j)| \delta(q_i + p_i\tau - q_j - p_j\tau) \\
&\qquad \times \big[f_N(q_1 + p_1\Delta t, p_1, \dots, q_i + p_i\tau + p_i^*(\Delta t - \tau), p_i^*, \dots, \\
&\qquad\qquad q_j + p_j\tau + p_j^*(\Delta t - \tau), p_j^*, \dots, q_N + p_N\Delta t, p_N) \\
&\qquad - f_N(q_1 + p_1\Delta t, p_1, \dots, q_i + p_i\Delta t, p_i, \dots, q_j + p_j\Delta t, p_j, \dots, \\
&\qquad\qquad q_N + p_N\Delta t, p_N)\big].
\end{aligned} \tag{2.6.12}
$$

The generalized function $\bar{S}_N(\Delta t) f_N(x_1, \ldots, x_N)$ is associated with the usual function $S_N(t) f_N(x_1, \ldots, x_N)$ given numerically by (2.6.9). $\bar{S}_N(\Delta t) f_N(x_1, \ldots, x_N)$ is used in the functional average of the observable $\varphi_N(x_1, \ldots, x_N)$ over the state $S_N(\Delta t) f_N(x_1, \ldots, x_N)$ and shows how to calculate this average.

The following formula is true for the derivative of functional (2.6.11) with respect to time:

$$\begin{aligned}
&\frac{d}{dt}(\bar{S}_N(t) f_N(x_1, \ldots, x_N))|_{t=o} \\
&\quad = \sum_{i=1}^{N} p_i \cdot \frac{\partial}{\partial q_i} f_N(x_1, \ldots, x_N) + \sum_{i<j=1}^{N} \frac{1}{4\pi} \int_{S^2_-} d\eta_{ij} |\eta_{ij} \cdot (p_i - p_j)| \\
&\qquad \times \big[f_N(q_1, p_1, \ldots, q_i, p_i^*, \ldots, q_j, p_j^*, \ldots, q_N, p_N) \\
&\qquad - f_N(q_1, p_1, \ldots, q_i, p_i, \ldots, q_j, p_j, \ldots, q_N, p_N) \big] \\
&\quad = \tilde{\mathcal{H}}_N^+ f_N(x_1, \ldots, x_N).
\end{aligned} \tag{2.6.13}$$

Consider also the following functional for infinitesimal Δt:

$$\begin{aligned}
&(\tilde{S}_N(\Delta t) f_N, \varphi_N) \\
&\quad = \int dq_1 dp_1 \ldots dq_N dp_N\, f_N(q_1 + p_1 \Delta t, p_1, \ldots, \\
&\qquad q_N + p_N \Delta t, p_N) \varphi_N(q_1, p_1, \ldots, q_N, p_N) \\
&\quad + \sum_{i<j=1}^{N} \int_0^{\Delta t} d\tau \int dq_1 dp_1 \ldots dq_N dp_N |\eta_{ij} \cdot (p_i - p_j)| \delta(q_i + p_i \tau - q_j - p_j \tau) \\
&\quad \times \big[f_N(q_1 + p_1 \Delta t, p_1, \ldots, q_i + p_i \tau + p_i^*(\Delta t - \tau), p_i^*, \ldots, \\
&\qquad q_j + p_j \tau + p_j^*(\Delta t - \tau), p_j^*, \ldots, q_N + p_N \Delta t, p_N) \\
&\quad - f_N(q_1 + p_1 \Delta t, p_1, \ldots, q_i + p_i \Delta t, \ldots, p_i, \ldots, q_j + p_j \Delta t, p_j, \ldots, \\
&\qquad q_N + p_N \Delta t, p_N) \big] \varphi_N(q_1, p_1, \ldots, q_i, p_i, \ldots, q_j, p_j, \ldots, q_N, p_N).
\end{aligned} \tag{2.6.14}$$

If follows from (2.6.14) that

$$\begin{aligned}
&\tilde{S}_N(\Delta t) f_N(x_1, \ldots, x_N) \\
&\quad = f_N(q_1 + p_1 \Delta t, p_1, \ldots, q_N + p_N \Delta t, p_N) \\
&\qquad + \sum_{i<j=1}^{N} \int_0^{\Delta t} d\tau |\eta_{ij} \cdot (p_i - p_j)| \Theta(-\eta_{ij} \cdot (p_i - p_j)) \delta(q_i + p_i \tau - q_j - p_j \tau)
\end{aligned}$$

$$\times\big[f_N(q_1+p_1\Delta t,p_1,\dots,q_i+p_i\tau+p_i^*(\Delta t-\tau),p_i^*,\dots,q_j$$
$$+p_j\tau+p_j^*(\Delta t-\tau),p_j^*,\dots,q_N+p_N\Delta t,p_N)-f_N(q_1+p_1\Delta t,p_1,\dots,$$
$$q_i+p_i\Delta t,p_i,\dots,q_j+p_j\Delta t,p_j,\dots,q_N+p_N\Delta t,p_N)\big]. \qquad (2.6.15)$$

We now explain the meaning of formulas (2.6.14) and (2.6.15). In the functional $(\tilde{S}_N(\Delta t)f_N,\varphi_N)$ and in the generalized function $\tilde{S}_N(\Delta t)f_N$ associated with the usual function $S_N(\Delta t)f_N$ we take into account the hypersurfaces where the stochastic particles interact, but do not average with respect to the random vectors η_{ij}. Note that in $(\bar{S}_N(\Delta t)f_N,\varphi_N)$ and $\bar{S}_N(\Delta t)f_N$ we average with respect to all random vectors η_{ij}, $i,j\in\{1,\dots,N\}$.

For fixed vectors η_{ij}, formula (2.6.15) yields

$$\frac{d}{dt}(\tilde{S}_N(t)f_N(x_1,\dots,x_N))|_{t=0}$$
$$=\sum_{i=1}^{N}p_i\cdot\frac{\partial}{\partial q_i}f_N(x_1,\dots,x_N)$$
$$+\sum_{i<j=1}^{N}|\eta_{ij}\cdot(p_i-p_j)|\Theta(-\eta_{ij}\cdot(p_i-p_j))\delta(q_i-q_j)$$
$$\times\big[f_N(q_1,p_1,\dots,q_i,p_i^*,\dots,q_j,p_j^*,\dots,q_N,p_N)$$
$$-f_N(q_1,p_1,\dots,q_i,p_i,\dots,q_j,p_j,\dots,q_N,p_N)\big]$$
$$=\tilde{\mathcal{H}}_N^+f_N(x_1,\dots,x_N). \qquad (2.6.16)$$

We have the boundary conditions and, according to them, one should replace (p_i,p_j) by (p_i^*,p_j^*) if $q_i=q_j$, $\eta_{ij}\in S^2$ in the Poisson bracket in (2.6.13), (2.6.16).

Formulas (2.6.9), (2.6.12), (2.6.14) define the state $S_N(\Delta t)f_N$ and associated with $S_N(\Delta t)f_N$ generalized functions $\bar{S}_N(\Delta t)f_N$, $\tilde{S}_N(\Delta t)f_N$ for infinitesimal Δt. For arbitrary time t, we define (formally) the state $S_N(t)f_N$ and associated generalized function $\bar{S}_N(t)f_N$, $\tilde{S}_N(t)f_N$ by the following formula:

$$S_N(t)f_N=\lim_{n\to\infty}\prod_{i=1}^{n}S_N(\Delta t_i)f_N,$$
$$\bar{S}_N(t)f_N=\lim_{n\to\infty}\prod_{i=1}^{n}\bar{S}_N(\Delta t_i)f_N. \qquad (2.6.17)$$
$$\tilde{S}_N(t)f_N=\lim_{n\to\infty}\prod_{i=1}^{n}\tilde{S}_N(\Delta t_i)f_N,\quad \sum_{i=1}^{N}\Delta t_i=t.$$

Suppose that we have the function

$$\bar{f}_N(t, x_1, \ldots, x_N) = \bar{S}_N(t) f_N(x_1, \ldots, x_N)$$

at time t and define the function

$$\bar{f}_N(t + \Delta t, x_1, \ldots, x_N) = \bar{S}_N(\Delta t) \bar{f}_N(t, x_1, \ldots, x_N)$$

by formula (2.6.12) with $\bar{f}_N(t, x_1, \ldots, x_N)$ instead of $f_N(x_1, \ldots, x_N)$. Then, according to (2.6.13), we have

$$\begin{aligned}
&\frac{d}{dt}\bar{f}_N(t, x_1, \ldots, x_N) \\
&\quad = \sum_{i=1}^{N} p_i \cdot \frac{\partial}{\partial q_i} \bar{f}_N(t, x_1, \ldots, x_N) \\
&\qquad + \sum_{i<j=1}^{N} \frac{1}{4\pi} \int_{S_-^2} d\eta_{ij} |\eta_{ij} \cdot (p_i - p_j)| \big[\bar{f}_N(t, x_1, \ldots, x_i^*, \ldots, x_j^*, \ldots, x_N) \\
&\qquad\qquad - \bar{f}_N(t, x_1, \ldots, x_i, \ldots, x_j, \ldots, x_N)\big]. \qquad (2.6.18)
\end{aligned}$$

Suppose that we have the function

$$\tilde{f}_N(t, x_1, \ldots, x_N) = \tilde{S}_N(t) f_N(x_1, \ldots, x_N)$$

at time t and define the function

$$\tilde{f}_N(t + \Delta t, x_1, \ldots, x_N) = \tilde{S}_N(\Delta t) \tilde{f}_N(t, x_1, \ldots, x_N)$$

by formula (2.6.15) with $\tilde{f}_N(t, x_1, \ldots, x_N)$ instead of $f_N(x_1, \ldots, x_N)$. Then it is easy to derive the following formulas for fixed vectors η_{ij}

$$\begin{aligned}
&\frac{d}{dt}\tilde{f}_N(t, x_1, \ldots, x_N) \\
&\quad = \sum_{i=1}^{N} p_i \cdot \frac{\partial}{\partial q_i} \tilde{f}_N(t, x_1, \ldots, x_N) + \sum_{i<j=1}^{N} |\eta_{ij} \cdot (p_i - p_j)| \Theta(-\eta_{ij} \cdot (p_i - p_j)) \\
&\qquad \times \big[\tilde{f}_N(t, x_1, \ldots, x_i^*, \ldots, x_j^*, \ldots, x_N) \\
&\qquad\qquad - \tilde{f}_N(t, x_1, \ldots, x_i, \ldots, x_j, \ldots, x_N)\big], \qquad (2.6.19)
\end{aligned}$$

$$x_i^* = (q_i, p_i^*), \quad x_j^* = (q_j, p_j^*).$$

2.7 Infinitesimal operator of the evolution operator of stochastic particles

2.7.1 Dynamics of finitely many particles

In this section, we calculate the infinitesimal operator of N stochastic particles in the sense of pointwise convergence and explain its connection with the infinitesimal operator and its action calculated in the previous sections in the weak sense as a certain generalized function. First, we summarize the definition of stochastic trajectories.

Consider N identical particles of unit mass located in a three-dimensional Euclidean space $\mathbb{R}^3$. Their positions and momenta are denoted by q_i and p_i, and $x_i = (q_i, p_i)$ is the phase point of the i-th particle, $i = 1, \ldots, N$. Denote by $Q_i(t)$, $P_i(t)$, and $X_i(t) = (P_i(t), Q_i(t))$ the position, momentum, and phase point of the i-th particle, respectively, at time $0 \leq t < \infty$ (this means that time changes from 0 to $+\infty$ or, more generally, that time increases), and let

$$Q_i(0) = q_i, \quad P_i(0) = p_i, \quad X_i(0) = x_i, \quad i = 1, \ldots, N.$$

The phase point of the entire system of N particles is denoted by $X(t) = (X_1(t), \ldots, X_N(t))$, $X(0) = (x_1, \ldots, x_N) = x = (q, p)$, $q = (q_1, \ldots, q_N)$, $p = (p_1, \ldots, p_N)$. To indicate the dependence of $X(t)$ on the initial value x, we write $X(t) = X(t, x) = X(t, x_1, \ldots, x_N)$, $X_i(t) = X_i(t, x) = X_i(t, x_1, \ldots, x_N)$. $X(t, x)$ is the trajectory of our system that passes through the point x at $t = 0$.

We assume that particles move freely until their positions coincide:

$$Q_i(t) = q_i + p_i t, \quad P_i(t) = p_i, \quad i = 1, \ldots, N.$$

If the positions of the i-th and j-th particles coincide at time t, i.e.,

$$Q_i(t) = Q_j(t), \tag{2.7.1}$$

then they collide instantaneously and their momenta change jumpwise.

After the collision, for time $t + 0$, the momenta have the following form:

$$\begin{aligned} P_i^*(t+0) &= P_i(t) + \eta_{ij}\, \eta_{ij} \cdot \big(P_i(t) - P_j(t)\big), \\ P_j^*(t+0) &= P_j(t) + \eta_{ij}\, \eta_{ij} \cdot \big(P_i(t) - P_j(t)\big), \end{aligned} \tag{2.7.2}$$

where η_{ij} is a unit vector, $|\eta_{ij}| = 1$, and $\eta_{ij} \cdot (P_i(t) - P_j(t))$ is the scalar product of the vector η_{ij} and $P_i(t) - P_j(t)$. We consider only vectors η_{ij} that satisfy the condition

$$\eta_{ij} \cdot (P_i(t) - P_j(t)) \leq 0 \tag{2.7.3}$$

for positive time $t \geq 0$. Denote the unit semisphere (2.7.3) by S^2_-.

If the vectors η_{ij} satisfy the condition

$$\eta_{ij} \cdot (P_i(t) - P_j(t)) \geq 0, \tag{2.7.3$'$}$$

then, after the collision, the momenta $P_i(t)$ and $P_j(t)$ do not change, i.e., they are the same as before the collision. Denote semisphere (2.7.3′) by S_+^2 and the sphere $|\eta_{ij}| = 1$ by S^2, $S^2 = S_-^2 \cup S_+^2$.

We assume that the vector η_{ij} is random with constant density of probability on the sphere S^2. Denote the density of probability by $\chi(\eta_{ij})$. Then

$$\int_{S^2} \chi(\eta_{ij}) d\eta_{ij} = 1, \quad \chi(\eta_{ij}) = \frac{1}{4\pi}. \tag{2.7.4}$$

It is obvious that two particles (the i-th and the j-th one) can collide only if the vector $p_i - p_j$ is parallel to the vector $q_i - q_j$.

We assume that, under a simultaneous collision of more than two particles, they move as free particles.

For negative time $-\infty < t \leq 0$ (this means that time changes from 0 to $-\infty$, or, more generally, that time decreases), all considerations remain the same as for $t \geq 0$, but the random vector η_{ij} satisfies the condition

$$\eta_{ij} \cdot (P_i(t) - P_j(t)) \geq 0, \tag{2.7.5}$$

i.e., the vector η_{ij} belongs to the unit semisphere S_+^2 (2.7.5) and, after the collision, for time $t - 0 \leq 0$, the momenta of the i-th and j-th particles have the following form:

$$\begin{aligned} P_i^*(t-0) &= P_i(t) - \eta_{ij}\, \eta_{ij} \cdot (P_i(t) - P_j(t)), \\ P_j^*(t-0) &= P_j(t) + \eta_{ij}\, \eta_{ij} \cdot (P_i(t) - P_j(t)). \end{aligned} \tag{2.7.2$'$}$$

If the vectors η_{ij} satisfy the condition

$$\eta_{ij} \cdot (P_i(t) - P_j(t)) \leq 0, \quad \eta_{ij} \in S_-^2, \tag{2.7.5$'$}$$

then, after the collision, the momenta $P_i(t)$ and $P_j(t)$ do not change. The vector η_{ij} is again random with constant density $\chi(\eta_{ij}) = 1/4\pi$.

We assume that random vectors η_{ij} related to different collisions are independent.

Note that the trajectories $X(t, x)$ constructed above are continuous functions of time t on the intervals of time between collisions. At the moment of collision, the trajectory $X(t, x)$ has a jump, and it is left continuous for $t > 0$ and right continuous for $t < 0$ (or for increasing and decreasing time, respectively). After collision particles move freely until the next collision and so on.

We call the above-defined evolution the *stochastic dynamics*.

Let us present shortly again a motivation for the definition of the concept of stochastic dynamics. For this purpose, we recall that the dynamics of hard spheres with diameter a is as follows:

They move as free particles if the distances between the centers of two spheres are greater than a: $|Q_i(t) - Q_j(t)| > a$. If $|Q_i(t) - Q_j(t)| = a$, they collide instantaneously and, after the collision, their momenta $P_i^*(t)$, $P_j^*(t)$ are given by (2.7.2), where the vector η_{ij} is determined by the formula

$$\eta_{ij} = \frac{Q_i(t) - Q_j(t)}{|Q_i(t) - Q_j(t)|}. \tag{2.7.6}$$

For fixed p_i, p_j and $a \neq 0$, the collision of the i-th and j-th particles may take place for all η_{ij} defined according to (2.7.6) and belonging to semisphere (2.7.3) for $t > 0$ or (2.7.5) for $t < 0$ (or for increasing and decreasing time, respectively).

Let the diameter a tend to zero, $a \to 0$, with fixed η_{ij}. In this limit, particles become point particles, and their dynamics coincides with the stochastic dynamics defined above with the same η_{ij} and with conditions (2.7.3) for $t > 0$ or (2.7.5) for $t < 0$. For the obtained point particles, momenta (2.7.2) after the collision are determined for arbitrary η_{ij} from the corresponding semisphere S_-^2 or S_+^2 with equal probability. (Details of the proof are presented in the previous sections.)

It is obvious that $X(t, x)$, regarded as a function of t for fixed x and arbitrary fixed vectors η_{ij} related to pair collisions, has the semigroup property

$$X(t_1 + t_2, x) = X(t_1, X(t_2, x)) = X(t_2, X(t_1, x)) \tag{2.7.7}$$

for arbitrary $t_1 > 0, t_2 > 0$ or $t_1 < 0, t_2 < 0$.

If $q_i = q_j$ for some pair $i, j \in \{1, \ldots, N\}$ at initial time $t = 0$, then the trajectory $X(t, x)$ is defined as follows:

$$X(t, x) = (q + p^* t, p^*), \quad p^* = (p_1, \ldots, p_i^*, \ldots, p_j^*, \ldots, p_N), \tag{2.7.8}$$

where p_i^*, p_j^* are defined as in (2.7.2) or (2.7.2′) with $\eta \subset S_-^2$ for $t > 0$ or $\eta \subset S_+^2$ for $t < 0$. If $\eta \subset S_+^2$ for $t > 0$ or $\eta \subset S_-^2$ for $t < 0$, then

$$X(t, x) = (q + pt, p). \tag{2.7.8′}$$

It is obvious that the above-mentioned properties of the trajectories are compatible with conditions (2.7.3), (2.7.3′) for increasing time because

$$(p_i^* - p_j^*) \cdot \eta = -\eta \cdot (p_i - p_j) < 0, \quad \eta \subset S_+^2,$$

and with conditions (2.7.5), (2.7.5′) for decreasing time because

$$(p_i^* - p_j^*) \cdot \eta = -\eta \cdot (p_i - p_j) > 0, \quad \eta \subset S_-^2.$$

It follows from this that $X(t, x)$ does possess the group property for fixed x and an arbitrary fixed vector $\eta_{i,j}$.

Note that we consider point particles that may interact only if the vectors $p_i - p_j$ are parallel to the vectors $q_i - q_j$. The points $(q_i + p_i^* t, p_i^*, q_i + p_j^* t, p_j^*)$ satisfy these conditions.

This means that if we consider the hypersurface all points of which satisfy the condition that the vectors of the difference of positions $q_1 - q_2$ are parallel to the vectors of the difference of momenta $p_1 - p_2$, then this hypersurface is invariant with respect to the stochastic dynamics.

The union of the hypersurfaces (with respect to time)

$$Q_i(t,x) = Q_j(t,x), \quad i,j \in \{1,\dots,N\}, \quad t > 0 \quad (t < 0), \tag{2.7.9}$$

that correspond to all pair collisions with fixed random vectors η_{ij} has lower dimension than the phase space.

In what follows, we denote by η the collection of random vectors η_{ij} related to all pair collisions in our system on the time interval $[0,t]$, $t > 0$, or $[t,0]$, $t < 0$. One can say that stochastic trajectories are defined by initial momenta and positions and by random vectors η, but depend explicitly on η only after collisions.

2.7.2 Evolution operator of finitely many particles and its infinitesimal operator

Consider a function $f_N(x_1,\dots,x_N) \equiv f(x)$, $x = (x_1,\dots,x_N)$ defined and continuous on the phase space of N particles. The operator of evolution of N particles $S_N(t)$ is formally defined by the formula

$$\begin{aligned}(S_N(t)f_N)(x_1,\dots,x_N) &= (S_N(t)f_N)(t) = f_N(X(t,x)) \\ &= f_N(X_1(t,x),\dots,X_N(t,x))\end{aligned} \tag{2.7.10}$$

for arbitrary time $t > 0$ or $t < 0$. The function $f(X(t,x))$ depends on time, initial phase point x, and random vectors η related to all pair collisions on the time interval $[0,t]$.

Consider the function $f_N(X(t,x))$ for an arbitrary fixed realization of random vectors η. This means that we also consider the operator $S_N(t)$ for some realization of η and, in what follows, we consider the properties of the operator $S_N(t)$ associated with some realization of η.

Outside hypersurfaces (2.7.9), the trajectory $X(t,x)$ coincides with the trajectory of free particles, i.e., it is a continuous function of time t and phase points x. Therefore, the function $f_N(X(t,x))$ is continuous outside hypersurface (2.7.9) , i.e., almost everywhere in the phase space (for a fixed realization of η).

The semigroup property of the operator $S_N(t)$

$$S_N(t_1+t_2) = S_N(t_1)S_N(t_2) = S_N(t_2)S_N(t_1) \tag{2.7.11}$$

for arbitrary $t_1 > 0, t_2 > 0$ or $t_1 < 0, t_2 < 0$ follows from the semigroup property of $X(t, x)$ with a fixed realization of η (2.7.7).

Let us proceed to the determination of the infinitesimal operator of the semigroup $S_N(t)$ on the set of differentiable functions f(x).

First, we consider the case $t \geq 0$ and differentiate the function $S_N(t)f_N(x) \equiv S(t)f(x)$ at time $t = 0$. (It is convenient to omit the sign N in this section.) We have

$$\frac{dS(t)f(x)}{dt}\bigg|_{t=0} = \lim_{\Delta t \to 0} \frac{f(X(\Delta t, x)) - f(x)}{\Delta t}. \tag{2.7.12}$$

If $q_i \neq q_j$ for arbitrary $i, j \in \{1, \dots, s\}$, then $X(\Delta t, x) = (q + p\Delta t, p)$ and

$$\frac{dS(t)f(x)}{dt}\bigg|_{t=0} = \sum_{i=1}^{N} p_i \cdot \frac{\partial}{\partial q_i} f(x). \tag{2.7.13}$$

If $q_i = q_j$ for some pair $i, j \in \{1, \dots, s\}$ and $\eta \subset S^2_-$, then

$$X(\Delta t, x) = (q + p^* \Delta t, p^*),$$

where

$$p^* = (p_1, \dots, p_i^*, \dots, p_j^*, \dots, p_N)$$

and p_i^*, p_j^* are defined by (2.7.2).

In order to calculate (2.7.12), we choose a coordinate system such that the first component of the vector (q_i, p_i, q_j, p_j) is directed along the fixed vector η_{ij}. In this coordinate system, we have

$$\begin{aligned} p_i^* &= (p_i^1 - (p_i^1 - p_j^1), p_i^2, p_i^3) = (p_j^1, p_i^2, p_i^3), \\ p_j^* &= (p_j^1 - (p_j^1 - p_i^1), p_j^2, p_j^3) = (p_i^1, p_j^2, p_j^3), \end{aligned} \tag{2.7.14}$$

and the function $f(X(t, x))$ has a jump at $t = 0$ which is connected with the jumps of $P_i^1(t)$ and $P_j^1(t)$ and is equal to $f(x^*) - f(x)$, $x^* = (q, p^*)$.

In this sense, the problem of the calculation of derivative (2.7.12) is reduced to the one-dimensional case, where the i-th and j-th particles change only the first components of their momenta. Taking this into account, we deduce from (2.7.12) that, for $q_i = q_j$, we have

$$\begin{aligned} \frac{dS(t)f(x)}{dt}\bigg|_{t=0} = \lim_{\Delta t \to 0} \Big\{ &(f(q + p^* \Delta t, p^*) - f(q + p^* O(\Delta t), p^*) \\ &+ f(q + p^* O(\Delta t), p^*) - f(q + pO(\Delta t), p)) \Big\} \Big/ \Delta t \end{aligned}$$

$$= \sum_{i=1}^{N} p_i \cdot \frac{\partial}{\partial q_i} f(x)\Big|_{p \to p_i^*} + \delta(t-\tau)[f(x^*) - f(x)]|_{t=0}, \tag{2.7.15}$$

where τ is the time of collision,

$$\tau = \frac{q_i^1 - q_j^1}{p_j^1 - p_i^1},$$

and $O(\Delta t)$ is an infinitesimal value of higher order.

Thus, in expression (2.7.15) for the infinitesimal operator of the evolution operator $S(t)$, we have the one-dimensional δ-function $\delta(t-\tau)|_{t=0}$, $q_i^1 = q_j^1$, $q_i^2 = q_j^2$, $q_i^3 = q_j^3$. As is known, the one-dimensional δ-function in the expression for the derivative $\frac{dS(t)}{dt} f(x)|_{t=0}$ is equivalent to the following boundary condition: For $q_i = q_j$ the momenta p_i, p_j should be replaced by p_i^*, p_j^* (this has been shown in the previous chapter for systems of hard spheres). Then we have the expression, equivalent to (2.7.15),

$$\frac{dS(t)}{dt} f(x)|_{t=0} = \sum_{i=1}^{N} p_i \cdot \frac{\partial}{\partial q_i} f(x) = \mathcal{H}_N^+ f(x) \tag{2.7.16}$$

with the boundary condition according to which, for $q_i = q_j$, $\eta \subset S_-^2$, the momenta on the right-hand side of (2.7.16) and in $f(x)$ should be replaced by p_i^*, p_j^*

On the set of functions $f(x)$ that are continuously differentiable and equal to zero in some neighborhood of the hyperplanes $q_i - q_j = 0$, $i, j \in \{1, \dots, N\}$, we have $f(x^*) - f(x) = 0$, the second term in (2.7.15) is equal to zero, and one directly obtains formulas (2.7.16).

By using the definition of the time of collision, one can represent the second term on the right-hand side of (2.7.15) as follows (for $q_i^2 = q_j^2$, $q_i^3 = q_j^3$):

$$\delta(t-\tau)[f(x^*) - f(x)]|_{t=0} = \delta(q_i^1 - q_j^1)|p_i^1 - p_j^1|[f(x^*) - f(x)]\Big|_{q_i^2 = q_j^2, q_i^3 = q_j^3}.$$

This expression, regarded as a generalized function in the three-dimensional space of the variables $q_i - q_j$, is concentrated on the first axis $q_i^1 - q_j^1$ (for $q_i^2 - q_j^2 = 0$, $q_i^3 - q_j^3 = 0$) and, therefore, is equal to

$$\begin{aligned} &\delta(q_i^1 - q_j^1)|p_i^1 - p_j^1|[f(x^*) - f(x)]\delta(q_i^2 - q_j^2)\delta(q_i^3 - q_j^3) \\ &\quad = \delta(q_i - q_j)|p_i^1 - p_j^1|[f(x^*) - f(x)] \end{aligned}$$

(for analogous calculations, see [GGV, p. 48]). Recalling that $p_i^1 - p_j^1 = \eta_{ij} \cdot (p_i - p_j)$, we obtain the following formula from (2.7.15):

$$\left.\frac{d\tilde{S}(t)f(x)}{dt}\right|_{t=0}$$
$$= \sum_{i=1}^{N} p_i \cdot \frac{\partial}{\partial q_i} f(x)\Big|_{p\to p^*} + \delta(q_i - q_j)|\eta_{ij}\cdot(p_i - p_j)|[f(x^*) - f(x)], \tag{2.7.17}$$

where $\left.\frac{d\tilde{S}(t)f(x)}{dt}\right|_{t=0}$ means the generalized function corresponding to $\left.\frac{dS(t)f(x)}{dt}\right|_{t=0}$ (2.7.15). Formula (2.7.17) does not depend on the choice of a coordinate system because the scalar product $\eta_{ij}\cdot(p_i - p_j)$ and $\delta(q_i - q_j)$ are invariant under rotation.

Let us explain from the physical point of view the replacement of the one-dimensional δ-function $\delta(q_i^1 - q_j^1)$ for $q_i^2 = q_j^2$, $q_i^3 = q_j^3$ by the three-dimensional δ-function $\delta(q_i - q_j) = \delta(q_i^1 - q_j^1)\delta(q_i^2 - q_j^2)\delta(q_i^3 - q_j^3)$. For this purpose, consider the following simple example: Let, in the three-dimensional Euclidean space, the mass m be continuously distributed on the first axis q^1 (for $q^2 = 0, q^3 = 0$) with density $m(q^1)$. Then

$$m = \int m(q^1)dq^1.$$

Considered as the distribution of mass in the three-dimensional space, it should be expressed as the following generalized function:

$$\tilde{m}(q) = m(q^1)\delta(q^2)\delta(q^3) \quad \text{and} \quad m = \int \tilde{m}(q)dq = \int m(q^1)\delta(q^2)\delta(q^3)dq.$$

Thus, numerically, the distribution of mass is given by the function $m(q^1)$ for $q^2 = 0, q^3 = 0$. If we calculate all masses in the three-dimensional space, then we should use the generalized function $m(q^1)\delta(q^2)\delta(q^3) = \tilde{m}(q)$.

In the infinitesimal operator (2.7.17), the second term is given numerically by the expression

$$\delta(q_i^1 - q_j^1)|p_i^1 - p_j^1|[f(x^*) - f(x)]\Big|_{q_i^2 = q_j^2,\, q_i^3 = q_j^3},$$

or, considered as the generalized function in the three-dimensional space, it is given by the expression $\delta(q_i - q_j)|\eta\cdot(p_i - p_j)|[f(x^*) - f(x)]$.

If $\eta \subset S_+^2$, $\eta_{ij}\cdot(p_i - p_j) > 0$, then $X(t,x) = (q + pt, p)$, and the second term is absent in (2.7.17). In the general case where $\eta_{ij} \subset S^2$, we obtain

$$\left.\frac{d\tilde{S}(t)f(x)}{dt}\right|_{t=0} = \sum_{i=1}^{N} p_i \cdot \frac{\partial}{\partial q_i} f(x) + \delta(q_i - q_j)$$
$$\times\, \Theta(-\eta_{ij}\cdot(p_i - p_j))|\eta_{ij}\cdot(p_i - p_j)|[f(x^*) - f(x)], \tag{2.7.18}$$

where $\Theta(a) = 1$ for $a > 0$, and $\Theta(a) = 0$ for $a < 0$.

In order to take into account all possible pair collisions of all particles, one should sum over all pairs $i, j \in \{1, \ldots, N\}$ in (2.7.17). After summation, we obtain the following final formula for arbitrary $q = (q_1, \ldots, q_N)$:

$$\begin{aligned}
&\frac{d\tilde{S}(t)f(x)}{dt}\Big|_{t=0} \\
&\quad = \sum_{i=1}^{N} p_i \cdot \frac{\partial}{\partial q_i} f(x) \\
&\qquad + \sum_{i<j=1}^{N} \delta(q_i - q_j)\Theta(-\eta_{ij} \cdot (p_i - p_j)) \times |\eta_{ij} \cdot (p_i - p_j)|[f(x^*) - f(x)] \\
&\quad = \tilde{\mathcal{H}}_N^+ f(x). \qquad (2.7.19)
\end{aligned}$$

We have obtained formula (2.6.14) deduced in the previous section in the weak sense.

After averaging (2.7.19) with respect to all random vectors η_{ij} with density $\chi(\eta_{ij}) = 1/4\pi$, one obtains

$$\begin{aligned}
&\frac{d\bar{S}(t)f(x)}{dt}\Big|_{t=0} \\
&\quad = \sum_{i=1}^{N} p_i \cdot \frac{\partial}{\partial q_i} f(x) \\
&\qquad + \sum_{i<j=1}^{N} \frac{1}{4\pi} \int_{S_-^2} d\eta_{ij} |\eta_{ij}(p_i - p_j)| \delta(q_i - q_j)[f(x^*) - f(x)] \\
&\quad = \bar{\mathcal{H}}_N^+ f(x), \qquad (2.7.20)
\end{aligned}$$

where $\frac{d\bar{S}(t)f(x)}{dt}\big|_{t=0}$ means the generalized function corresponding to $\frac{dS(t)f(x)}{dt}\big|_{t=0}$ and averaged with respect to all η_{ij}.

We have obtained from formula (2.7.15) three equivalent representations $\mathcal{H}^+$ (2.7.16), $\tilde{\mathcal{H}}^+$ (2.7.19), and $\bar{\mathcal{H}}^+$ (2.7.20) of the infinitesimal operator of the evolution operator of N stochastic point particles.

The first one, $\mathcal{H}^+$, is obtained is the sense of pointwise convergence, the second one, $\tilde{\mathcal{H}}^+$, is obtained in the weak sense and shows how to integrate $\frac{dS(t)f(x)}{dt}\big|_{t=0}$ with test functions $\varphi(x)$ for fixed random vectors η_{ij}, $i, j \in \{1, \ldots, N\}$, and the third one is obtained in the weak sense and shows how to integrate $\frac{dS(t)f(x)}{dt}\big|_{t=0}$ with test functions $\varphi(x)$ and to average with respect to all random vectors η_{ij}, $i, j \in \{1, \ldots, N\}$. Recall that formulas (2.7.19), (2.7.20) have been obtained in the previous section from functionals $(\tilde{S}_N(t) f_N, \varphi_N)$ and $(\bar{S}_N(t) f_N, \varphi_N)$, respectively.

By using the semigroup property (2.7.11), one can prove the following formula for the derivative of $S(t)f(x)$ at arbitrary time $t > 0$:

$$\frac{dS(t)f(x)}{dt} = \mathcal{H}^+ S(t)f(x) = S(t)\mathcal{H}^+ f(x),$$

$$\frac{d\tilde{S}(t)f(x)}{dt} = \tilde{\mathcal{H}}_N^+ \tilde{S}(t)f(x) = \tilde{S}(t)\tilde{\mathcal{H}}_N^+ f(x), \tag{2.7.21}$$

$$\frac{d\bar{S}(t)f(x)}{dt} = \bar{\mathcal{H}}_N^+ \bar{S}(t)f(x) = \bar{S}(t)\bar{\mathcal{H}}_N^+ f(x).$$

Indeed, formulas (2.7.18), (2.7.19) follow directly from the identity

$$S(t+\Delta t) - S(t) = (S(\Delta t) - I)S(t) = S(t)(S(\Delta t) - I).$$

and the operators $S(t)$, $\tilde{S}(t)$ $\bar{S}(t)$ have been obtained by formulas (2.6.17). We stress that, in the expressions for $S(t)f(x^*)$ and $S(t)f(x)$ in (2.7.21), the initial phase points x^* and x are considered as the states after collision at time $t = +0$. This means that, for $t > 0$, the particles have the momenta p^* in the first expression $S(t)f(x^*) = f(X(t,x^*))$ and p in the second expression $S(t)f(x) = f(X(t,x))$ until new collisions take place.

2.7.3 Evolution operator for negative time

The case of negative time can be considered in a completely analogous way.

Consider the function $S_N(-t)f(x)$ for $t > 0$ and define its derivative at $t = 0$ by the formula

$$\left.\frac{dS(-t)f(x)}{dt}\right|_{t=0} = \lim_{\Delta t\to 0} \frac{f(X(-\Delta t,x)) - f(x)}{\Delta t}.$$

Repeating the calculation performed above almost word for word, we get

$$\begin{aligned}
&\left.\frac{d\tilde{S}(-t)f(x)}{dt}\right|_{t=0} \\
&\quad= -\sum_{i=1}^{N} p_i \cdot \frac{\partial}{\partial q_i} f(x) \\
&\qquad + \sum_{i<j=1}^{N} \delta(q_i - q_j)\Theta(\eta_{ij}(p_i - p_j))\eta_{ij}(p_i - p_j)[f(x^*) - f(x)] \\
&\quad= \tilde{\mathcal{H}}_N^- f(x).
\end{aligned} \tag{2.7.22}$$

Note that we again have an equivalent representation of the infinitesimal operator (2.7.22):

$$\frac{dS(-t)f(x)}{dt}\Big|_{t=0} = \sum_{i=1}^{N} p_i \cdot \frac{\partial}{\partial q_i} f(x) = \mathcal{H}_N^- f(x). \tag{2.7.23}$$

We have the boundary condition according to which, at $q_i = q_j$, $i, j \in \{1, \dots, N\}$, $\eta \subset S_+^2$, the momenta p_i, p_j should be replaced by p_i^*, p_j^* in $\sum_{i=1}^N p_i \cdot \frac{\partial}{\partial q_i} f(x)$ and in $f(x)$. By averaging formula (2.7.22) with respect to η_{ij}, $i, j \in \{1, \dots, N\}$, one obtains

$$\begin{aligned}\frac{d\bar{S}(-t)f(x)}{dt}\Big|_{t=0} &\\ &= -\sum_{i=1}^{N} p_i \cdot \frac{\partial}{\partial q_i} f(x)\\ &\quad + \sum_{i<j=1}^{N} \int_{S_+^2} d\eta_{ij} \delta(q_i - q_j)\eta_{ij} \cdot (p_i - p_j)\big[f(x^*) - f(x)\big]\\ &= \bar{\mathcal{H}}_N^- f(x).\end{aligned} \tag{2.7.24}$$

We now present the definition of the operators $S(-t)$, $\bar{S}(-t)$, $\tilde{S}(-t)$ for an arbitrary time. If suffices to make obvious changes in the corresponding formulas for $S(t)$, $\bar{S}(t)$, $\tilde{S}(t)$.

Namely, for infinitesimal Δt, we have

$$\begin{aligned}S_N(-\Delta t) f_N(x_1, \dots, x_N) &= f_N(x_1(-\Delta t), \dots, x_N(-\Delta t))\\ &= f_N(q_1 - p_1\Delta t, p_1, \dots, q_N - p_N\Delta t, p_N)\end{aligned}$$

if $q_i - p_i\tau \neq q_j - p_j$ for all $i, j \in \{1, \dots, N\}$ and $0 \le \tau \le \Delta t$,

$$\begin{aligned}S_N(-\Delta t) f_N(x_1, \dots, x_N) &= f_N(x_1(-\Delta t), \dots, x_N(-\Delta t))\\ &= f_N(q_1 - p_1\Delta t, p_1, \dots, q_i - p_i\tau - p_i^*(\Delta t - \tau), p_i^*, \dots,\\ &\qquad q_j - p_j\tau - p_j^*(\Delta t - \tau), p_j^*, \dots, q_N - p_N\Delta t, p_N)\end{aligned} \tag{2.7.25}$$

if $q_i - p_i\tau = q_j - p_j\tau$, $0 \le \tau \le \Delta t$, $\eta_{ij} \subset S_2^+$,

$$\begin{aligned}S_N(-\Delta t) f_N(x_1, \dots, x_N) &= f_N(x_1(-\Delta t), \dots, x_N(-\Delta t))\\ &= f_N(q_1 - p_1\Delta t, p_1, \dots, q_i - p_i\Delta t, p_i, \dots, q_j - p_j\Delta t,\\ &\qquad p_j, \dots, q_N - p_N\Delta t, p_N)\end{aligned}$$

if $q_i - p_i\tau = q_j - p_j\tau, 0 \le \tau \le \Delta t, \eta_{ij} \subset S^2_-, i, j \in \{1, \dots, N\}$,

$$\begin{aligned}
&\tilde{S}_N(-\Delta t) f_N(x_1, \dots, x_N) \\
&\quad = f_N(q_1 - p_1\Delta t, p_1, \dots, q_N - p_N\Delta t, p_N) \\
&\qquad + \sum_{i<j=1}^{N} \int_0^{\Delta t} d\tau \eta_{ij} \cdot (p_i - p_j)\Theta(\eta_{ij} \cdot (p_1 - p_2))\delta(q_i - p_i\tau - q_j + p_j\tau) \\
&\qquad \times \big[f_N(q_1 - p_1\Delta t, p_1, \dots, q_i - p_i\tau - p_i^*(\Delta t - \tau), p_i^*, \dots, \\
&\qquad\qquad q_j - p_j\tau + p_j^*(\Delta t - \tau), p_j^*, \dots, q_N - p_N\Delta t, p_N) \\
&\qquad - f_N(q_1 - p_1\Delta t, p_1, \dots, q_i - p_i\Delta t, \dots, p_i, \dots, q_j - p_j\Delta t, p_j, \dots, \\
&\qquad\qquad q_N - p_N\Delta t, p_N)\big],
\end{aligned} \tag{2.7.26}$$

and

$$\begin{aligned}
&\bar{S}_N(-\Delta t) f_N(x_1, \dots, x_N) \\
&\quad = f_N(q_1 - p_1\Delta t, p_1, \dots, q_N - p_N(\Delta t), p_N) \\
&\qquad + \sum_{i<j=1}^{N} \int_0^{\Delta t} d\tau \int_{S^2_+} d\eta_{ij}\, \eta_{ij} \cdot (p_i - p_j)\delta(q_i - p_i\tau - q_j + p_j\tau) \\
&\qquad \times \big[f_N(q_1 - p_1\Delta t, p_1, \dots, q_i - p_i\tau - p_i^*(\Delta t - \tau), p_i^*, \dots, \\
&\qquad\qquad q_j - p_j\tau + p_j^*(\Delta t - \tau), p_j^*, \dots, q_N - p_N\Delta t, p_N) \\
&\qquad - f_N(q_1 - p_1\Delta t, p_1, \dots, q_i - p_i\Delta t, \dots, p_i, \dots, q_j - p_j\Delta t, p_j, \dots, \\
&\qquad\qquad q_N - p_N\Delta t, p_N)\big].
\end{aligned} \tag{2.7.27}$$

For arbitrary t, the operators $S(-t)$, $\tilde{S}(-t)$, $\bar{S}(-t)$ are defined by analogy with the definition of $S(t)$, $\tilde{S}(t)$, $\bar{S}(t)$ (2.6.17). Namely,

$$S(-t) = \lim_{n\to\infty} \prod_{i=1}^{n} S(-\Delta t_i), \quad \tilde{S}(-t) = \lim_{n\to\infty} \prod_{i=1}^{n} \tilde{S}(-\Delta t_i),$$

$$\bar{S}(-t) = \lim_{n\to\infty} \prod_{i=1}^{n} \bar{S}(-\Delta t_i), \quad \sum_{i=1}^{n} \Delta t_i = t. \tag{2.7.28}$$

If we have the functions

$$\tilde{f}_N(t, x_1, \dots, x_N) = \tilde{S}_N(-t) f_N(x_1, \dots, x_N),$$

$$\bar{f}_N(t, x_1, \ldots, x_N) = \bar{S}_N(-t) f_N(x_1, \ldots, x_N),$$

then the functions $\tilde{f}_N(t + \Delta t, x_1, \ldots, x_N)$, $\bar{f}_N(t + \Delta t, x_1, \ldots, x_N)$ are defined by formulas (2.7.26), (2.7.27) with $\tilde{f}_N(t, x_1, \ldots, x_N)$, $\bar{f}_N(t, x_1, \ldots, x_N)$ instead of $f_N(x_1, \ldots, x_N)$.

Namely,

$$\begin{aligned}
&\tilde{f}_N(t + \Delta t, x_1, \ldots, x_N) \\
&\quad = \tilde{f}_N(t, q_1 - p_1\Delta t, p_1, \ldots, q_N - p_N\Delta t, p_N) \\
&\qquad + \sum_{i<j=1}^{N} \int_0^{\Delta t} d\tau \eta_{ij} \cdot (p_i - p_j)\Theta(\eta_{ij} \cdot (p_i - p_j))\delta(q_i - p_i\tau - q_j + p_j\tau) \\
&\qquad \times \big[\tilde{f}_N(t, q_1 - p_1\Delta t, p_1, \ldots, q_i - p_i\tau - p_i^*(\Delta t - \tau), p_i^*, \ldots, \\
&\qquad\qquad q_j - p_j\tau + p_j^*(\Delta t - \tau), p_j^*, \ldots, q_N - p_N\Delta t, p_N) \\
&\qquad - \tilde{f}_N(t, q_1 - p_1\Delta t, p_1, \ldots, q_i - p_i\Delta t, p_i, \ldots, q_j - p_j\Delta t, p_j, \ldots, \\
&\qquad\qquad q_N - p_N\Delta t, p_N)\big],
\end{aligned} \tag{2.7.29}$$

$$\begin{aligned}
&\bar{f}_N(t + \Delta t, x_1, \ldots, x_N) \\
&\quad = \bar{f}_N(t, q_1 - p_1\Delta t, p_1, \ldots, q_N - p_N\Delta t, p_N) \\
&\qquad + \sum_{i<j=1}^{N} \int_0^{\Delta t} d\tau \int_{S_+^2} d\eta_{ij}\, \eta_{ij} \cdot (p_i - p_j)\delta(q_i - p_i\tau - q_j + p_j\tau) \\
&\qquad \times \big[\bar{f}_N(t, q_1 - p_1\Delta t, p_1, \ldots, q_i - p_i\tau - p_i^*(\Delta t - \tau), p_i^*, \ldots, \\
&\qquad\qquad q_j - p_j\tau + p_j^*(\Delta t - \tau), p_j^*, \ldots, q_N - p_N\Delta t, p_N) \\
&\qquad - \bar{f}_N(t, q_1 - p_1\Delta t, p_1, \ldots, q_i - p_i\Delta t, p_i, \ldots, q_j - p_j\Delta t, p_j, \ldots, \\
&\qquad\qquad q_N - p_N\Delta t, p_N)\big].
\end{aligned} \tag{2.7.30}$$

From (2.7.29) and (2.7.30), one obtains formulas (2.7.22) and (2.7.24), respectively. We need these formulas in terms of the functions $\tilde{f}_N(t, x_1, \ldots, x_N)$ and

$\bar{f}_N(t, x_1, \ldots, x_N)$. They have the form

$$\begin{aligned}
\frac{d\tilde{f}_N(t, x_1, \ldots, x_N)}{dt} &= -\sum_{i=1}^{N} p_i \cdot \frac{\partial}{\partial q_i} \tilde{f}_N(t, x_1, \ldots, x_N) \\
&\quad + \sum_{i<j=1}^{N} \eta_{ij} \cdot (p_i - p_j)\Theta(\eta_{ij} \cdot (p_i - p_j))\delta(q_i - q_j) \\
&\quad \times \big[\tilde{f}_N(t, q_1, p_1, \ldots, q_i, p_i^*, \ldots, q_j, p_j^*, \ldots, q_N, p_N) \\
&\quad - \tilde{f}_N(t, q_1, p_1, \ldots, q_i, p_i, \ldots, q_j, p_j, \ldots, q_N, p_N)\big] \\
&= \tilde{\mathcal{H}}_N^- \tilde{f}_N(t, x_1, \ldots, x_N),
\end{aligned} \tag{2.7.31}$$

$$\begin{aligned}
\frac{d\bar{f}_N(t, x_1, \ldots, x_N)}{dt} &= -\sum_{i=1}^{N} p_i \cdot \frac{\partial}{\partial q_i} \bar{f}_N(t, x_1, \ldots, x_N) \\
&\quad + \sum_{i<j=1}^{N} \int_{S_+^2} d\eta_{ij}\, \eta_{ij} \cdot (p_i - p_j)\delta(q_i - q_j) \\
&\quad \times \big[\bar{f}_N(t, q_1, p_1, \ldots, q_i, p_i^*, \ldots, q_j, p_j^*, \ldots, q_N, p_N) \\
&\quad - \bar{f}_N(t, q_1, p_1, \ldots, q_i, p_i, \ldots, q_j, p_j, \ldots, q_N, p_N)\big] \\
&= \bar{\mathcal{H}}_N^- \bar{f}_N(t, x_1, \ldots, x_N).
\end{aligned} \tag{2.7.32}$$

One also has an analog of formulas (2.7.21):

$$\begin{aligned}
\frac{dS(-t)f(x)}{dt} &= \mathcal{H}^+ S(-t)f(x) = S(-t)\mathcal{H}^+ f(x), \\
\frac{d\tilde{S}(-t)f(x)}{dt} &= \tilde{\mathcal{H}}_N^- \tilde{S}(-t)f(x) = \tilde{S}(-t)\tilde{\mathcal{H}}_N^- f(x), \\
\frac{dS(-t)f(x)}{dt} &= \bar{\mathcal{H}}_N^- \bar{S}(-t)f(x) = \bar{S}(-t)\bar{\mathcal{H}}_N^- f(x).
\end{aligned} \tag{2.7.33}$$

As is known in statistical mechanics, the state of N particles at time $t > 0$ is given by formula $S(-t)f(x)$, where the initial state at $t = 0$ is equal to $f(x)$. As usual, we suppose that $f(x) = f_N(x_1, \ldots, x_N)$ is a real symmetric, i.e. invariant with respect to permutation, function normalized to unit: $\int f(x)dx = 1$.

In what follows, we use the states

$$\bar{S}(-t)f(x) \equiv \bar{S}_N(-t)f_N(x_1,\dots,x_N) = \bar{f}_N(t,x_1,\dots,x_N),$$
$$\tilde{S}(-t)f(x) \equiv \tilde{S}_N(-t)f_N(x_1,\dots,x_N) = \tilde{f}_N(t,x_1,\dots,x_N), \tag{2.7.34}$$
$$S(-t)f(x) \equiv S_N(-t)f_N(x_1,\dots,x_N) = f_N(t,x_1,\dots,x_N)$$

and equations (2.7.31), (2.7.32), respectively, for the derivation of the corresponding hierarchies for reduced correlation functions.

Note that states (2.7.29), (2.7.30) and equations (2.7.31), (2.7.32) are equivalent. Details about their equivalence have been explained in the previous subsections.

2.7.4 Equivalence of the infinitesimal operators

Remark 2.4. We explain that both forms of the representation of the infinitesimal operator $\mathcal{H}_N^+$ (without and with δ-function) are equivalent. Their equivalence directly follows from two equivalent representations of the differential equation associated with the stochastic dynamics. The stochastic dynamics is defined by pair collisions and, therefore, it is sufficient to consider a system of two stochastic particles. Moreover, since we use the coordinate system with the first coordinates parallel to the vector η, we will consider a one-dimensional system because only the momenta parallel to η have jumps at the instant of collision.

The one-dimensional dynamics of two point particles consists of the following: Let $(q_1, p_1), (q_2, p_2)$ be the initial data in the phase space at $t = 0$. Then, at time $t > 0$, we have $Q_1(t) = q_1 + p_1 t$, $P_1(t) = p_1$, $Q_2(t) = q_2 + p_2 t$, $P_2(t) = p_2$ at $t < \tau$, where τ is the instant of collision, i.e., $Q_1(\tau) = Q_2(\tau)$, $\tau = \frac{q_1 - q_2}{p_2 - p_1} > 0$. At time $t > \tau$, we have

$$\begin{gathered} Q_1(t) = q_1 + p_1\tau + p_1^*(t-\tau), \quad P_1(t) = p_1^*, \\ Q_2(t) = q_2 + p_2\tau + p_2^*(t-\tau), \quad P_2(t) = p_2^*, \quad p_1^* = p_2, \quad p_2^* = p_1, \\ p_1^* = p_1 - \eta\cdot(p_1 - p_2) = p_1 - p_1 + p_2 = p_2, \\ p_2^* = p_2 + \eta\cdot(p_1 - p_2) = p_2 + p_1 - p_2 = p_1, \quad \eta = 1. \end{gathered} \tag{2.7.35}$$

The above-defined $\big(Q_1(t), P_1(t)\big), \big(Q_2(t), P_2(t)\big)$ satisfy the differential equations

$$\frac{dP_1(t)}{dt} = 0, \quad \frac{dQ_1(t)}{dt} = P_1(t), \quad \frac{dP_2(t)}{dt} = 0, \quad \frac{dQ_2(t)}{dt} = P_2(t) \tag{2.7.36}$$

with the following initial and boundary (at $t = \tau$) conditions:

$$\begin{gathered} P_1(0) = p_1, \quad P_2(0) = p_2, \quad Q_1(0) = q_1, \quad Q_2(0) = q_2, \\ P_1(\tau + 0) = p_1^*, \quad P_2(\tau + 0) = p_2^*. \end{gathered} \tag{2.7.37}$$

It is easy to check that solutions of the differential equation (2.7.36) with initial and boundary conditions (2.7.37) coincide with (2.7.35). Consider the operator of evolution $S_2(t)$ on a continuously differentiable function $f_2(x_1, x_2)$:

$$S_2(t) f_2(x_1, x_2) = f_2(X_1(t, (x)_2)), \quad X_2(t, (x_2)) = f_2(t, x_1, x_2),$$

$$X_1(t) = (Q_1(t), P_1(t)), \quad X_2(t) = (Q_2(t), P_2(t)),$$

$$f_2(t, x_1, x_2)|_{t=0} = f_2(x_1, x_2).$$

Using the group property of the operators $S_2(t)$, we get

$$\begin{aligned} \frac{\partial}{\partial t} f_2(t, x_1, x_2) &= \Big(P_1(t)\frac{\partial}{\partial Q_1(t)} + P_2(t)\frac{\partial}{\partial Q_2(t)}\Big) f_2(t, x_1, x_2) \\ &= S_2(t)\Big(p_1\frac{\partial}{\partial q_1} + p_2\frac{\partial}{\partial q_2}\Big) f_2(x_1, x_2) \\ &= \Big(p_1\frac{\partial}{\partial q_1} + p_2\frac{\partial}{\partial q_2}\Big) S_2(t) f_2(x_1, x_2) \\ &= \Big(p_1\frac{\partial}{\partial q_1} + p_2\frac{\partial}{\partial q_2}\Big) f_2(t, x_1, x_2) \end{aligned} \tag{2.7.38}$$

with the boundary condition according to which, at $q_1 = q_2$, the momenta (p_1, p_2) should be replaced by (p_1^*, p_2^*), or $(P_1(\tau), P_2(\tau))$ should be replaced by $(P_1^*(\tau), P_2^*(\tau))$ at $t = \tau$ or $Q_1(\tau) = Q_2(\tau)$.

The relation

$$\frac{\partial}{\partial t} f_2(t, x_1, x_2) = \Big(p_1\frac{\partial}{\partial q_1} + p_2\frac{\partial}{\partial q_2}\Big) f_2(t, x_1, x_2) \tag{2.7.39}$$

with the above-formulated boundary condition can be considered as the Liouville equation.

We obtain equation (2.7.36) from (2.7.38)–(2.7.39) as the corresponding equations of characteristics or if we take $f(x_1, x_2)$ equal to (q_1, p_1, q_2, p_2) and use the first equality from (2.7.38) and the corresponding boundary condition.

The results above obtained mean that the infinitesimal operator $\mathcal{H}_2$ of the group of evolution operators $S_2(t)$ is given on functions $f_2(x_1, x_2)$ as follows:

$$\mathcal{H}_2 f_2(x_1, x_2) = \Big(p_1\frac{\partial}{\partial q_1} + p_2\frac{\partial}{\partial q_2}\Big) f_2(x_1, x_2) \tag{2.7.40}$$

with the corresponding boundary condition at $q_1 = q_2$.

From another point of view, equations (2.7.36) with the initial and boundary conditions (2.7.37) are equivalent to the following equations:

$$\frac{dP_1(t)}{dt} = \delta(t-\tau)(P_1^*(t) - P_1(t)), \quad \frac{dQ_1(t)}{dt} = P_1(t),$$
$$\frac{dP_2(t)}{dt} = \delta(t-\tau)(P_2^*(t) - P_2(t)), \quad \frac{dQ_2(t)}{dt} = P_2(t), \tag{2.7.41}$$
$$P_1(0) = p_1, \quad Q_1(0) = q_1, \quad P_2(0) = p_2, \quad Q_2(0) = q_2.$$

The instant of collision is again defined from the equation

$$Q_1(t_1) = Q_2(t_1), \quad \tau = \frac{q_1 - q_2}{p_2 - p_1}.$$

Integrating equation (2.7.41) with initial condition, one obtains solution (2.7.35). Note that, in the equation (2.7.41), the boundary conditions (2.7.37) are absent. Instead of them, one has the δ-function.

The difference between equations (2.7.36) and (2.7.41) consists of the following: In (2.7.36), one differentiates on the intervals $[0,\tau)$ and $(\tau, t]$ and uses the boundary condition at $t = \tau$, whereas, in (2.7.41), one differentiates on the interval $[0,t)$ without boundary condition at $t = \tau$.

The infinitesimal operator $\mathcal{H}_2$ of the evolution operator $S_2(t)$ corresponding to equations (2.7.41) is defined on functions $f_2(x_1, x_2)$ as follows:

$$\begin{aligned}\frac{\partial f_2(t,x_1,x_2)}{\partial t} &= \Big(P_1(t)\frac{\partial}{\partial Q_1(t)} + P_2(t)\frac{\partial}{\partial Q_2(t)}\Big) f_2(t,x_1,x_2) \\ &\quad + \delta(t-\tau)[f_2(t,x_1^*,x_2^*) - f_2(t,x_1,x_2)] \\ &= \Big(p_1\frac{\partial}{\partial q_1} + p_2\frac{\partial}{\partial q_2}\Big) f_2(t,x_1,x_2) + \delta(q_1-q_2)|p_1-p_2| \\ &\quad \times [f_2(t,x_1^*,x_2^*) - f_2(t,x_1,x_2)],\end{aligned} \tag{2.7.42}$$

$$x_1^* = (q_1, p_1^*), \quad x_2^* = (q_2, p_2^*).$$

It is easy to check that equations (2.7.41) can be obtained from the first equality in (2.7.42) if one puts functions (q_1, p_1, q_2, p_2) instead of $f_2(x_1, x_2)$, i.e., equations (2.7.41) are the equations of characteristics corresponding to (2.7.42).

The difference between equations (2.7.38) and (2.7.42) again consists of the following: In (2.7.38), one differentiates on the intervals $[0,\tau)$, $(\tau, t]$, whereas, in (2.7.42), one differentiates on the entire interval $[0,t]$.

Thus, both forms of the infinitesimal operator $\mathcal{H}_2$,

$$\mathcal{H}_2 f_2(x_1,x_2) = \Big(p_1\frac{\partial}{\partial q_1} + p_2\frac{\partial}{\partial q_2}\Big) f_2(t,x_1,x_2)$$

with the corresponding boundary condition or

$$\mathcal{H}_2 f_2(x_1, x_2) = \left(p_1 \frac{\partial}{\partial q_1} + p_2 \frac{\partial}{\partial q_2}\right) f_2(x_1, x_2)$$
$$+ \delta(q_1 - q_2)|p_1 - p_2|[f_2(x_1^*, x_2^*) - f_2(x_1, x_2)]$$

without the boundary condition, are equivalent because equations (2.7.38), (2.7.39) and (2.7.42) have the same solutions $f_2(t, x_1, x_2) = f_2(X_1(t, x_1, x_2), X_2(t, x_1, x_2))$, where $X_1(t), X_2(t)$ are equivalent solutions of (2.7.36)–(2.7.37) or (2.7.41).

Chapter 3

Stochastic Boltzmann hierarchy

3.1 Introduction

In Chapter 2, we have introduced the stochastic dynamics of point particles, which is obtained from the Hamilton dynamics of a system of hard spheres in the Boltzmann–Grad limit. According to this stochastic dynamics, point particles move as free ones until their positions coincide, and then they undergo elastic scattering. The unit vector that determines elastic scattering is a random vector uniformly distributed on the unit sphere. Then particles move as free ones until the next collision.

In the present chapter, we derive a hierarchy of equations for the sequence of correlation functions. We call this hierarchy the *stochastic Boltzmann hierarchy*. The *ordinary Boltzmann hierarchy* is also derived for correlation functions that are equal to the product of solutions of the Boltzmann equation. The connection between both hierarchies is explained.

As is customary in classical statistical mechanics [PGM3, CGP], the initial state of a system is defined by a distribution function on the phase space. The state of the system at arbitrary time is defined as the result of the action of an evolution operator, i.e., the operator of transition along the trajectory, on the initial distribution function. The distribution function thus defined differs from the distribution function of the free system of particles at arbitrary time only on the hyperplanes of lower dimension where the point particles interact.

From the viewpoint of traditional classical statistical mechanics, the system of point particles moving according to the stochastic dynamics should be regarded as a free system. Indeed, in traditional statistical mechanics, averages are calculated via the Lebesgue integral, and the behavior of distribution functions on hyperplanes of lower dimension is not taken into account. In this connection, in a series of papers, a new concept of averages of observables over distribution functions was introduced. This concept takes into account, in a special way described in Chapter 2, the contribution of the hyperplanes where the particles interact. With the use of averages thus introduced, correlation functions are defined that also take into account, in a special way, the contribution of the hyperplanes where the particles interact.

The hierarchy of equations derived for the sequence of correlation functions has the form of the ordinary Boltzmann hierarchy [Gra1, Gra2, Ger1, Cer2, Cer3, Cer4] but takes into account the boundary conditions where the positions of particles coincide. If the hierarchy is considered in the weak sense, then it contains δ-functions that differ from zero when the positions of particles coincide. The hierarchy obtained was called the stochastic Boltzmann hierarchy.

Note that the stochastic Boltzmann hierarchy is obtained from the stochastic dynamics in the same way as the BBGKY hierarchy is obtained from the Hamilton dynamics. The ordinary Boltzmann hierarchy is obtained directly from the Boltzmann equations [Pet5] or from the BBGKY hierarchy for a system of hard spheres in the Boltzmann–Grad limit where the boundary conditions are not taken into account [Cer1, Cer2, Cer3, Cer4, CIP], and it is likely that there is no dynamics corresponding to it.

The stochastic Boltzmann hierarchy was also obtained in the Boltzmann–Grad limit directly from the BBGKY hierarchy of system of hard spheres. And again it differs from the ordinary Boltzmann hierarchy by the boundary condition or, in the weak sense, by additional δ-functions.

We show that the solutions of the Boltzmann equation and the Boltzmann hierarchies are represented by integrals over the hyperplanes of lower dimension where the stochastic particles interact with one another. Surprisingly, this property of solutions of the Boltzmann equation and the Boltzmann hierarchies has not been observed earlier.

3.2 Average of observables over state

3.2.1 Stochastic dynamics

For convenience, in this subsection we summarize the definition of the stochastic dynamics for negative time. Consider point particles with unit mass in the three-dimensional space $\mathbb{R}^3$ and denote by $x_1 = (q_1, p_1), \dots, x_N = (q_N, p_N)$ their phase points; $(x)_N = (x_1, \dots, x_N) = x$ at initial time $t = 0$.

We define their stochastic dynamics for negative time $-t$, $t > 0$ as follows: Particles move as free ones until $q_i - p_i\tau = q_j - p_j\tau$, $0 \leq \tau \leq t$, $i, j \in \{1, \dots, N\}$. Then these two particles collide, and their momenta become

$$
\begin{gathered}
p_i^* = p_i - \eta_{ij}\eta_{ij} \cdot (p_i - p_j), \quad p_j^* = p_j + \eta_{ij}\eta_{ij} \cdot (p_i - p_j), \\
\eta_{ij} \subset S_+^2(\eta_{ij} | \eta_{ij} \cdot (p_i - p_j) \geq 0), \quad |\eta_{ij}| = 1;
\end{gathered}
\tag{3.2.1}
$$

if $\eta_{ij} \subset S_-^2(\eta_{ij} \mid \eta_{ij} \cdot (p_i - p_j) \leq 0)$, then $p_i^* = p_i$ and $p_j^* = p_j$; here, $\eta_{ij} \cdot (p_i - p_j)$ is the scalar product of the vectors η_{ij} and $p_i - p_j$.

At time $-t$, their phase points are

$$
\begin{gathered}
x_i(-t) = (q_i - p_i\tau - p_i^*(t - \tau), p_i^*), \\
x_j(-t) = (q_j - p_j\tau - p_j^*(t - \tau), p_j^*), \quad \eta_{ij} \subset S_+^2;
\end{gathered}
\tag{3.2.2}
$$

if $\eta_{ij} \subset S_-^2$, then $x_i(-t) = (q_i - p_i t, p_i)$ and $x_j(-t) = (q_j - p_j t, p_j)$.

Particles scatter elastically, but the vectors η_{ij} are random ones with constant density of probability on the unit sphere.

If $\eta_{ij} \subset S^2_-(\eta_{ij} \mid \eta_{ij} \cdot (p_i - p_j) \leq 0)$, then particles continue to move freely even in the case where $q_i - p_i\tau = q_j - p_j\tau$.

For positive time $t > 0$, it is necessary to put in (3.2.1) $(+\tau)$, $+(t-\tau)$ instead of $(-\tau)$, $-(t-\tau)$ and S^2_- and S^2_+ instead of S^2_+ and S^2_-, respectively. We neglect the case where three or more particles collide at the same point, or suppose that they continue to move freely.

The above-introduced stochastic dynamics defines the trajectory in phase space with fixed random vectors η_{ij}:

$$\begin{aligned} X(-t) &= (x(-t))_N = (x_1(-t), \dots, x_N(-t)) \\ &= (x_1(-t, (x)_N), \dots, x_N(-t, (x)_N)) \\ &= X(-t, (x)_N) = X(-t, x). \end{aligned}$$

Obviously, the trajectory $X(-t)$ possesses the group property

$$X(-t_1 - t_2, x) = X(-t_1, X(-t_2, x)) = X(-t_2, X(-t_1, x)).$$

We define the operator $S_N(-t)$ as the operator of shift along the trajectory:

$$\begin{aligned} S_N(-t) f_N(x_1, \dots, x_N) &= f_N(x_1(-t), \dots, x_N(-t)) \\ &= f_N(x_1(-t, (x)_N), \dots, x_N(-t, (x)_N)). \end{aligned} \tag{3.2.3}$$

Let $f_N(x_1, \dots, x_N)$ be a real symmetric continuously differentiable normalized function and let $\varphi_N(x_1, \dots, x_N)$ be a real symmetric test function. The function $f_N(x_1, \dots, x_N)$ is the initial state, $S_N(-t) f_N(x_1, \dots, x_N)$ is the state at time t for systems of N particles with some fixed realization of the random vectors η_{ij}. The function $\varphi_N(x_1, \dots, x_N)$ is an observable.

3.2.2 Average for infinitesimal time

Consider an infinitesimal time $-\Delta t$ and introduce the following functional average of the observable $\varphi_N(x_1, \dots, x_N)$ over the state $S_N(-\Delta t) f_N(x_1, \dots, x_N)$:

$$\begin{aligned} &(S_N(-\Delta t) f_N, \varphi_N) \\ &\quad = \int dx_1 \dots dx_N \, f_N(q_1 - p_1 \Delta t, p_1, \dots, \\ &\qquad\qquad q_N - p_N \Delta t, p_N) \varphi_N(q_1, p_1, \dots, q_N, p_N) \\ &\qquad + \sum_{i<j=1}^{N} \int dx_1 \dots dx_N \int_0^{\Delta t} d\tau \int_{S^2_+} d\eta_{ij} \, \eta_{ij} \cdot (p_i - p_j) \end{aligned}$$

$$
\begin{aligned}
&\times \delta(q_i - p_i\tau - q_j + p_j\tau) \\
&\times \big[f_N(q_1 - p_1\Delta t, p_1, \dots, q_i - p_i\tau - p_i^*(\Delta t - \tau), p_i^*, \dots, \\
&\qquad q_j - p_j\tau - p_j^*(\Delta t - \tau), p_j^*, \dots, q_N - p_N\Delta t, p_N) \\
&- f_N(q_1 - p_1\Delta t, p_1, \dots, q_i - p_i\Delta t, p_i, \dots, q_j - p_j\Delta t, p_j, \dots, \\
&\qquad q_N - p_N\Delta t, p_N)\big]\varphi_N(q_1, p_1, \dots, q_N, p_N) \\
= &\int dx_1 \dots dx_N \big[f_N(x_1, \dots, x_N) \\
&- \Delta t \sum_{i=1}^N p_i \cdot \frac{\partial}{\partial q_i} f_N(x_1, \dots, x_N)\big]\varphi_N(x_1, \dots, x_N) \\
&+ \int dx_1 \dots dx_N \Delta t \int_{S_+^2} d\eta_{ij}\delta(q_i - q_j)\eta_{ij} \cdot (p_i - p_j) \\
&\quad \times \Big[f_N(q_1, p_1, \dots, q_i, p_i^*, \dots, q_j, p_j^*, \dots, q_N, p_N) \\
&\qquad - f_N(q_1, p_1, \dots, q_i, p_i, \dots, q_j, p_j, \dots, q_N, p_N)\Big] \\
&\quad \times \varphi_N(q_1, p_1, \dots, q_N, p_N).
\end{aligned} \tag{3.2.4}
$$

For the sake of simplicity, in what follows we denote by $d\eta_{ij}$ the measure

$$\chi(\eta_{ij})d\eta_{ij} = \frac{1}{4\pi}d\eta_{ij}.$$

Here the operator $S_N(-\Delta t)$ is defined, as in the previous chapter, according to the stochastic dynamics as follows:

For $q_i - p_i\tau = q_j - p_j\tau$, one has

$$
\begin{aligned}
&S_N(-\Delta t) f_N(x_1, \dots, x_i, \dots, x_j, \dots, x_N)|_{q_i - p_i\tau = q_j - p_j\tau} \\
&\quad = f_N(q_1 - p_1\Delta t, p_1, \dots, q_i - p_i\tau - p_i^*(\Delta t - \tau), p_i^*, \dots, \\
&\qquad q_j - p_j\tau - p_j^*(\Delta t - \tau), p_j^*, \dots, q_N - p_N\Delta t, p_N)
\end{aligned}
$$

for $\eta_{ij} \in S_+^2$ and

$$
\begin{aligned}
&S_N(-\Delta t) f_N(x_1, \dots, x_i, \dots, x_j, \dots, x_N)|_{q_i - p_i\tau = q_j - p_j\tau} \\
&\quad = f_N(q_1 - p_1\Delta t, p_1, \dots, q_i - p_i\Delta t, p_i, \dots, q_j \\
&\qquad - p_j\Delta t, p_j, \dots, q_N - p_N\Delta t, p_N)
\end{aligned} \tag{3.2.5}
$$

for $\eta_{ij} \in S^2_-$, and for $q_i - \tau p_i \neq q_j - p_j\tau$ for all $i, j \in \{1, \dots, N\}$ one has

$$S_N(-\Delta t) f_N(x_1, \dots, x_N) = f_N(q_1 - p_1\Delta t, p_1, \dots, q_N - p_N\Delta t, p_N)$$

and $0 \leq \tau \leq \Delta t$.

We now represent the functional average as follows:

$$\begin{aligned}&\big(S_N(-\Delta t) f_N, \varphi_N\big)\\ &\quad = \int dx_1 \dots dx_N \bar{S}_N(-\Delta t) f_N(x_1, \dots, x_N)\varphi_N(x_1, \dots, x_N). \qquad (3.2.6)\end{aligned}$$

The introduced operator $\bar{S}_N(-\Delta t)$ is the usual operator of evolution in the theory of Markov processes and is obtained as a result of a specific averaging procedure with respect to random vectors η_{ij} that takes into account the contribution of the hypersurfaces of lower dimension $q_i - p_i\tau = q_j - p_j\tau$, $0 \leq \tau \leq t$, $i, j \in \{1, \dots, N\}$, where the stochastic particles interact.

The functional average (3.2.6) is defined for arbitrary test function and it defines the result of the action of the operator $\bar{S}_N(-\Delta t)$ on the function $f_N(x_1, \dots, x_N)$, i.e., $\bar{S}_N(-\Delta t) f_N(x_1, \dots, x_N)$, as a generalized function:

$$\begin{aligned}
&\bar{S}_N(-\Delta t) f_N(x_1, \dots, x_N)\\
&\quad = f_N(q_1 - p_1\Delta t, p_1, \dots, q_N - p_N\Delta t, p_N)\\
&\quad\quad + \sum_{i<j=1}^{N} \int_0^{\Delta t} d\tau \int_{S^2_+} d\eta_{ij}\, \eta_{ij} \cdot (p_i - p_j)\delta(q_i - p_i\tau - q_j + p_j\tau)\\
&\quad\quad \times \big[f_N(q_1 - p_1\Delta t, p_1, \dots, q_i - p_i\tau - p_i^*(\Delta t - \tau), p_i^*, \dots,\\
&\quad\quad\quad\quad q_j - p_j\tau - p_j^*(\Delta t - \tau), p_j^*, \dots, q_N - p_N\Delta t, p_N)\\
&\quad\quad - f_N(q_1 - p_1\Delta t, p_1, \dots, q_i - p_i\Delta t, p_i, \dots, q_j - p_j\Delta t, p_j, \dots,\\
&\quad\quad\quad\quad q_N - p_N\Delta t, p_N)\big]\\
&\quad = f_N(q_1 - p_1\Delta t, p_1, \dots, q_N - p_N\Delta t, p_N)\\
&\quad\quad + \sum_{i<j=1}^{N} \int_0^{\Delta t} d\tau \int_{S^2} d\eta_{ij}\, \eta_{ij} \cdot (p_i - p_j)\delta(q_i - p_i\tau - q_j + p_j\tau)\\
&\quad\quad \times S_N(-\Delta t) f_N(x_1, \dots, x_N) = \bar{f}_N(\Delta t, x_1, \dots, x_N). \qquad (3.2.7)
\end{aligned}$$

We use in (3.2.7) the fact that $\eta_{ij} \cdot (p_i - p_j)|_{\eta_{ij} \in S^2_-} = -\eta_{ij} \cdot (p_i - p_j)|_{\eta_{ij} \in S^2_+}$ and relation (3.2.5).

Note that there is no contradiction between the definition (3.2.3), (3.2.5) of

$$S_N(-\Delta t) f_N(x_1,\dots,x_N) = f_N(x_1(-\Delta t),\dots,x_N(-\Delta t))$$

and (3.2.7). Formula (3.2.7) simply defines the function $\bar{S}_N(-\Delta t) f_N(x_1,\dots,x_N)$ as a generalized function, and the averages of the function $\bar{S}_N(-\Delta t) f_N(x_1,\dots,x_N)$ (3.2.7) over the observable $\varphi_N(x_1,\dots,x_N)$ should be calculated as the following functional:

$$\begin{aligned}(\bar{f}_N(\Delta t),\varphi_N) &= (\bar{S}_N(-\Delta t) f_N,\varphi_N)\\ &= \int dx_1\dots dx_N \bar{S}_N(-\Delta t) f_N(x_1,\dots,x_N)\varphi_N(x_1,\dots,x_N)\\ &= (S_N(-\Delta t) f_N,\varphi_N). \end{aligned} \tag{3.2.8}$$

Thus, numerically, the state $S_N(-\Delta t) f_N(x_1,\dots,x_N)$ is given by formulas (3.2.3), (3.2.5):

$$\begin{aligned}S_N(-\Delta t) f_N(x_1,\dots,x_N) &= f_N(x_1(-\Delta t),\dots,x_2(-\Delta t))\\ &= f_N(x_1(-\Delta t,(x)_N),\dots,x_1(-\Delta t,(x)_N)).\end{aligned}$$

When we calculate the average of $S_N(-\Delta t) f_N(x_1,\dots,x_N)$ over the observable $\varphi_N(x_1,\dots,x_N)$, we use the generalized function $\bar{S}_N(-\Delta t) f_N(x_1,\dots,x_N) = \bar{f}_N(\Delta t, x_1,\dots,x_N)$ given by formula (3.2.7) and calculate the average $(\bar{S}_N(-\Delta t) f_N,\varphi_N)$ as functional (3.2.8) that coincides with (3.2.4).

Note again that functional (3.2.4) is the average of the observable $\varphi_N(x_1,\dots,x_N)$ over the state

$$S_N(-\Delta t) f_N(x_1,\dots,x_N) = f_N(x_1(-\Delta t),\dots,x_N(-\Delta t)),$$

where φ_N is a real symmetric test function and $f_N \geq 0$ is also a real symmetric continuously differentiable function normalized so that

$$\int f_N(x_1,\dots,x_N) dx_1\dots dx_N = 1. \tag{3.2.9}$$

We stress that, in functionals (3.2.4), (3.2.6), and (3.2.8), the contributions of the hyperplanes of lower dimension $q_i - p_i\tau = q_j - p_j\tau$, $0 \leq \tau \leq t$, $1 \leq i < j \leq N$, where stochastic particles interact are taken into account; they are equal to the second term on the right-hand side of (3.2.4).

3.2.3. Infinitesimal operator

It follows from (3.2.7) that

$$\frac{\partial \bar{S}_N(-t)}{\partial t}\Big|_{t=0} f_N(x_1,\dots,x_N)$$

$$= -\sum_{i=1}^{N} p_i \cdot \frac{\partial}{\partial q_i} f_N(x_1, \ldots, x_N)$$

$$+ \sum_{i<j=1}^{N} \int_{S_+^2} d\eta_{ij} \eta_{ij} \cdot (p_i - p_j) \delta(q_i - q_j)$$

$$\times \left[f_N(x_1, \ldots, x_i^*, \ldots, x_j^*, \ldots, x_N) - f_N(x_1, \ldots, x_i, \ldots, x_j, \ldots, x_N) \right]$$

$$= \bar{\mathcal{H}}_N f_N(x_1, \ldots, x_N), \quad x_i^* = (q_i, p_i^*), \quad x_j^* = (q_j, p_j^*). \tag{3.2.10}$$

We formally define the group of operators $\bar{S}_N(-t)$ at arbitrary time $-t$ as follows:

$$\bar{S}_N(-t) = \lim_{n \to \infty} \prod_{i=1}^{n} \bar{S}_N(-\Delta t_i), \quad \sum_{i=1}^{n} \Delta t_i = t, \tag{3.2.11}$$

where the operator $\bar{S}_N(-\Delta t)$ for infinitesimal Δt is defined according to (3.2.7), and the infinitesimal generator of the group $\bar{S}_N(-t)$ is defined by (3.2.10) and is equal to $\bar{\mathcal{H}}_N$.

We now formally define the state $\bar{f}_N(t, x_1, \ldots, x_N)$ at arbitrary time $t > 0$ as follows:

$$\bar{f}_N(t) = \bar{f}_N(t, x_1, \ldots, x_N) = \bar{S}_N(-t) f_N(x_1, \ldots, x_N)$$

$$= \lim_{n \to \infty} \prod_{i=1}^{n} \bar{S}_N(-\Delta t_i) f_N(x_1, \ldots, x_N), \tag{3.2.12}$$

$$\sum_{i=1}^{n} \Delta t_i = t, \quad \bar{f}_N(t + \Delta t) = \bar{S}_N(-\Delta t) \bar{f}_N(t),$$

i.e., $\bar{S}_N(-\Delta t) \bar{f}_N(t)$ is defined by formula (3.2.7) with $\bar{f}_N(t, x_1, \ldots, x_N)$ instead of $f_N(x_1, \ldots, x_N)$.

We define the functional average (with infinitesimal Δt) of the state $\bar{f}_N(t + \Delta t)$:

$$(\bar{f}_N(t + \Delta t), \varphi_N)$$

$$= (\bar{S}_N(-\Delta t) \bar{f}_N(t), \varphi_N)$$

$$= \int dx_1 \ldots dx_N \, \bar{f}_N(t, q_1 - p_1 \Delta t, p_1, \ldots, q_N - p_N \Delta t, p_N)$$

$$\times \varphi_N(q_1, p_1, \ldots, q_N, p_N)$$

$$+ \sum_{i<j=1}^{N} \int dx_1 \ldots dx_N \int_0^{\Delta t} d\tau \int_{S_+^2} d\eta_{ij} \eta_{ij} \cdot (p_i - p_j) \delta(q_i - p_i \tau - q_j + p_j \tau)$$

$$\times \big[\bar{f}_N(t, q_1 - p_1\Delta t, p_1, \dots, q_i - p_i\tau - p_i^*(\Delta t - \tau), p_i^*, \dots,$$
$$q_j - p_j\tau - p_j^*(\Delta t - \tau), p_j^*, \dots, q_N - p_N\Delta t, p_N)$$
$$- \bar{f}_N(t, q_1 - p_1\Delta t, p_1, \dots, q_i - p_i\Delta t, p_i, \dots, q_j - p_j\Delta t, p_j, \dots,$$
$$q_N - p_N\Delta t, p_N)\varphi_N(x_1, \dots, x_N)\big]$$
$$= \int \bar{S}_N(-\Delta t)\bar{f}_N(t, x_1, \dots, x_N)\varphi_N(x_1, \dots, x_N)dx_1 \dots dx_N$$
$$= \int \bar{f}_N(t + \Delta t, x_1, \dots, x_N)\varphi_N(x_1, \dots, x_N)dx_1 \dots dx_N. \tag{3.2.13}$$

It follows from (3.2.13) that

$$\bar{S}_N(-\Delta t)\bar{f}_N(t, x_1, \dots, x_N)$$
$$= \bar{f}_N(t + \Delta t, x_1, \dots, x_N)$$
$$= \bar{f}_N(t, q_1 - p_1\Delta t, p_1, \dots, q_N - p_N\Delta t, p_N)$$
$$+ \sum_{i<j=1}^{N} \int_0^{\Delta t} d\tau \int_{S_+^2} d\eta_{ij}\eta_{ij} \cdot (p_i - p_j)\delta(q_i - p_i\tau - q_j + p_j\tau)$$
$$\times \big[\bar{f}_N(t, q_1 - p_1\Delta t, p_1, \dots, q_i - p_i\tau - p_i^*(\Delta t - \tau),$$
$$p_i^*, \dots, q_j - p_j\tau - p_j^*(\Delta t - \tau), p_j^*, \dots, q_N - p_N\Delta t, p_N)$$
$$- \bar{f}_N(t, q_1 - p_1\Delta t, p_1, \dots, q_i - p_i\Delta t, p_i, \dots, q_j - p_j\Delta t,$$
$$p_j, \dots, q_N - p_N\Delta t, p_N)\big]. \tag{3.2.14}$$

Using (3.2.14), we obtain the following differential equation for $\bar{f}_N(t, x_1, \dots, x_N)$:

$$\frac{\partial}{\partial t}\bar{f}_N(t, x_1, \dots, x_N)$$
$$= -\sum_{i=1}^{N} p_i \cdot \frac{\partial}{\partial q_i}\bar{f}_N(t, x_1, \dots, x_N) + \sum_{i<j=1}^{N} \int_{S_+^2} d\eta_{ij}\eta_{ij} \cdot (p_i - p_j)\delta(q_i - q_j)$$
$$\times \big[\bar{f}_N(t, x_1, \dots, x_i^*, \dots, x_j^*, \dots, x_N) - \bar{f}_N(t, x_1, \dots, x_i, \dots, x_j, \dots, x_N)\big]$$
$$= \bar{\mathcal{H}}_N \bar{f}_N(t, x_1, \dots, x_N), \quad x_i^* = (q_i, p_i^*), \quad x_j^* = (q_j, p_j^*) \tag{3.2.15}$$

with the initial condition

$$\bar{f}_N(t, x_1, \dots, x_N)|_{t=0} = f_N(x_1, \dots, x_N).$$

Equation (3.2.15) is the *Itô–Liouville equation* for $\bar{f}_N(t, x_1, \ldots, x_N)$. It was derived from functional average (3.2.4), (3.2.13) or from formulas (3.2.7), (3.2.14). We stress again that in functional average (3.2.4), (3.2.13), the contribution of the hyperplanes $q_i - p_i\tau = q_j - p_j\tau$, $1 \le i < j \le N$, are taken into account. These contributions are expressed in (3.2.4) and (3.2.13) by the second terms. Note that equation (3.2.15) defines the derivative of the function $\bar{f}_N(t, x_1, \ldots, x_N)$ in the sense of generalized functions.

We will also call $\bar{f}_N(t, x_1, \ldots, x_N)$ the distribution function of N particles at time t.

3.2.3 Infinitesimal operator with fixed random vectors

We can also differentiate the function $f_N(x_1(-t), \ldots, x_N(-t))$ in the sense of pointwise convergence, i.e., differentiate $f_N(x_1(-t), \ldots, x_N(-t))$ with respect to time along the trajectory $(x_1(-t), \ldots, x_N(-t))$ with fixed parameters η_{ij} and then introduce an associated generalized function with δ-functions $\delta(q_i - q_j)$, $i, j \in \{1, \ldots, N\}$. Denote the generalized function corresponding to the function $f_N(x_1(-t), \ldots, x_N(-t))$ with fixed parameters η_{ij} by $\tilde{f}_N(t, x_1, \ldots, x_N)$. Repeating our calculation from Chapter 2 word for word, we obtain the equation

$$\begin{aligned}
&\frac{\partial \tilde{f}_N}{\partial t}(t, x_1, \ldots, x_N) \\
&\quad = -\sum_{i=1}^{N} p_i \cdot \frac{\partial}{\partial q_i} \tilde{f}_N(t, x_1, \ldots, x_N) \\
&\quad\quad + \sum_{i<j=1}^{N} \Theta(\eta_{ij} \cdot (p_i - p_j))\eta_{ij} \cdot (p_i - p_j)\delta(q_i - q_j) \\
&\quad\quad \times \left[\tilde{f}_N(t, x_1, \ldots, x_i^*, \ldots, x_j^*, \ldots, x_N) - \tilde{f}_N(t, x_1, \ldots, x_i, \ldots, x_j, \ldots, x_N)\right] \\
&\quad = \tilde{\mathcal{H}}_N \tilde{f}_N(t, x_1, \ldots, x_N),
\end{aligned} \tag{3.2.16}$$

$$f_N(t, x_1, \ldots, x_N)_{|t=0} = f_N(x_1, \ldots, x_N)$$

with a boundary condition according to which, for $q_i = q_j$ momenta (p_i, p_j) in the first term of (3.2.16) should be replaced by (p_i^*, p_j^*) with $\eta_{ij} \cdot (p_i - p_j) \ge 0$ $\Theta(\alpha) = 1$, $\alpha > 0$, $\Theta(\alpha) = 0$, $\alpha < 0$. Note that this boundary condition is satisfied automatically due the second term with δ-functions.

We now present a new derivation of equation (3.2.16). Equation (3.2.16) and the infinitesimal operator $\tilde{\mathcal{H}}_N$ can be obtained from the following functional average:

$$\int dx_1 \ldots dx_N \, f_N(q_1 - p_1\Delta t, p_1, \ldots, q_N - p_N\Delta t, p_N)\varphi_N(q_1, p_1, \ldots, q_N, p_N)$$

$$
\begin{aligned}
&+ \sum_{i<j=1}^{N} \int dx_1 \dots dx_N \int_0^{\Delta t} d\tau (\eta_{ij} \cdot (p_i - p_j)) \\
&\quad \times \Theta(\eta_{ij} \cdot (p_i - p_j)) \delta(q_i - p_i\tau - q_j + p_j\tau) \\
&\quad \times \big[f_N(q_1 - p_1\Delta t, p_1, \dots, q_i - p_i\tau - p_i^*(\Delta t - \tau), p_i^*, \dots, \\
&\qquad q_j - p_j\tau - p_j^*(\Delta t - \tau), p_j^*, \dots, q_N - p_N\Delta t, p_N) \\
&\quad - f_N(q_1 - p_1\Delta t, p_1, \dots, q_i - p_i\Delta t, p_i, \dots, q_j - p_j\Delta t, p_j, \dots, \\
&\qquad q_N - p_N\Delta t, p_N) \big] \varphi_N(x_1, \dots, x_N) \\
&= \int dx_1 \dots dx_N \tilde{S}_N(-\Delta t) f_N(x_1, \dots, x_N) \varphi_N(x_1, \dots, x_N) \\
&= (\tilde{f}_N(\Delta t), \varphi_N). \qquad (3.2.17)
\end{aligned}
$$

Equality (3.2.17) holds for an arbitrary test function $\varphi_N(x_1, \dots, x_N)$, and $\tilde{S}_N(-\Delta t) f_N(x_1, \dots, x_N)$ is determined from (3.2.17) as follows:

$$
\begin{aligned}
&\tilde{S}_N(-\Delta t) f_N(x_1, \dots, x_N) \\
&\quad = \tilde{f}_N(\Delta t, x_1, \dots, x_N) \\
&\quad = f_N(q_1 - p_1\Delta t, p_1, \dots, q_N - p_N\Delta t, p_N) \\
&\qquad + \sum_{i<j=1}^{N} \int_0^{\Delta t} d\tau \Theta(\eta_{ij} \cdot (p_i - p_j)) \eta_{ij} \cdot (p_i - p_j) \delta(q_i - p_i\tau - q_j + p_j\tau) \\
&\qquad \times \big[f_N(q_1 - p_1\Delta t, p_1, \dots, q_i - p_i\tau - p_i^*(\Delta t - \tau), p_i^*, \dots, \\
&\qquad\quad q_j - p_j\tau - p_j^*(\Delta t - \tau), p_j^*, \dots, q_N - p_N\Delta t, p_N) \\
&\qquad - f_N(q_1 - p_1\Delta t, p_1, \dots, q_i - p_i\Delta t, p_i, \dots, q_j - p_j\Delta t, p_j, \dots, \\
&\qquad\quad q_N - p_N\Delta t, p_N) \big], \qquad (3.2.18)
\end{aligned}
$$

$$
\frac{\partial \tilde{S}_N(-t)}{\partial t}_{|t=0} f_N(x_1, \dots, x_N) = \tilde{\mathcal{H}}_N f_N(x_1, \dots, x_N).
$$

We formally define a group of operators $\tilde{S}_N(-t)$ at arbitrary time $-t$ as follows:

$$
\tilde{S}_N(-t) = \lim_{n \to \infty} \prod_{i=1}^{n} \bar{S}_N(-\Delta t_i), \quad \sum_{i=1}^{n} \Delta t_i = t,
$$

where the operator $\bar{S}_N(-\Delta t)$ for infinitesimal Δt is defined according to (3.2.18), and the infinitesimal generator of the group $\tilde{S}_N(-t)$ is equal to $\tilde{\mathcal{H}}_N$. Using (3.2.18) and the definition of $\tilde{S}_N(-t)$, one can obtain the distribution function at arbitrary time t, and it is $\tilde{f}_N(t, x_1, \ldots, x_N) = \tilde{S}_N(-t) f_N(x_1, \ldots, x_N)$. If depends on random vectors. Assume that the distribution function is already obtained at time t. Then $\tilde{f}_N(t + \Delta t, x_1, \ldots, x_N)$ is defined through $\tilde{f}_N(t, x_1, \ldots, x_N)$ as follows:

$$\begin{aligned}
&\tilde{f}_N(t + \Delta t, x_1, \ldots, x_N) = \tilde{S}_N(-\Delta t)\tilde{f}_N(t, x_1, \ldots, x_N) \\
&\quad = \tilde{f}_N(t, q_1 - p_1\Delta t, p_1, \ldots, q_N - p_N\Delta t, p_N) \\
&\qquad + \sum_{i<j=1}^{N} \int_0^{\Delta t} d\tau \Theta(\eta_{ij} \cdot (p_i - p_j))\eta_{ij} \cdot (p_i - p_j)\delta(q_i - p_i\tau - q_j + p_j\tau) \\
&\qquad\qquad \times \big[\tilde{f}_N(t, q_1 - p_1\Delta t, p_1, \ldots, q_i - p_i\tau - p_i^*(\Delta t - \tau), p_i^*, \ldots, \\
&\qquad\qquad\qquad q_j - p_j\tau - p_j^*(\Delta t - \tau), p_j^*, \ldots, q_N - p_N\Delta t, p_N) \\
&\qquad\qquad - \tilde{f}_N(t, q_1 - p_1\Delta t, p_1, \ldots, q_i - p_i\Delta t, p_i, \ldots, \\
&\qquad\qquad\qquad q_j - p_j\Delta t, p_j, \ldots, q_N - p_N\Delta t, p_N)\big], \qquad (3.2.19)
\end{aligned}$$

or, in terms of averages with test functions $\varphi_N(x_1, \ldots, x_N)$,

$$\begin{aligned}
&(\tilde{f}_N(t + \Delta t), \varphi_N) \\
&\quad = \int dx_1 \ldots dx_N \tilde{f}_N(t, q_1 - p_1\Delta t, p_1, \ldots, q_N - p_N\Delta t, p_N)\varphi_N(x_1, \ldots, x_N) \\
&\qquad + \sum_{i<j=1}^{N} \int dx_1 \ldots dx_N \Big\{ \int_0^{\Delta t} d\tau \Theta(\eta_{ij} \cdot (p_i - p_j))\eta_{ij} \cdot (p_i - p_j) \\
&\qquad\qquad \times \delta(q_i - p_i\tau - q_j + p_j\tau) \\
&\qquad\qquad \times \big[\tilde{f}_N(t, q_1 - p_1\Delta t, p_1, \ldots, q_i - p_i\tau - p_i^*(\Delta t - \tau), p_i^*, \ldots, \\
&\qquad\qquad\qquad q_j - p_j\tau - p_j^*(\Delta t - \tau), p_j^*, \ldots, q_N - p_N\Delta t, p_N) \\
&\qquad\qquad - \tilde{f}_N(t, q_1 - p_1\Delta t, p_1, \ldots, q_i - p_i\Delta t, p_i, \ldots, q_j - p_j\Delta t, p_j, \ldots, \\
&\qquad\qquad\qquad q_N - p_N\Delta t, p_N)\big]\Big\}\varphi_N(x_1, \ldots, x_N). \qquad (3.2.20)
\end{aligned}$$

The differential equation (3.2.16) follows from (3.2.19), (3.2.20).

It was shown in the previous chapter that the operator $\tilde{\mathcal{H}}_N$, is equivalent to the operator $\mathcal{H}_N$, and equation (3.2.16) can be represented in the form

$$\frac{\partial f_N}{\partial t}(t, x_1, \ldots, x_N) = -\sum_{i=1}^{N} p_i \cdot \frac{\partial}{\partial q_i} f_N(t, x_1, \ldots, x_N)$$

$$= \mathcal{H}_N f_N(t, x_1, \dots, x_N) \tag{3.2.21}$$

with the boundary condition according to which, at $q_i = q_j$, $i, j \in \{1, \dots, N\}$, $\eta \cdot (p_i - p_j) \geq 0$, momenta p_i, p_j should be replaced by p_i^*, p_j^* on the right-hand side of (3.2.21) and in $f_N(t, x_1, \dots, x_N)$.

Equation (3.2.16) shows how to integrate equation (3.2.21) with test functions and take into account the hyperplanes where the stochastic particles interact, but without averaging (3.2.16) with respect to the random vectors η_{ij}. Equation (3.2.15) also shows how to average with respect to the random vectors η_{ij}. Both of them correspond to equation (3.2.21), considered as an equation for usual function defined numerically, and are equivalent to it as generalized functions.

3.2.4 Duality principle

We now explain in what sense the functions $\tilde{S}_N(-\Delta t) f_N(x_1, \dots, x_N)$ given by formulas (3.2.18) are equivalent to the function $S_N(-\Delta t) f_N(x_1, \dots, x_N)$ given by formulas (3.2.3), (3.2.5):

$$\begin{aligned} S_N(-\Delta t) f_N(x_1, \dots, x_N) &= f_N(x_1(-\Delta t), \dots, x_N(-\Delta t)) \\ &= f_N(q_1 - p_1 \Delta t, p_1, \dots, q_N - p_N \Delta t, p_N) \end{aligned}$$

if $q_i - p_i \tau \neq q_j - p_j \tau$ for all $i, j \in \{1, \dots, N\}$ and $0 \leq \tau \leq \Delta t$,

$$\begin{aligned} &S_N(-\Delta t) f_N(x_1, \dots, x_N) \\ &\quad = f_N(x_1(-\Delta t), \dots, x_N(-\Delta t)) \\ &\quad = f_N(q_1 - p_1 \Delta t, p_1, \dots, q_i - p_i \tau - p_i^*(\Delta t - \tau), p_i^*, \dots, \\ &\qquad\quad q_j - p_j \tau - p_j^*(\Delta t - \tau), p_j^*, \dots, q_N - p_N \Delta t, p_N) \end{aligned} \tag{3.2.22}$$

if $q_i - p_i \tau = q_j - p_j \tau$, $0 \leq \tau \leq \Delta t$, $\eta_{ij} \subset S_+^2$, and

$$\begin{aligned} &S_N(-\Delta t) f_N(x_1, \dots, x_N) \\ &\quad = f_N(x_1(-\Delta t), \dots, x_N(-\Delta t)) \\ &\quad = f_N(q_1 - p_1 \Delta t, p_1, \dots, q_i - p_i \Delta t, p_i, \dots, q_j - p_j \Delta t, \\ &\qquad\quad p_j, \dots, q_N - p_N \Delta t, p_N) \end{aligned} \tag{3.2.23}$$

if $q_i - p_i \tau = q_j - p_j \tau$, $0 \leq \tau \leq \Delta t$, $\eta_{ij} \subset S_-^2$, $i, j \in \{1, \dots, N\}$. The definition of $S_N(-\Delta t) f_N(t, x_1, \dots, x_N)$ is obvious.

Thus, the functions $S_N(-\Delta t) f_N(x_1, \dots, x_N)$ and $S_N(-\Delta t) f_N(t, x_1, \dots, x_N)$ are numerically given by relations (3.2.22) and (3.2.23) in the entire phase space $(x_1, \dots, x_N)$. Expressions (3.2.18) and (3.2.19) for $\tilde{S}_N(-\Delta t) f_N(x_1, \dots, x_N)$

and $\tilde{S}_N(-\Delta t)\tilde{f}_N(t, x_1, \ldots, x_N)$ determine how to integrate the functions $S_N(-\Delta t)f_N(x_1, \ldots, x_N)$ and $S_N(-\Delta t)f_N(t, x_1, \ldots, x_N)$ with the test function $\varphi_N(x_1, \ldots, x_N)$, but with the fixed random vectors η_{ij}, $1 \leq i < j \leq N$, and how to take into account the contributions of the hyperplanes $q_i - p_i\tau = q_j - p_j\tau$, $1 \leq i < j \leq N$.

Expressions (3.2.7) and (3.2.14) for $\bar{S}_N(-\Delta t)f_N(x_1, \ldots, x_N)$ and $\bar{S}_N(-\Delta t)\bar{f}_N(t, x_1, \ldots, x_N)$ determine how to integrate these functions with test functions $\varphi_N(x_1, \ldots, x_N)$, to average with respect to the random vectors η_{ij}, and to take into account the contribution of the hyperplanes $q_i - p_i\tau = q_j - p_j\tau$.

We can now formulate the *principle of duality for the distribution function* $f_N(t, x_1, \ldots, x_N)$. Assume that the initial distribution function $f_N(0, x_1, \ldots, x_N) \equiv f_N(x_1, \ldots, x_N)$ is real, symmetric, continuously differentiable, and normalized in the phase space. Then the distribution function (3.2.3) $f_N(t, x_1, \ldots, x_N) = S_N(-t)f_N(x_1, \ldots, x_N)$ is a well-defined continuously differentiable function everywhere outside the hyperplanes of lower dimensions where particles interact. But in the functional average with some observable $\varphi_N(x_1, \ldots, x_N)$, which is a real symmetric smooth test function, we consider $S(-t)f_N(x_1, \ldots, x_N) = f_N(t, x_1, \ldots, x_N)$ as some definite generalized function and calculate the contribution of these hyperplanes of lower dimension where particles interact. Calculating the functional averages $(\tilde{f}(t), \varphi_N)$ or $(\bar{f}_N(t), \varphi_N)$, we use $\tilde{f}_N(t)$ or $\bar{f}_N(t)$ instead of $f_N(t)$. We cannot calculate $(\bar{f}_N(t), \varphi_N)$ or $(\tilde{f}_N(t), \varphi_N)$ directly for arbitrary finite time t; we have explicit formulas (3.2.4)–(3.2.8), (3.2.17)–(3.2.18) only for infinitesimal Δt. For arbitrary time t we use formulas (3.2.14) and (3.2.19) for the definition of $\bar{f}_N(t+\Delta t)$ and $\tilde{f}_N(t+\Delta t)$ through already defined $\bar{f}_N(t)$, $\tilde{f}_N(t)$, and then calculate $(\bar{f}_N(t+\Delta t), \varphi_N)$ according to (3.2.13), and $(\tilde{f}_N(t + \Delta t), \varphi_N)$ according to (3.2.20).

The duality principle defines the generalized functions $\bar{f}_N(t, x_1, \ldots, x_N) = \bar{S}_N(-t)f_N(x_1, \ldots, x_N)$ or $\tilde{f}_N(t, x_1, \ldots, x_N) = \tilde{S}_N(-t) \times f_N(x_1, \ldots, x_N)$ through the usual function $f_N(t, x_1, \ldots, x_N) = S_N(-t)f_N(x_1, \ldots, x_N)$ by formulas (3.2.7), (3.2.14) or (3.2.18), (3.2.19). If one has to consider the distribution function as a usual function (numerically), then one should take $f_N(t, x_1, \ldots, x_N) = S_N(-t)f_N(x_1, \ldots, x_N)$; if one has to calculate the functional average (3.2.13) or (3.2.20) with observable $\varphi_N(x_1, \ldots, x_N)$, then one should take the generalized function $\bar{f}_N(t, x_1, \ldots, x_N) = \bar{S}_N(-t)f_N(x_1, \ldots, x_N)$ or the generalized function $\tilde{f}_N(t, x_1, \ldots, x_N) = \tilde{S}_N(-t)f_N(x_1, \ldots, x_N)$.

It has been shown in Chapter 2 that a differential equation for $f_N(t, x_1, \ldots, x_N)$ in the sense of pointwise convergence has the form

$$\frac{\partial f_N}{\partial t}(t, x_1, \ldots, x_N)$$

$$= -\sum_{i=1}^{N} p_i \cdot \frac{\partial}{\partial q_i} f_N(t, x_1, \ldots, x_N)$$

$$+ \sum_{i<j=1}^{S} \Theta(\eta_{ij} \cdot (p_i - p_j))\delta(\Delta t - \tau_{ij})_{|\Delta t=0}$$
$$\times \big[f_N(t, x_1, \ldots, x_i^*, \ldots, x_j^*, \ldots, x_N)$$
$$- f_N(t, x_1, \ldots, x_i, \ldots, x_j, \ldots, x_N) \big]|_{q_i=q} \tag{3.2.24}$$

with the above-described boundary condition in the Poisson bracket according to which (p_i, p_j) should be replaced by (p_i^*, p_j^*) if $q_i - q_j = 0$,

$$\big[f_N(t+0, x_1, \ldots, x_N) = f_N(t, x_1, \ldots, x_i^*, \ldots, x_j^*, \ldots, x_N) \big],$$
$$q_i = q_j, \quad \eta_{ij} \cdot (p_i - p_j) > 0,$$

and

$$\big[f_N(t+0, x_1, \ldots, x_N) = f_N(t, x_1, \ldots, x_N) \big],$$
$$q_i = q_j, \quad (\eta_{ij} \cdot (p_i - p_j) \leq 0).$$

Here τ_{ij} is the time of collision. In the coordinate system where the first components of the vector (q_i, p_i, q_j, p_j) are directed along the vector η_{ij}, the time of collision τ_{ij} is defined as follows:

$$\tau_{ij} = \frac{q_i^1 - q_j^1}{p_i^1 - p_j^1}.$$

Then the (i, j)-th term in (3.2.24) can be expressed as follows:

$$\Theta(p_i^1 - p_j^1)\delta(q_i^1 - q_j^1)(p_i^1 - p_j^1)\big[f_N(t, x_1, \ldots, x_i^*, \ldots, x_j^*, \ldots, x_N)$$
$$- f_N(t, x_1, \ldots, x_i, \ldots, x_j, \ldots, x_N) \big]|_{(q_i^2=q_j^2),(q_i^3=q_j^3)}.$$

This term is different from zero on the first axis $q_i^1 - q_j^1$ (with respect to the vector $q_i - q_j$, i.e., for $q_i^2 - q_j^2 = 0$, $q_i^3 - q_j^3 = 0$), and, regarded as a generalized function in the three-dimensional space, it is equal to

$$\Theta(p_i^1 - p_j^1)\delta(q_i^1 - q_j^1)\delta(q_i^2 - q_j^2)\delta(q_i^3 - q_j^3)(p_i^1 - p_j^1)$$
$$\times \big[f_N(t, x_1, \ldots, x_i^*, \ldots, x_j^*, \ldots, x_N)$$
$$- f_N(t, x_1, \ldots, x_i, \ldots, x_j, \ldots, x_N) \big]$$
$$= \Theta(\eta_{ij} \cdot (p_i - p_j))\delta(q_i - q_j)\eta_{ij} \cdot (p_i - p_j)$$
$$\times \big[f_N(t, x_1, \ldots, x_i^*, \ldots, x_j^*, \ldots, x_N)$$
$$- f_N(t, x_1, \ldots, x_i, \ldots, x_j, \ldots, x_N) \big]. \tag{3.2.25}$$

(For analogous calculations, see [GGV].) The expression obtained does not depend on the choice of a coordinate system because $\delta(q_i - q_j)$ and $\eta_{ij} \cdot (p_i - p_j)$ are invariant under rotation. Substituting (3.2.25) in (3.2.24), we obtain (3.2.16).

Recall that the operator

$$
\begin{aligned}
-\sum_{i=1}^{N} p_i \cdot \frac{\partial}{\partial q_i} f_N(t, x_1, \dots, x_N) + \sum_{i<j=1}^{N} \delta(\Delta t - \tau_{ij})|_{\Delta t=0} \\
\times \Theta(\eta_{ij}(p_i - p_j)\big[f_N(t, x_1, \dots, x_i^*, \dots, x_j^*, \dots, x_N) \\
- f_N(t, x_1, \dots, x_i, \dots, x_j, \dots, x_N)\big]|_{q_i=q}
\end{aligned}
\tag{3.2.26}
$$

with one-dimensional δ-functions $\delta(\Delta t - \tau_{ij})|_{\Delta t=0}$ is equivalent to the operator

$$
-\sum_{i=1}^{N} p_i \cdot \frac{\partial}{\partial q_i} f_N(t, x_1, \dots, x_N) = \mathcal{H}_N f_N(t, x_1, \dots, x_N) \tag{3.2.27}
$$

with the boundary condition according to which, at points $q_i = q_j, i, j \in \{1, \dots, N\}$, momenta p_i and p_j should be replaced by p_i^* and p_j^* if $\eta_{ij} \cdot (p_i - p_j) \geq 0$, and (p_i, p_j) do not change if $\eta_{ij} \cdot (p_i - p_j) \leq 0$.

Thus, we have three expressions for the infinitesimal operator $\mathcal{H}_N$ of the group $S_N(-t)$. The first one $\bar{\mathcal{H}}_N$ (3.2.15) was obtained in the weak sense and shows how to take into account, in the functional average, the hypersurfaces of lower dimensions $q_i = q_j, i, j \in \{1, \dots, N\}$ where particles interact. In the second one $\tilde{\mathcal{H}}_N$ (3.2.16) (also calculated in the weak sense), the average with respect to the random vectors η_{ij} was not performed, but the hypersurfaces $q_i = q_j, i, j \in \{1, \dots, N\}$, were taken into account. In the third one $\mathcal{H}_N$ (3.2.24)–(3.2.27) calculated pointwise, the infinitesimal operator $\mathcal{H}_N$ of the group $S_N(-t)$ is equal to the operator $-\sum_{i=1}^{N} p_i \cdot \frac{\partial}{\partial q_i}$ with the boundary conditions at $q_i = q_j, i, j \in \{1, \dots, N\}$, according to which momenta p_i, p_j should be replaced by p_i^*, p_j^* in (3.2.27) and in $f_N(t, x_1, \dots, x_N)$.

All these expressions for the infinitesimal operator $\mathcal{H}_N$ are equivalent, but the first one (3.2.15) and the second one (3.2.16) show how to calculate the average of $\frac{\partial \bar{f}}{\partial t}(t, x_1, \dots, x_N)$ and $\frac{\partial \tilde{f}_N(t, x_1, \dots, x_N)}{\partial t}$ with observable $\varphi_N(x_1, \dots, x_N)$, or they define $\frac{\partial \bar{f}_N(t, x_1, \dots, x_N)}{\partial t}$ and $\frac{\partial \tilde{f}_N(t, x_1, \dots, x_N)}{\partial t}$ as generalized functions. The third expression for the infinitesimal operator (3.2.26), (3.2.27) defines

$$
\frac{\partial f_N(t, x_1, \dots, x_N)}{\partial t} = \mathcal{H}_N f_N(t, x_1, \dots, x_N)
$$

in the sense of pointwise differentiation and defines it as a usual function with jumps at $q_i = q_j, i, j \in \{1, \dots, N\}$, which is expressed in the boundary conditions.

Thus, for the derivative $\frac{\partial f_N(t, x_1, \dots, x_N)}{\partial t}$, the principle of duality is also formulated as for $f_N(t, x_1, \dots, x_N)$, and, according to it, the same $\frac{\partial f_N(t, x_1, \dots, x_N)}{\partial t}$ is considered as a usual function or as special generalized functions in the functional average.

3.2.5 Generalized function

It will be useful to consider the action of the operator of evolution $S_N(-\Delta t)$ on the function $f_N(x_1,\dots,x_s,x_{s+1},\dots,x_N)$ as a usual function with respect to the variables $(x_1,\dots,x_s)$ and a generalized function with respect to $(x_{s+1},\dots,x_N)$. Namely, with the usual function $S_N(-\Delta t)f_N(x_1,\dots,x_s,x_{s+1},\dots,x_N)$, we associate the following expression with infinitesimal Δt:

$$\begin{aligned}
&\bar{f}_{s,N-s}(\Delta t, x_1,\dots,x_s,x_{s+1},\dots,x_N)\\
&= S_s(-\Delta t,(x)_s)S^0_{N-s}(-\Delta t,(x)^s_N)f_N(x_1,\dots,x_s,x_{s+1},\dots,x_N)\\
&\quad + \sum_{i<j=1}^{N}{}' \int_0^{\Delta t} d\tau \int_{S^2_+} d\eta_{ij}\,\eta_{ij}\cdot(p_i-p_j)\delta(q_i-p_i\tau-q_j+p_j\tau)\\
&\qquad \times \big[f_N(q_1-p_1\Delta t, p_1,\dots,q_i-p_i\tau-p_i^*(\Delta t-\tau),p_i^*,\dots,\\
&\qquad\qquad q_j-p_j\tau-p_j^*(\Delta t-\tau),p_j^*,\dots,q_N-p_N\Delta t,p_N)\\
&\qquad\quad - f_N(q_1-p_1\Delta t,p_1,\dots,q_i-p_i\Delta t,p_i,\dots,q_j-p_j\Delta t,\\
&\qquad\qquad p_j,\dots,q_N-p_N\Delta t,p_N)\big]\\
&= \bar{S}_{s,N-s}(-\Delta t)f_N(x_1,\dots,x_s,x_{s+1},\dots,x_N). \qquad (3.2.28)
\end{aligned}$$

In (3.2.28) we have used the following notation: $S^0_{N-s}(-\Delta t,(x)^s_N)$ is the operator of the free evolution of $N-s$ particles, $(x)^s_N=(x_{s+1},\dots,x_N)$, and $\sum'^{N}_{i<j=1}$ means that summation with respect to $1\le i<j\le s$ is excluded.

For arbitrary time t the function $f_{s,N-s}(t,x_1,\dots,x_s,x_{s+1},\dots,x_N)$ is formally defined as follows:

$$\begin{aligned}
&\bar{f}_{s,N-s}(t,x_1,\dots,x_s,x_{s+1},\dots,x_N)\\
&\qquad = \lim_{n\to\infty}\prod_{i=1}^{n}\bar{S}_{s,N-s}(-\Delta t_i)f_N(x_1,\dots,x_s,x_{s+1},\dots,x_N), \qquad (3.2.29)
\end{aligned}$$

$$\sum_{i=1}^{n}\Delta t_i = t.$$

It is obvious that $\bar{f}_{s,N-s}(t,x_1,\dots,x_s,x_{s+1},\dots,x_N)$ satisfies the following equation:

$$\begin{aligned}
&\frac{\partial f_{s,N-s}(t,x_1,\dots,x_s,x_{s+1},\dots,x_N)}{\partial t}\\
&\qquad = \Big(\mathcal{H}_s - \sum_{i=s+1}^{N} p_i\cdot\frac{\partial}{\partial q_i}\Big)\bar{f}_{s,N-s}(t,x_1,\dots,x_s,x_{s+1},\dots,x_N)
\end{aligned}$$

$$+\sum_{i<j=1}^{N}{}' \int_{S_+^2} d\eta_{ij}\, \eta_{ij}\cdot(p_i-p_j)\delta(q_i-q_j)$$
$$\times\big[\bar{f}_{s,N-s}(t,x_1,\dots,x_i^*,\dots,x_j^*,\dots,x_N)$$
$$-\bar{f}_{s,N-s}(t,x_1,\dots,x_i,\dots,x_j,\dots,x_N)\big] \qquad (3.2.30)$$

where the operator $\mathcal{H}_s$ is equal to $-\sum_{i=1}^{N} p_i\cdot\frac{\partial}{\partial q_i}$ with the known boundary condition at $q_i=q_j$, $i,j\in\{1,\dots,s\}$. The boundary conditions in $\sum_{i=s+1}^{N} p_i\cdot\frac{\partial}{\partial q_i}$ are absent; they are taken into account in the second term in (3.2.30).

3.3 Hierarchy for correlation functions

3.3.1 Derivation of hierarchy from equation for distribution function

We define the following sequence of correlation functions:

$$\bar{F}_s^{(N)}(t,x_1,\dots,x_s)$$
$$= N(N-1)\dots(N-s+1)\int dx_{s+1}\dots dx_N$$
$$\times\bar{f}_N(t,x_1,\dots,x_s,\dots,x_{s+1},\dots,x_N),\quad 1\le s\le N. \qquad (3.3.1)$$

We denote the sequence of correlation functions (3.3.1) by

$$\bar{F}^{(N)}(t)=\Big(\bar{F}_1^{(N)}(t,x_1),\dots,\bar{F}_s^{(N)}(t,x_1,\dots,x_s),\dots,\bar{F}_N^{(N)}(t,x_1,\dots,x_N)\Big).$$

The sequence $\bar{F}^{(N)}(t)$ is the state of a system of N particles at time t within the framework of canonical ensemble.

Let us proceed to the derivation of an equation for the state $\bar{F}^{(N)}(t)$. For this purpose, we differentiate the left- and right-hand sides of (3.3.1) with respect to time. Differentiating under the integral sign and using (3.2.15), we obtain

$$\frac{\partial\bar{F}_s^{(N)}(t,x_1,\dots,x_N)}{\partial t}$$
$$= N(N-1)\dots(N-s+1)$$
$$\times\int\frac{\partial}{\partial t}\bar{f}_N(t,x_1,\dots,x_s,x_{s+1},\dots,x_N)dx_{s+1}\dots dx_N$$
$$=\frac{N!}{(N-s)!}\int\Big\{-\sum_{i=1}^{N} p_i\cdot\frac{\partial}{\partial q_i}\bar{f}_N(t,x_1,\dots,x_s,x_{s+1},\dots,x_N)$$

$$+\sum_{i<j=1}^{N}\int_{S_+^2} d\eta_{ij}\delta(q_i-q_j)\eta_{ij}\cdot(p_i-p_j)$$
$$\times\big[\bar f_N(t,x_1,\dots,x_i^*,\dots,x_j^*,\dots,x_N)$$
$$-\bar f_N(t,x_1,\dots,x_i,\dots,x_j,\dots,x_N)\big]\Big\}dx_{s+1}\dots dx_N. \quad (3.3.2)$$

It is easy to see that the integration of the expressions

$$-\sum_{i=1}^{s} p_i\cdot\frac{\partial}{\partial q_i}\bar f_N(t,x_1,\dots,x_s,x_{s+1},\dots,x_N)$$
$$+\sum_{i<j=1}^{s}\int_{S_+^2} d\eta_{ij}\delta(q_i-q_j)\eta_{ij}\cdot(p_i-p_j)$$
$$\times\big[\bar f_N(t,x_1,\dots,x_i^*,\dots,x_j^*,\dots,x_s,x_{s+1},\dots,x_N)$$
$$-\bar f_N(t,x_1,\dots,x_s,x_{s+1},\dots,x_N)\big]$$

on the right-hand side of (3.3.2) yields

$$-\sum_{i-1}^{s} p_i\cdot\frac{\partial}{\partial q_i}\bar F_s^{(N)}(t,x_1,\dots,x_s)$$
$$+\sum_{i<j=1}^{s}\int_{S_+^2} d\eta_{ij}\delta(q_i-q_j)\eta_{ij}\cdot(p_i-p_j)$$
$$\times\big[\bar F_s^{(N)}(t,x_1,\dots,x_i^*,\dots,x_j^*,\dots,x_s)-\bar F_s^{(N)}(t,x_1,\dots,x_s)\big]. \quad (3.3.3)$$

Further, the integration of the expression

$$\sum_{i=1}^{s}\sum_{j=s+1}^{N}\int_{S_+^2} d\eta_{ij}\delta(q_i-q_j)\eta_{ij}\cdot(p_i-p_j)$$
$$\times\big[\bar f_N(t,x_1,\dots,x_i^*,\dots,x_s,x_{s+1},\dots,x_j^*,\dots,x_N)$$
$$-\bar f_N(t,x_1,\dots,x_s,x_{s+1},\dots,x_N)\big]$$

yields

$$\sum_{i=1}^{s}\int dp_{s+1}\int_{S_+^2} d\eta_{is+1}\eta_{is+1}\cdot(p_i-p_{s+1})$$

$$\times \big[\bar{F}_{s+1}^{(N)}(t, x_1, \dots, x_i^*, \dots, x_s, x_{s+1}) \\ - \bar{F}_{s+1}^{(N)}(t, x_1, \dots, x_i, \dots, x_s, x_{s+1}, \dots, x_N)\big]_{q_{s+1}=q_i}.$$

Here, we have used the invariance of the function $\bar{f}_N(t, x_1, \dots, x_N)$ under the permutation.

The contribution of the term

$$-\sum_{i=s+1}^{N} p_i \cdot \frac{\partial}{\partial q_i} \bar{f}_N(t, x_1, \dots, x_N) \tag{3.3.4}$$

is zero because the function $\bar{f}_N(t, x_1, \dots, x_N)$ tends to zero as $|q_i| \to \infty$ and the boundary conditions have been taken into account by δ-functions.

The contribution of the term

$$\sum_{i<j=s+1}^{N} \int_{S_+^2} d\eta_{ij} \delta(q_i - q_j) \eta_{ij} \cdot (p_i - p_j) \\ \times \big[\bar{f}_N(t, x_1, \dots, x_i^*, \dots, x_j^*, \dots, x_N) - \bar{f}_N(t, x_1, \dots, x_N)\big] \tag{3.3.5}$$

is equal to the integral

$$\frac{(N-s)(N-s-1)}{2} \int dx_{s+1} dx_{s+2} \\ \times \int_{S_+^2} d\eta \delta(q_{s+1} - q_{s+2}) \eta \cdot (p_{s+1} - p_{s+2}) \\ \times \big[\bar{F}_{s+2}^{(N)}(t, x_1, \dots, x_s, x_{s+1}^*, x_{s+2}^*) - \bar{F}_{s+2}^{(N)}(t, x_1, \dots, x_s, x_{s+1}, x_{s+2})\big], \tag{3.3.6}$$

which is zero. This fact can be proved by using new variables $p_{s+1}^* = p'_{s+1}$, $p_{s+2}^* = p'_{s+2}$ and taking into account that the Jacobian of this transformation is equal to unity.

Summarizing the calculations performed above, we obtain the following chain of equations:

$$\frac{\partial \bar{F}_s^{(N)}(t, x_1, \dots, x_N)}{\partial t} \\ = -\sum_{i=1}^{s} p_i \cdot \frac{\partial}{\partial q_i} \bar{F}_s^{(N)}(t, x_1, \dots, x_s) + \sum_{i<j=1}^{s} \int_{S_+^2} d\eta_{ij} \delta(q_i - q_j) \eta_{ij} \cdot (p_i - p_j) \\ \times \big[\bar{F}_s^{(N)}(t, x_1, \dots, x_i^*, \dots, x_j^*, \dots, x_s) - \bar{F}_s^{(N)}(t, x_1, \dots, x_s)\big] \\ + \sum_{i=1}^{s} \int dp_{s+1} \int_{S_+^2} d\eta_{is+1} \eta_{is+1} \cdot (p_i - p_{s+1})$$

$$\times \big[\bar{F}^{(N)}_{s+1}(t, x_1, \dots, x_i^*, \dots, x_s, x_{s+1}^*) \\ - \bar{F}^{(N)}_{s+1}(t, x_1, \dots, x_s, x_{s+1})\big]_{q_{s+1}=q_i}, \quad s = 1, 2, \dots, N. \tag{3.3.7}$$

We call equations (3.3.7) the *stochastic hierarchy*, or the *stochastic Boltzmann hierarchy in the weak sense*.

The obtained stochastic hierarchy (3.3.7) is equivalent to the Itô–Liouville equation (3.2.15); moreover, the last equation for $\bar{F}_N(t, x_1, \dots, x_N)$ completely coincides with the Itô–Liouville equation (3.2.15).

Hierarchy (3.3.7) describes the evolution of the state $\bar{F}^{(N)}(t)$ of a finite system. In order to describe the state of an infinite system within the framework of canonical ensemble, it is necessary to perform the thermodynamic limit procedure and let the number of particles tend to infinity ($N \to \infty$). Or, more exactly, one must consider the system of N particles located at the initial time $t = 0$ in a bounded domain Λ of the configurational space (this means that $\bar{F}^{(N)}_s(0, x_1, \dots, x_s) = 0$ if $q_i \notin \Lambda$ for at least one particle) and then let the number of particles and the volume $V(\Lambda)$ of the domain Λ tend to infinity ($N \to \infty$, $V(\Lambda) \to \infty$) so that the density is constant:

$$\frac{1}{V} = \frac{N}{V(\Lambda)} = \text{const.}$$

Performing the formal thermodynamic limit transition for sequence (3.3.1) and supposing that one can also perform this limit transition in hierarchy (3.3.7), we obtain the limiting hierarchy

$$\frac{\partial \bar{F}_s(t, x_1, \dots, x_s)}{\partial t} \\ = -\sum_{i=1}^{s} p_i \cdot \frac{\partial}{\partial q_i} \bar{F}_s(t, x_1, \dots, x_s) + \sum_{i<j=1}^{s} \int_{S^2_+} d\eta_{ij}\, \eta_{ij} \cdot (p_i - p_j)\delta(q_i - q_j) \\ \times \big[\bar{F}_s(t, x_1, \dots, x_i^*, \dots, x_j^*, \dots, x_s) - \bar{F}_s(t, x_1, \dots, x_i, \dots, x_j, \dots, x_s)\big] \\ + \sum_{i=1}^{s} \int dx_{s+1} \int_{S_2^+} d\eta_{is+1}\, \eta_{is+1} \cdot (p_i - p_{s+1})\delta(q_i - q_{s+1}) \\ \times \big[\bar{F}_{s+1}(t, x_1, \dots, x_i^*, \dots, x_{s+1}^*) - \bar{F}_{s+1}(t, x_1, \dots, x_i, \dots, x_{s+1})\big], \quad s \ge 1,$$

$$\bar{F}_s(t, x_1, \dots, x_s) = \lim_{N \to \infty} \bar{F}^{(N)}_s(t, x_1, \dots, x_s). \tag{3.3.8}$$

Note that the correlation functions $\bar{F}_s(t, x_1, \dots, x_s)$ do not depend on any random vectors η_{ij}.

The sequence of correlation functions $\bar{F}_s(t, x_1, \dots, x_s)$, $s \ge 1$, and hierarchy (3.3.8) for it were obtained through the distribution function $\bar{f}_N(t, x_1, \dots, x_N)$ considered as a definite generalized function.

The first two terms on the right-hand side of hierarchy (3.3.8) are the result of the action of the infinitesimal operator $\bar{\mathcal{H}}_s$ (of the group $\bar{S}_s(-t)$), which is useful for the functional average. We want to deduce a hierarchy for the sequence of correlation function considered as a usual (not generalized) function. To do this we replace the operator $\bar{\mathcal{H}}_s$ in (3.3.8) by the equivalent operator

$$\mathcal{H}_s = -\sum_{i=1}^{s} p_i \cdot \frac{\partial}{\partial q_i}$$

with known boundary condition. One obtains the hierarchy

$$\begin{aligned}
&\frac{\partial F_s(t, x_1, \ldots, x_s)}{\partial t} \\
&\quad = -\sum_{i=1}^{s} p_i \cdot \frac{\partial}{\partial q_i} F_s(t, x_1, \ldots, x_s) \\
&\qquad + \sum_{i=1}^{s} \int dx_{s+1} \int_{S_+^2} d\eta_{is+1} \eta_{is+1} \cdot (p_i - p_{s+1}) \delta(q_i - q_{s+1}) \\
&\qquad\quad \times \big[F_{s+1}(t, x_1, \ldots, x_i^*, \ldots, x_s, x_{s+1}^*) \\
&\qquad\qquad - F_{s+1}(t, x_1, \ldots, x_i, \ldots, x_s, x_{s+1}^*)\big], \quad s \geq 1, \qquad (3.3.9)
\end{aligned}$$

with boundary conditions according to which, at $q_i = q_j$, $\eta_{ij} \in S_+^2$, $i, j \in \{1, \ldots, s\}$, the momenta (p_i, p_j) in the first term on the right-hand side of (3.3.9) should be replaced by (p_i^*, p_j^*).

This hierarchy coincides with the stochastic Boltzmann hierarchy (1.5.4) derived in Chapter 1 from the BBGKY hierarchy for a system of hard spheres in the Boltzmann–Grad limit (see also the next section).

Hierarchy (3.3.2) can be written in equivalent form:

$$\begin{aligned}
&\frac{\partial \tilde{F}_s(t, x_1, \ldots, x_s)}{\partial t} \\
&\quad = -\sum_{i=1}^{s} p_i \cdot \frac{\partial}{\partial q_i} \tilde{F}_s(t, x_1, \ldots, x_s) \\
&\qquad + \sum_{i<j=1}^{s} \Theta(\eta_{ij}(p_i - p_j)) \eta_{ij}(p_i - p_j) \delta(q_i - q_j) \\
&\qquad\quad \times \big[\tilde{F}_s(t, x_1, \ldots, x_i^*, \ldots, x_j^*, \ldots, x_s) - \tilde{F}_s(t, x_1, \ldots, x_i, \ldots, x_j, \ldots, x_s)\big] \\
&\qquad + \sum_{i=1}^{s} \int dx_{s+1} \int_{S_+^2} d\eta_{is+1} \eta_{is+1} \cdot (p_i - p_{s+1})
\end{aligned}$$

$$\times \big[\tilde{F}_{s+1}(t, x_1, \dots, x_i^*, \dots, x_s, x_{s+1}^*) - \tilde{F}_{s+1}(t, x_1, \dots, x_i, \dots, x_s, x_{s+1})\big], \quad s \geq 1, \tag{3.3.10}$$

without boundary conditions.

In (3.3.9) and (3.3.10), we have used an equivalent representation of the infinitesimal operator $\mathcal{H}_s$. Note that the correlation functions in (3.3.9), (3.3.10) depend on the random vectors η_{ij}, $i, j \in \{1, \dots, s\}$.

In what follows we use the hierarchy for a sequence of $F_s(t, q_1, \dots, q_s)$, $s \geq 1$, with the infinitesimal operator $\mathcal{H}_s$ and its well-known boundary condition, namely

$$\frac{\partial F_s(t, x_1, \dots, x_s)}{\partial t} = \mathcal{H}_s F_s(t, x_1, \dots, x_s) + \sum_{i=1}^{s} \int dx_{s+1} \int_{S_+^2} d\eta_{is+1} \eta_{is+1} \cdot (p_i - p_j) \delta(q_i - q_{s+1}) \times \big[F_{s+1}(t, x_1, \dots, x_i^*, \dots, x_s, x_{s+1}^*) - F_{s+1}(t, x_1, \dots, x_i, \dots, x_s, x_{s+1})\big] \tag{3.3.11}$$

with the initial condition

$$F_s(t, x_1, \dots, x_s)|_{t=0} = F_s(x_1, \dots, x_s).$$

Remark 3.1. Hierarchies (3.3.9) and (3.3.10) can be obtained from equation (3.2.16) by integrating over $x_{s+1}, \dots, x_N$ and averaging over all η_{ij} excluding those η_{ij} with $i, j \in \{1, \dots, s\}$. Namely, define the following correlation functions

$$\tilde{F}_s^{(N)}(t, x_1, \dots, x_s) = \int \tilde{f}_N(t, x_1, \dots, x_s, x_{s+1}, \dots, x_N) dx_{s+1} \dots dx_N \prod_{i,j}{}' d\eta_{ij}, \tag{3.3.12}$$

$$1 \leq s \leq N,$$

where the symbol $\prod'_{i,j} d\eta_{ij}$ means that one integrates with respect to all random vectors η_{ij} that correspond to all collisions of N particles, excluding collisions of s-particles with themselves, and pass to the thermodynamic limit. One obtains hierarchy (3.3.10) with $\tilde{\mathcal{H}}_s$ instead of $\mathcal{H}_s$. Then one uses the duality principle and replaces $\tilde{\mathcal{H}}_s$ by $\mathcal{H}_s$. The duality principle for hierarchies (3.3.8)–(3.3.10) consists of exactly this replacement.

Note that the correlation functions $F_s(t, x_1, \dots, x_s)$ depend on all random vectors η_{ij} with $i, j \in \{1, \dots, s\}$ that correspond to collisions of particles with numbers $(1, \dots, s)$ and appear on the entire interval $[-t, 0]$.

The stochastic hierarchy (3.3.11) with the operator

$$\mathcal{H}_s = -\sum_{i=1}^{s} p_i \cdot \frac{\partial}{\partial q_i}$$

and the well-known boundary conditions at $q_i = q_j,\ i, j \in \{1, \ldots, s\}$ can also be obtained from the modification of the Itô–Liouville equation (3.2.30)

$$\begin{aligned}
&\frac{\partial \bar{f}_{s,N-s}(t, x_1, \ldots, x_s, x_{s+1}, \ldots, x_N)}{\partial t} \\
&\quad = -\sum_{i=1}^{N} p_i \cdot \frac{\partial}{\partial q_i} \bar{f}_{s,N-s}(t, x_1, \ldots, x_s, x_{s+1}, \ldots, x_N) \\
&\quad + \sum_{i<j=1}^{N}{}' \int_{S^2_+} d\eta_{ij} \delta(q_i - q_j)\eta_{ij} \cdot (p_i - p_j))\delta(q_i - q_j)\eta_{ij} \cdot (p_i - p_j) \\
&\qquad \times \big[\bar{f}_{s,N-s}(t, x_1, \ldots, x_i^*, \ldots, x_j^*, \ldots, x_N) \\
&\qquad\quad - \bar{f}_{s,N-s}(t, x_1, \ldots, x_i, \ldots, x_j, \ldots, x_N)\big]
\end{aligned}$$

with boundary condition at $q_i - q_j,\ i, j \in \{1, \ldots, s\}$, by integration with respect to $x_{s+1}, \ldots, x_N$. The corresponding N-particle distribution function $\bar{f}_{s,N-s}(t, x_1, \ldots, x_s, x_{s+1}, \ldots, x_N)$ is defined by formulas (3.2.29). Then the correlation functions $F_s^{(N)}(t, (x)_s)$ can be obtained from the function $\bar{f}_{s,N-s}(t, x_1, \ldots, x_s, x_{s+1}, \ldots, x_N)$.

3.3.2 Stochastic hierarchy in grand canonical ensemble

Let us introduce a sequence of distribution functions within the framework of grand canonical ensemble where a system consists of an arbitrary number of particles with certain probability. The s-particle distribution functions are defined according to the formula

$$\begin{aligned}
&\bar{F}_s(t, x_1, \ldots, x_s) \\
&\quad = \frac{1}{\Xi} \sum_{n=0}^{\infty} \frac{1}{n!} \int \bar{f}_{s+n}(t, x_1, \ldots, x_s, x_{s+1}, \ldots, x_{s+n}) dx_{s+1} \ldots dx_{s+n},
\end{aligned} \tag{3.3.13}$$

where the functions $\bar{f}_{s+n}(t, x_1, \ldots, x_{s+n})$ are defined by (3.2.13)–(3.2.14). Note that, in this case, the functions $\bar{f}_{s+n}(0, x_1, \ldots, x_{s+n})$ are not normalized to unity.

In (3.3.12), Ξ is the grand partition function:

$$\Xi = \sum_{n=0}^{\infty} \frac{1}{n!} \int \bar{f}_n(t, x_1, \dots, x_n) dx_1 \dots dx_n$$
$$= \sum_{n=0}^{\infty} \frac{1}{n!} \int \bar{f}_n(0, x_1, \dots, x_n) dx_1 \dots dx_n,$$

$$\bar{f}_n(t, x) = \bar{S}_n(-t) f_n(x).$$

Here, we have used the Liouville theorem, according to which

$$\int \bar{f}_n(t, x_1, \dots, x_n) dx_1 \dots dx_n = \int \bar{f}_n(0, x_1, \dots, x_n) dx_1 \dots dx_n.$$

The proof follows directly from equation (3.2.3) and the same tricks as in (3.3.6). In order to perform the thermodynamic limit procedure within the framework of grand canonical ensemble, it is necessary to let the average number of particles $\bar{N}$ tend to infinity or, if the system of particles is located in a bounded domain Λ of the configurational space, let the volume $V(\Lambda)$ also tend to infinity so that so that the density remains constant:

$$\frac{1}{V} = \frac{\bar{N}}{V(\Lambda)} = \text{const.}$$

It is easy to prove that the sequence of functions (3.3.13)

$$\bar{F}(t) = (\bar{F}_1(t, x_1), \dots, \bar{F}_s(t, x_1, \dots, x_s), \dots)$$

satisfies the chain of equations (3.3.8). By using the duality principle, one obtains the chain of equations (3.3.9) or (3.3.10).

3.3.3 Duality principle for correlation functions

The principle of duality for correlation functions consists of the following: The correlation functions $F_s(t, (x)_s)$ are defined numerically as usual functions; the correlation functions $\bar{F}_s(t, (x)_s)$ are generalized functions associated with the usual functions $F_s(t, (x)_s)$, and they show how to integrate the usual functions $F_s(t, (x)_s)$ with test functions $\varphi_s((x)_s)$ and average with respect to the random vectors η_{ij}. The generalized correlation functions $\tilde{F}_s(t, (x)_s)$ play the some role, but without averaging with respect to η_{ij}.

We have introduced three kinds of correlation functions: $\bar{F}_s(t, x_1, \dots, x_s)$, $\tilde{F}_s(t, x_1, \dots, x_s)$, and $F_s(t, x_1, \dots, x_s)$. For infinitesimal time Δt, they are defined as follows:

$$\bar{F}_s(\Delta t, x_1, \dots, x_s)$$

$$= \int \bar{S}_N(-\Delta t) f_N(x_1, \dots, x_s, x_{s+1}, \dots, x_N) dx_{s+1} \dots dx_N$$

$$= \int \Big\{ f_N(q_1 - p_1\Delta t, p_1, \dots, q_s - p_s\Delta t, p_s,$$

$$q_{s+1} - p_{s+1}\Delta t, p_{s+1}, \dots, q_N - p_N\Delta t, p_N)$$

$$+ \int_0^{\Delta t} d\tau \sum_{i<j=1}^{N} \int_{S_+^2} d\eta_{ij} \eta_{ij} \cdot (p_i - p_j) \delta(q_i - p_i\tau - q_j + p_j\tau)$$

$$\times \big[f_N(q_1 - p_1\Delta t, p_1, \dots, q_i - p_i\tau - p_i^*(\Delta t - \tau), p_i^*, \dots,$$

$$q_j - p_j\tau - p_j^*(\Delta t - \tau), p_j^*, \dots, q_N - p_N\Delta t, p_N)$$

$$- f_N(q_1 - p_1\Delta t, p_1, \dots, q_i - p_i\Delta t, p_i, \dots,$$

$$q_j - p_j\Delta t, p_j, \dots, q_N - p_N\Delta t, p_N) \big] \Big\} dx_{s+1} \dots dx_N, \tag{3.3.14}$$

$$\tilde{F}_s(\Delta t, x_1, \dots, x_s)$$

$$= \int \tilde{S}_N(-\Delta t) f_N(x_1, \dots, x_s, x_{s+1}, \dots, x_N) dx_{s+1} \dots dx_N \prod_{i<j=1}^{N}{}' d\eta_{ij}$$

$$= \int \Big\{ f_N(q_1 - p_1\Delta t, p_1, \dots, q_s - p_s\Delta t, p_s,$$

$$q_{s+1} - p_{s+1}\Delta t, p_{s+1}, \dots, q_N - p_N\Delta t, p_N)$$

$$+ \int_0^{\Delta t} d\tau \sum_{i<j=1}^{N} \Theta(\eta_{ij}(p_i - p_j)) \eta_{ij} \cdot (p_i - p_j) \delta(q_i - p_i\tau - q_j + p_j\tau)$$

$$\times \big[f_N(q_1 - p_1\Delta t, p_1, \dots, q_i - p_i\tau - p_i^*(\Delta t - \tau), p_i^*, \dots,$$

$$q_j - p_j\tau - p_j^*(\Delta t - \tau), p_j^*, \dots, q_N - p_N\Delta t, p_N)$$

$$- f_N(q_1 - p_1\Delta t, p_1, \dots, q_i - p_i\Delta t, p_i, \dots, q_j - p_j\Delta t, p_j, \dots,$$

$$q_N - p_N\Delta t, p_N) \big] \Big\} dx_{s+1} \dots dx_N \prod_{i<j=1}^{N}{}' d\eta_{ij}, \tag{3.3.15}$$

$$
\begin{aligned}
&F_s(\Delta t, x_1, \dots, x_s)\\
&\quad = \int \bar{S}_{s,N-s}(-\Delta t) f_N(x_1, \dots, x_s, x_{s+1}, \dots, x_N) dx_{s+1} \dots dx_N\\
&\quad = \int \Big\{ S_s(-\Delta t, x_1, \dots, x_s) S^0_{N-s}(-\Delta t, x_{s+1}, \dots, x_N) f_N(x_1, \dots, x_N)\\
&\qquad + \int_0^{\Delta t} d\tau \sum_{i<j}{}' \int_{S^2_+} \eta_{ij}\, \eta_{ij} \cdot (p_i - p_j) \delta(q_i - p_i\tau - q_j + p_j\tau)\\
&\qquad \times \big[f_N(q_1 - p_1\Delta t, p_1, \dots, q_i - p_i\tau - p_i^*(\Delta t - \tau), p_i^*, \dots,\\
&\qquad\qquad q_j - p_j\tau - p_j^*(\Delta t - \tau), p_j^*, \dots, q_N - p_N\Delta t, p_N)\\
&\qquad - f_N(q_1 - p_1\Delta t, p_1, \dots, q_i - p_i\Delta t, p_i, \dots,\\
&\qquad\qquad q_j - p_j\Delta t, p_j, \dots, q_N - p_N\Delta t, p_N) \big] \Big\} dx_{s+1} \dots dx_N .
\end{aligned}
\tag{3.3.16}
$$

Note that, in (3.3.16), the boundary conditions with respect to the variables $x_{s+1}, \dots, x_{s+N}$ are taken into account in the second term.

For arbitrary t one can use the functions $\bar{f}_N(t, x_1, \dots, x_N)$, $\tilde{f}_N(t, x_1, \dots, x_N)$ and $f_{s,N-s}(t, x_1, \dots, x_N)$ defined by using

$$
\prod_{i=1}^n \bar{S}_N(-\Delta t_i), \quad \prod_{i=1}^n \tilde{S}_N(-\Delta t_i), \quad \prod_{i=1}^n \bar{S}_{s,N-s}(-\Delta t_i), \quad \text{and} \quad \sum_{i=1}^n \Delta t_i = t
$$

(see (3.2.14), (3.2.19), (3.2.28), (3.2.29)).

We now explain the relationship between $\bar{F}_s(\Delta t, (x)_s)$, $\tilde{F}_s(\Delta t, (x)_s)$ and $F_s(\Delta t, (x)_s)$. It is obvious that the following assertions are true:

1. Every correlation function can be obtained from another one by replacing the operator of evolution. For example, $\bar{F}_s(\Delta t, (x)_s)$ and $\tilde{F}_s(\Delta t, (x)_s)$ can be obtained from $F_s(\Delta t, (x)_s)$ by replacing the operator $\bar{S}_{s,N-s}(-\Delta t)$ by $\bar{S}_N(-\Delta t)$ and $\tilde{S}_N(-\Delta t)$, respectively. The same replacement should be made in
$$
\lim_{n\to\infty} \prod_{i=1}^n \bar{S}_{s,N-s}(-\Delta t_i), \quad \sum_{i=1}^n \Delta t_i = t.
$$
This is the duality principle for the correlation function (3.3.14)–(3.3.16). The duality principle for hierarchies has already been formulated in Subsection 3.3.1.
2. All correlation functions $\bar{F}_s(\Delta t, (x)_s)$, $\tilde{F}_s(\Delta t, (x)_s)$ and $F_s(\Delta t, (x)_s)$, $s \ge 2$, coincide outside all hyperplanes $q_i - p_i\tau = q_j - p_j\tau$, $0 \le \tau \le \Delta t$, $i, j \in \{1, \dots, s\}$, $\bar{F}_1(\Delta t, x_1) = \tilde{F}_1(\Delta t, (x)_1) = F_1(\Delta t, (x)_1)$ in the entire phase space. It follows directly from formulas (3.3.14)–(3.3.16).

Both properties are also true for arbitrary time t.

3.4 Derivation of hierarchy from functional average

3.4.1 Functional average for s-particle observable

We now define the functional average for the following s-particle observable of the N-particle system:

$$\varphi_N(x_1,\dots,x_N) = \sum_{i_1<\dots<i_s} \varphi_s(x_{i_1},\dots,x_{i_s}), \tag{3.4.1}$$

where the summation is carried out over all $(i_1 < \dots < i_s) \subset \{1,\dots,N\}$.

Using (3.2.13) and the symmetry of the functions $\bar{f}_N(t+\Delta t, x_1,\dots,x_N)$ and $\bar{f}_N(t, x_1,\dots,x_N)$ with respect to the variables $(x_1,\dots,x_N)$, we get

$$\begin{aligned}
&\big(\bar{f}_N(t+\Delta t), \varphi_N\big)\\
&= \int \bar{f}_N(t+\Delta t, x_1,\dots,x_N)\varphi_N(x_1,\dots,x_N)dx_1\dots x_N\\
&= \Big[\frac{N!}{s!(N-s)!}\int \bar{f}_N(t+\Delta t, x_1,\dots,x_s,x_{s+1},\dots,x_N)\\
&\qquad \times dx_{s+1}\dots dx_N)\Big]\varphi_s(x_1,\dots,x_s)dx_1\dots dx_s\\
&= \Big[\frac{N!}{s!(N-s)!}\int \bar{f}_N(t, q_i - p_1\Delta t, p_1,\dots,q_N - p_N\Delta t, p_N)\\
&\qquad \times dx_{s+1}\dots dx_N\Big]\varphi_s(x_1,\dots,x_s)dx_1\dots dx_s\\
&\quad + \frac{N!}{s!(N-s)!}\int\Big\{\sum_{i<j=1}^{s}\int_0^{\Delta t} d\tau \int_{S_+^2} d\eta_{ij}\eta_{ij}\cdot(p_i - p_j)\\
&\qquad \times \delta(q_i - p_i\tau - q_j + p_j\tau)\\
&\qquad \times \int \big[\bar{f}_N(t, q_1 - p_1\Delta t, p_1,\dots,q_i - p_i\tau - p_i^*(\Delta t-\tau), p_i^*,\dots,\\
&\qquad\qquad q_j - p_j\tau - p_j^*(\Delta t-\tau), p_j^*,\dots,q_N - p_N\Delta t, p_N)\\
&\qquad - \bar{f}_N(t, q_1 - p_1\Delta t, p_1,\dots,q_i - p_i\Delta t, p_i,\dots,q_j - p_j\Delta t, p_j,\dots,\\
&\qquad\qquad q_N - p_N\Delta t, p_N)dx_{s+1}\dots dx_N\big]\Big\}\\
&\quad \times \varphi_s(x_1,\dots,x_s)dx_1\dots dx_s + \frac{N!(N-s)}{s!(N-s)!}\\
&\quad \times \int\Big\{\sum_{i=1}^{s}\int_0^{\Delta t} d\tau\int_{S_+^2} d\eta_{is+1}\eta_{is+1}\cdot(p_i - p_{s+1})
\end{aligned}$$

$$\times \delta(q_i - p_i\tau - q_{s+1} + p_{s+1}\tau)$$
$$\times \int \big[\bar{f}_N(t, q_1 - p_1\Delta t, p_1, \ldots, q_i - p_i\tau - p_i^*(\Delta t - \tau), p_i^*, \ldots,$$
$$q_s - p_s\Delta t, p_s, q_{s+1} - p_{s+1}\tau - p_{s+1}^*(\Delta t - \tau), p_{s+1}^*, q_{s+2} - p_{s+2}\Delta t,$$
$$p_{s+2}, \ldots, q_N - p_N\Delta t, p_N) - \bar{f}_N(t, q_1 - p_1\Delta t, p_1, \ldots, q_s - p_s\Delta t, p_s,$$
$$q_{s+1} - p_{s+1}\Delta t, p_{s+1}, q_{s+2} - p_{s+2}\Delta t, p_{s+2}, \ldots, q_N - p_N\Delta t, p_N)\big]$$
$$\times dx_{s+1} \ldots dx_N \Big\} \varphi_s(x_1, \ldots, x_s) dx_1 \ldots dx_s. \tag{3.4.2}$$

Note that the terms with $i, j \in \{1, \ldots, N\}$ are equal to zero. The proof is the same as in (3.3.6). We suppose that the functional average on the right-hand side of (3.4.2) are already defined through the correlation functions $\bar{F}_s^{(N)}(t, x_1, \ldots, x_s)$. Then the above-obtained formula has the following form:

$$\frac{1}{s!} \int \bar{F}_s^{(N)}(t + \Delta t, x_1, \ldots, x_s) \varphi_s(x_1, \ldots, x_s) dx_1 \ldots dx_s$$
$$= \frac{1}{s!} \int \bar{F}_s^{(N)}(t, q_1 - p_1\Delta t, p_1, \ldots, q_s - p_s\Delta t, p_s)$$
$$\times \varphi_s(x_1, \ldots, x_s) dx_1 \ldots dx_s$$
$$+ \frac{1}{s!} \int \Big\{ \sum_{i<j=1}^{s} \int_0^{\Delta t} d\tau \int_{S_+^2} d\eta_{ij} \eta_{ij} \cdot (p_i - p_j) \delta(q_i - p_i\tau - q_j + p_j\tau)$$
$$\times \big[\bar{F}_s^{(N)}(t, q_1 - p_1\Delta t, p_1, \ldots, q_i - p_i\tau - p_i^*(\Delta t - \tau), p_i^*, \ldots,$$
$$q_j - p_j\tau - p_j^*(\Delta t - \tau), p_j^*, \ldots, q_s - p_s\Delta t, p_s)$$
$$- \bar{F}_s^{(N)}(t, q_1 - p_1\Delta t, p_1, \ldots, q_i - p_i\Delta t, p_i, \ldots, q_j - p_j\Delta t, p_j, \ldots,$$
$$q_s - p_s\Delta t, p_s)\big]\Big\} \varphi_s(x_1, \ldots, x_s) dx_1 \ldots dx_s$$
$$+ \frac{1}{s!} \int dx_{s+1} \Big\{ \sum_{i=1}^{s} \int_0^{\Delta t} d\tau \int_{S_+^2} d\eta_{is+1} \eta_{is+1} \cdot (p_i - p_{s+1})$$
$$\times \delta(q_i - p_i\tau - q_{s+1} + p_{s+1}\tau)$$
$$\times \int \big[\bar{F}_{s+1}^{(N)}(t, q_1 - p_1\Delta t, p_1, \ldots, q_i - p_i\tau - p_i^*(\Delta t - \tau), p_i^*, \ldots,$$
$$q_s - p_s\Delta t, p_s, q_{s+1} - p_{s+1}\tau - p_{s+1}^*(\Delta t - \tau), p_{s+1}^*)$$
$$- \bar{F}_{s+1}^{(N)}(t, q_1 - p_1\Delta t, p_1, \ldots, q_i - p_i\Delta t, p_i, \ldots, q_s - p_s\Delta t, p_s,$$

$$q_{s+1} - p_{s+1}\Delta t, p_{s+1})\big]\Big\}\varphi_s(x_1,\dots,x_s)dx_1\dots dx_s. \qquad (3.4.3)$$

By differentiating this recurrence relation with respect to time, one obtains the following equation (in the weak sense):

$$\int \frac{\partial \bar{F}_s^{(N)}(t,x_1,\dots,x_s)}{\partial t}\varphi_s(x_1,\dots,x_s)dx_1\dots dx_s$$
$$= \int\Big\{-\sum_{i=1}^{s} p_i\cdot\frac{\partial}{\partial q_i}\bar{F}_s^{(N)}(t,x_1,\dots,x_s) + \sum_{i<j=1}^{s}\int_{S_+^2} d\eta_{ij}\,\eta_{ij}\cdot(p_i-p_j)$$
$$\times\,\delta(q_i-q_j)\big[\bar{F}_s^{(N)}(t,x_1,\dots,x_i^*,\dots,x_j^*,\dots,x_s)$$
$$-\,\bar{F}_s^{(N)}(t,x_1,\dots,x_i,\dots,x_j,\dots,x_s)\big]\Big\}\varphi_s(x_1,\dots,x_s)dx_1\dots dx_s$$
$$+\int\Big\{\sum_{i=1}^{s}\int dx_{s+1}\int_{S_+^2} d\eta_{ij}\,\eta_{is+1}\cdot(p_i-p_{s+1})\delta(q_i-q_{s+1})$$
$$\times\big[\bar{F}_{s+1}^{(N)}(t,x_1,\dots,x_i^*,\dots,x_s,x_{s+1}^*) - \bar{F}_{s+1}^{(N)}(t,x_1,\dots,x_i,\dots,$$
$$x_s,x_{s+1})\big]\Big\}\varphi_s(x_1,\dots,x_s)dx_1\dots dx_s,\quad 1\le s\le N. \qquad (3.4.4)$$

In the obtained hierarchy of equations, the derivative of $\bar{F}_s^{(N)}(t,x_1,\dots,x_s)$ with respect to time is expressed through $\bar{F}_s^{(N)}(t,x_1,\dots,x_s)$ and $\bar{F}_{s+1}^{(N)}(t,x_1,\dots,x_s,x_{s+1})$, and the contribution of the hyperplanes of lower dimension, where particles interact is taken into account. Obviously, hierarchy (3.3.7) directly follows from (3.4.4).

According to the duality principle, the obtained hierarchy is equivalent to the following hierarchy for correlation functions considered as usual functions at every point of the phase space $(x_1,\dots,x_s)$:

$$\frac{\partial F_s^{(N)}(t,x_1,\dots,x_s)}{\partial t}$$
$$= \mathcal{H}_s F_s^{(N)}(t,x_1,\dots,x_s)$$
$$+\sum_{i=1}^{s}\int dx_{s+1}\int_{S_+^2} d\eta_{is+1}\,\eta_{is+1}\cdot(p_i-p_{s+1})\delta(q_i-q_{s+1})$$
$$\times\big[F_{s+1}^{(N)}(t,x_1,\dots,x_i^*,\dots,x_s,x_{s+1}^*)$$
$$-\,F_{s+1}^{(N)}(t,x_1,\dots,x_i,\dots,x_s,x_{s+1})\big],\quad s\ge 1, \qquad (3.4.5)$$

where $\mathcal{H}_s$ is the infinitesimal operator of the group of operators $S_s(-t)$, $-\infty < t < \infty$, calculated in the sense of pointwise convergence. Recall that $\mathcal{H}_s$ can be represented on differentiable functions $f_s(x_1, \dots, x_s)$ in the form

$$\mathcal{H}_s f_s(x_1, \dots, x_s) = -\sum_{i=1}^{s} p_i \cdot \frac{\partial}{\partial q_i} f_s(x_1, \dots, x_s) \tag{3.4.6}$$

with the boundary condition according to which, at the points $q_i = q_j$, $i, j \in \{1, \dots, s\}$, the momenta p_i and p_j should be replaced by p_i^* and p_j^* in $\mathcal{H}_s$ if $\eta_{ij} \cdot (p_i - p_j) \geq 0$, and (p_i, p_j) do not change if $\eta_{ij} \cdot (p_i - p_j) \leq 0$.

One can repeat the above-performed calculation with the functional average for $\tilde{f}_N(t, x_1, \dots, x_N)$ (3.2.20), obtain hierarchy (3.4.5) with the operator $\tilde{\mathcal{H}}_s$ instead of $\mathcal{H}_s$, and then, using the duality principle, replace $\tilde{\mathcal{H}}_s$ by $\mathcal{H}_s$.

Passing formally to the thermodynamic limit as $N \to \infty$ in hierarchy (3.4.5) and supposing that the limit correlation functions exist, one obtains the limit hierarchy

$$\begin{aligned} &\frac{\partial F_s(t, x_1, \dots, x_s)}{\partial t} \\ &\quad = \mathcal{H}_s F_s(t, x_1, \dots, x_s) \\ &\quad\quad + \sum_{i=1}^{s} \int dx_{s+1} \int_{S_+^2} d\eta_{is+1} \eta_{is+1} \cdot (p_i - p_{s+1}) \delta(q_i - q_{s+1}) \\ &\quad\quad\quad \times \big[F_{s+1}(t, x_1, \dots, x_i^*, \dots, x_{s+1}^*) - F_{s+1}(t, x_1, \dots, x_i, \dots, x_{s+1}^*) \big], \\ &F_s(t, x_1, \dots, x_s) = \lim_{N \to \infty} F_s^{(N)}(t, x_1, \dots, x_s), \quad s \geq 1, \end{aligned} \tag{3.4.7}$$

with boundary condition in $\mathcal{H}_s$ and initial data

$$F_s(t, x_1, \dots, x_s)|_{t=0} = F_s(x_1, \dots, x_s). \tag{3.4.8}$$

Note that the correlation functions $F_s(t, (x)_s)$ depend on random vectors $\eta_{ij}, i, j \in \{1, \dots, s\}$.

Hierarchy (3.4.7) is known as the *stochastic Boltzmann hierarchy*.

Remark 3.2. It is easy to show that hierarchy (3.4.7) can also be derived in the framework of grand canonical ensemble.

3.4.2 Derivation of the stochastic boltzmann hierarchy from the Itô–Liouville equation

Consider the Itô–Liouville equation (3.2.21) for the N-particle distribution function

$$\frac{\partial f_N(t, x_1, \dots, x_N)}{\partial t} = -\sum_{i=1}^{N} p_i \cdot \frac{\partial}{\partial q_i} f_N(t, x_1, \dots, x_N)$$

with the boundary condition according to which, at $q_i = q_j$, $i, j \in \{1, \dots, N\}$, $\eta \cdot (p_i - p_j) \geq 0$, the momenta p_i, p_j should be replaced by p_i^*, p_j^* on the right-hand side of (3.2.21) and in $f_N(t, x_1, \dots, x_N)$.

We now show that hierarchy (3.4.7) can be obtained from equation (3.2.21) for the correlation functions $F_s(t, (x)_s)$. For the convenience of the reader, we perform a detailed derivation for a two-particle system. We have the equation

$$\frac{\partial f_2(t, x_1, x_2)}{\partial t} = \Big(- p_1 \frac{\partial}{\partial q_1} - p_2 \frac{\partial}{\partial q_2} \Big) f_2(t, x_1, x_2) \tag{3.4.9}$$

with the above-formulated boundary condition and without the δ-function.

The sequence of correlation functions consists of the two functions

$$F_1(t, x_1) = 2 \int dx_2 d\eta f_2(t, x_1, x_2),$$

$$F_2(t, x_1, x_2) = 2 f_2(t, x_1, x_2).$$

In order to deduce, the corresponding hierarchy, we integrate the Itô–Liouville equation (3.4.9) with respect to x_2 and take into account the boundary condition. For this purpose, we use the coordinate system in which the vector η is directed along the first coordinate q_1^1, q_2^1, p_1^1, p_2^1. First, we perform integration with respect to q_2^1, p_2^1 with $q_2^2 = q_1^2$, $q_2^3 = q_1^3$ and take into account the boundary condition.

One obtains

$$- \Big(\int_{-\infty}^{q_1^1} + \int_{q_1^1}^{\infty} \Big) p_2^1 \frac{\partial}{\partial q_2^1} f_2(t, x_1, x_2) \Big|_{q_1^2 = q_2^2,\, q_1^3 = q_2^3} dq_2^1$$

$$= (-p_2^1 f_2(t, x_1, x_2))|_{q_2^1 = q_1^1 - 0} + p_2^1 f_2(t, x_1, x_2)|_{q_2^1 = q_1^1 + 0}\Big)\Big|_{q_1^2 = q_2^2,\, q_1^3 = q_2^3},$$

$$- \Big(\int_{-\infty}^{q_1^1} + \int_{q_1^1}^{\infty} \Big) p_1^1 \frac{\partial}{\partial q_1^1} f_2(t, x_1, x_2) \Big|_{q_1^2 = q_2^2,\, q_1^3 = q_2^3} dq_2^1$$

$$= - p_1^1 \frac{\partial}{\partial q_1^1} \int_{-\infty}^{\infty} f_2(t, x_1, x_2) \Big|_{q_1^2 = q_2^2,\, q_1^3 = q_2^3} dq_2^1$$

$$+ \Big(p_1^1 f_2(t, x_1, x_2)|_{q_2^1 = q_1^1 - 0} - p_1 f_2(t, x_1, x_2)|_{q_2^1 = q_1^1 + 0} \Big)\Big|_{q_1^2 = q_2^2,\, q_1^3 = q_2^3}.$$

Using the boundary condition and taking into account that $p_1^1 - p_2^1 = \eta \cdot (p_1 - p_2)$, one establishes that the sum of these expressions is equal to

$$\eta \cdot (p_1 - p_2) \Theta(\eta \cdot (p_1 - p_2)) \Big[f_2(t, q_1, p_1^*, q_1, p_2^*) - f_2(t, q_1, p_1, q_1, p_2) \Big]$$

$$- p_1^1 \frac{\partial}{\partial q_1^1} \int_{-\infty}^{\infty} f_2(t, x_1, x_2) \Big|_{q_1^2 = q_2^2,\, q_1^3 = q_2^3} dq_2^1.$$

Note that the integral

$$2\int\Big(-p_1\frac{\partial}{\partial q_1}-p_2\frac{\partial}{\partial q_2}\Big)f_2(t,x_1,x_2)dx_2d\eta,$$

calculated as a usual Lebesgue integral, is equal to

$$-p_1\frac{\partial}{\partial q_1}F_1(t,x_1)$$

because

$$\int p_2\frac{\partial}{\partial q_2}f_2(t,x_1,x_2)dx_2=0.$$

Thus, we (i) integrate the right-hand side of (3.4.9) as a usual function and (ii) take into account, in a special way, the boundary condition.

Summing up all terms, integrating with respect to p_2, and averaging with respect to η, one finally obtains

$$\begin{aligned}\frac{\partial F_1(t,x)}{\partial t}&=-p_1\frac{\partial}{\partial q_1}F_1(t,x_1)+\int dp_2\int_{S_2^+}d\eta\,\eta\cdot(p_1-p_2)\\&\quad\times\big[F_2(t,q_1,p_1^*,q_1,p_2^*)-F_2(t,q_1,p_1,q_2,p_2)\big],\qquad(3.4.10)\\\frac{\partial F_2(t,x_1,x_2)}{\partial t}&=-\Big(p_1\frac{\partial}{\partial q_1}+p_2\frac{\partial}{\partial q_2}\Big)F_2(t,x_1,x_2),\\F_2(t,x_1,x_2)&=f_2(t,x_1,x_2)\end{aligned}$$

with the boundary condition for the equation for $F_2(t,x_1,x_2)$.

Note that the term

$$-p_1^1\frac{\partial}{\partial q_1}\int f_2(t,x_1,x_2)\Big|_{q_1^2=q_2^2,\,q_1^3=q_2^3}dq_2^1dp_2$$

is included in $-p_1^1\frac{\partial}{\partial q_1}F_1(t,x_1)$ because it is defined on a hyperplane of lower dimension and the one-particle correlation function is considered as a usual summable function if the initial function $f_2(0,x_1,x_2)$ is summable.

The general case of an N-particle system can be performed analogously by summing up the contributions of all pair collisions and using the same tricks as in (3.3.6).

Thus, hierarchy (3.3.9) can also be derived from equation (3.2.21) with operator $\mathcal{H}_N$ without δ-functions but taking into account the boundary condition and calculating in a special way contribution of hyperplanes of lower dimension. If one uses the operators $\tilde{\mathcal{H}}_N$ or $\bar{\mathcal{H}}_N$ with δ-functions in the Itô–Liouville equation, then the boundary conditions are taken into account automatically and one obtains the same integral term in the stochastic hierarchy as above.

3.4.3 Derivation of ordinary Boltzmann hierarchy from Boltzmann equation

There exist two methods for the derivation of the ordinary Boltzmann hierarchy for hard spheres: from the Boltzmann equation and from the BBGKY hierarchy for systems of hard spheres in the Boltzmann–Grad limit without the known boundary condition. Consider the first method.

We have the Boltzmann equation for hard spheres:

$$\begin{aligned}\frac{\partial F_1(t,x_1)}{\partial t}&\\ &= -p_1\frac{\partial}{\partial q_1}F_1(t,x_1) + \int_{S^2_+} d\eta \int dp_2 \eta\cdot(p_1-p_2)\\ &\quad\times\Big[F_1(t,q_1,p_1^*)F_1(t,q_1,p_2^*) - F_1(t,q_1,p_1)F_1(t,q_1,p_2)\Big],\quad t\geq 0,\end{aligned} \tag{3.4.11}$$

where, as usual,

$$x=(q,p),\ |\eta|=1,\quad S^2_+(\eta|\eta\cdot(p_1-p_2)\geq 0),$$

$$p_1^* = p_1-\eta\eta\cdot(p_1-p_2),\quad p_2^* = p_2+\eta\eta\cdot(p_1-p_2).$$

We consider the Cauchy problem for equation (3.4.11) with the initial condition

$$F_1(t,x_1)|_{t=0} = F_1(0,x_1) \tag{3.4.12}$$

in a certain functional space and suppose that a solution of (3.4.11)–(3.4.12) exists.

We define the sequence of s-particle correlation functions associated with the solution of the Boltzmann equation as follows:

$$F_s(t,x_1,\dots,x_s) = F_1(t,x_1)F_1(t,x_2)\dots F_1(t,x_s),\quad s\geq 1. \tag{3.4.13}$$

It follows from (3.4.11)–(3.4.12) that sequence (3.4.13) satisfies the chain of integro-differential equations

$$\begin{aligned}\frac{\partial F_s(t,x_1,\dots,x_s)}{dt}&\\ &= -\sum_{i=1}^{s} p_i\cdot\frac{\partial}{\partial q_i}F_s(t,x_1,\dots,x_s)\\ &\quad+\sum_{i=1}^{s}\int_{S^2_+} d\eta\int dp_{s+1}\eta\cdot(p_i-p_{s+1})\\ &\quad\times\big[F_{s+1}(t,x_1,\dots,q_i,p_i^*,\dots,x_s,q_i,p_{s+1}^*)\\ &\qquad - F_{s+1}(t,x_1,\dots,q_i,p_i,\dots,x_s,q_i,p_{s+1})\big],\ s\geq 1,\ t\geq 0\end{aligned} \tag{3.4.14}$$

with the initial conditions

$$F_s(t, x_1, \dots, x_s)|_{t=0} = F_s(0, x_1, \dots, x_s) = F_1(0, x_1) \dots F_1(0, x_s). \tag{3.4.15}$$

We say that the initial conditions (3.4.15) possess the chaos property.

The chain of equations (3.4.14) is known as the *ordinary Boltzmann hierarchy*.

It is obvious that the ordinary Boltzmann hierarchy (3.4.14) with the initial conditions (3.4.15) is equivalent to the Boltzmann equation (3.4.11) in the following sense:

We have shown that the s-particle distribution functions (3.4.13) are solutions of the ordinary Boltzmann hierarchy (3.4.14) with the initial conditions (3.4.15). Conversely, the solutions of the ordinary Boltzmann hierarchy (3.4.14) with the initial conditions (3.4.15) also possess the chaos property, i.e., the s-particle distribution functions are the products of one-particle distribution functions:

$$F_s(t, x_1, \dots, x_s) = F_1(t, x_1) \dots F_1(t, x_s),$$

where the one-particle distribution functions satisfy the ordinary Boltzmann equation (3.4.11).

The last assertion follows directly from (3.4.14). Indeed, the ordinary Boltzmann hierarchy admits the separation of variables because, on the right-hand side of (3.4.14), we have the sum of s operators that act with respect to each s-variables x_i, $i = 1, \dots, s$.

3.5 Derivation of stochastic Boltzmann hierarchy from BBGKY hierarchy for hard spheres

3.5.1 Stochastic Boltzmann hierarchy

We now consider the second method for the derivation of the stochastic Boltzmann hierarchy from the BBGKY hierarchy for hard spheres.

We have the BBGKY hierarchy for systems of hard spheres with diameter a

$$\begin{aligned}
&\frac{\partial F_s^a(t, x_1, \dots, x_s)}{\partial t} \\
&= -\sum_{i=1}^{s} p_i \cdot \frac{\partial}{\partial q_i} F_s^a(t, x_1, \dots, x_s) \\
&\quad + \sum_{i=1}^{s} a^2 \int_{S_+^2} d\eta \int dp_{s+1} \eta \cdot (p_i - p_{s+1}) \\
&\quad \times \big[F_{s+1}^a(t, x_1, \dots, q_i, p_i^*, \dots, x_s, q_i - a\eta, p_{s+1}^*) \\
&\quad - F_{s+1}^a(t, x_1, \dots, q_i, p_i, \dots, x_s, q_i + a\eta, p_{s+1})\big], \quad t \geq 0, \qquad (3.5.1)
\end{aligned}$$

with the initial conditions

$$F_s(t, x_1, \dots, x_s)|_{t=0} = F_s(0, x_1, \dots, x_s).$$

The correlation functions $F_s^a(x_1, \dots, x_s)$ are symmetric and satisfy the following condition: $F_s^a(t, x_1, \dots, x_s) = 0$ if $|q_i - q_j| < a$ for at least one pair $i, j \in \{1, \dots, s\}$. This means that the particles under consideration are hard spheres and the distances between their centers should be greater than or equal to a, i.e., hard spheres occupy admissible configurations.

The correlation functions $F_s^a(t, x_1, \dots, x_s)$ satisfy the following boundary conditions:

$$F_s^a(t+0, x_1, \dots, x_i, \dots, x_j, \dots, x_s) = F_s(t, x_1, \dots, x_i^*, \dots, x_j^*, \dots, x_s)$$

if $q_i - q_j = a\eta_{ij}$ and $\eta_{ij} \cdot (p_i - p_j) \ge 0$, and

$$F_s^a(t+0, x_1, \dots, x_i, \dots, x_j, \dots, x_s) = F_s(t, x_1, \dots, x_i, \dots, x_j, \dots, x_s) \quad (3.5.2)$$

if $q_i - q_j = a\eta_{ij}$ and $\eta_{ij} \cdot (p_i - p_j) \le 0$. Here,

$$x_i^* = (q_i, p_i^*), \quad x_j^* = (q_j, p_j^*).$$

We also have the following boundary conditions for the Poisson bracket $-\sum_{i=1}^s p_i \cdot \frac{\partial}{\partial q_i} F_s(t, x_1, \dots, x_s)$: If $q_i - q_j = a\eta_{ij}$ and $\eta_{ij} \cdot (p_i - p_j) \ge 0$, i.e., $\eta_{ij} \in S_+^2$, then the momenta (p_i, p_j) should be replaced by (p_i^*, p_j^*). The momenta (p_i, p_j) do not change if $\eta_{ij} \cdot (p_i - p_j) \ge 0$, i.e., $\eta_{ij} \in S_-^2$.

We stress that the boundary conditions in the Poisson bracket are of great importance because these boundary conditions are associated with the dynamics of hard spheres. Namely, they are responsible for the jumps of momenta $(p_i, p_j) \to (p_i^*, p_j^*)$ when hard spheres touch one another ($q_i - q_j = a\eta_{ij}$) and elastically collide. The Poisson bracket with the boundary conditions is the infinitesimal operator of the evolution operator of s hard spheres. Detailed information about these boundary conditions can be found in our books [PGM3, CGP]. Unfortunately, in all available books of other authors devoted to statistical mechanics of systems of hard spheres, the information about boundary conditions for the BBGKY hierarchy (3.5.1) is absent, and this led to serious mistakes.

We introduce the sequence of renormalized distribution functions

$$\tilde{F}_s^a(t, x_1, \dots, x_s) = a^{2s} F_s^a(t, x_1, \dots, x_s), \quad s \ge 1, \quad (3.5.3)$$

and let the diameter a tend to zero: $a \to 0$. We assume that the Boltzmann–Grad limit of the renormalized sequence (3.5.3) exists in a certain sense (see [PGM3, CGP]):

$$\lim_{a\to 0} \tilde{F}_s^a(t, x_1, \dots, x_s) = F_s(t, x_1, \dots, x_s), \quad s \ge 1, \quad (3.5.4)$$

Performing formally the limit procedure (3.5.4) in (3.5.1), we obtain

$$\frac{\partial F_s(t, x_1, \dots, x_s)}{\partial t}$$
$$= -\sum_{i=1}^{s} p_i \cdot \frac{\partial}{\partial q_i} F_s(t, x_1, \dots, x_s)$$
$$+ \sum_{i=1}^{s} \int_{S_+^2} d\eta \int dp_{s+1} \eta \cdot (p_i - p_{s+1}) \big[F_{s+1}(t, x_1, \dots, q_i, p_i^*, \dots, x_s, q_i, p_{s+1}^*)$$
$$- F_{s+1}(t, x_1, \dots, q_i, p_i, \dots, x_s, q_i, p_{s+1}) \big]. \quad (3.5.5)$$

At first sight, hierarchy (3.5.5) obtained from the BBGKY hierarchy in the Boltzmann–Grad limit resembles the ordinary Boltzmann hierarchy (3.4.14) obtained from the Boltzmann equation (3.4.11). In fact, there are principal differences between them.

Namely, the correlation functions $F_s(t, x_1, \dots, x_s)$ (3.5.4) satisfy the following boundary conditions:

$$F_s(t+0, x_1, \dots, x_i, \dots, x_j, \dots, x_s) = F_s(t, x_1, \dots, x_i^*, \dots, x_j^*, \dots, x_s)$$

if $q_i = q_j$ and $\eta_{ij} \cdot (p_i - p_j) \geq 0$, and

$$F_s(t+0, x_1, \dots, x_i, \dots, x_j, \dots, x_s) = F_s(t, x_1, \dots, x_i, \dots, x_j, \dots, x_s) \quad (3.5.6)$$

if $q_i = q_j$ and $\eta_{ij} \cdot (p_i - p_j) \leq 0$.

In hierarchy (3.5.5), we also have the Poisson bracket

$$-\sum_{i=1}^{s} p_i \cdot \frac{\partial}{\partial q_i} F_s(t, x_1, \dots, x_s) = \mathcal{H}_s F_s(t, x_1, \dots, x_s)$$

with boundary conditions according to which one should replace (p_i, p_j) by (p_i^*, p_j^*) if $q_i = q_j$ and $\eta_{ij} \cdot (p_i - p_j) \geq 0$, i.e., $\eta_{ij} \in S_+^2$. The momenta (p_i, p_j) do not change if $\eta_{ij} \cdot (p_i - p_j) \leq 0$, i.e., $\eta_{ij} \in S_-^2$. These boundary conditions (3.5.6) (in the correlation functions and the Poisson bracket) follow from the corresponding boundary conditions (3.5.2) for the BBGKY hierarchy (3.5.1) for hard spheres. For hard spheres, the boundary conditions are given on the spheres $q_i - q_j = a\eta_{ij}$, which are reduced to the hyperplanes $q_i - q_j = 0$ in the Boltzmann–Grad limit. In hierarchy (3.4.14), these boundary conditions are absent because they are also absent in the Boltzmann equation (3.4.11).

The second principal difference is the random nature of the unit vector η in (3.5.5). In the BBGKY hierarchy (3.5.1), the vector η is directed between the centers of colliding spheres and defines the relative shift of the centers of the spheres. In hierarchy (3.5.5), we have point particles and η is a random vector with constant density of probability on the sphere S^2. It characterizes how the point particles were obtained from the colliding hard spheres. The details on the mechanism of obtaining the stochastic point particles from hard spheres were presented in Chapter 2.

Thus, we have derived two different hierarchies, namely, hierarchy (3.4.14) from the Boltzmann equation and hierarchy (3.5.5) from the BBGKY hierarchy for hard spheres. They differ by the boundary conditions for the correlation functions and the Poisson bracket in hierarchy (3.5.5), which are obtained from the BBGKY hierarchy for hard spheres. These boundary conditions are absent for hierarchy (3.4.14) obtained from the Boltzmann equation.

We call hierarchy (3.5.5) with boundary conditions (3.5.6) the *stochastic Boltzmann hierarchy*.

Thus, the ordinary Boltzmann hierarchy (3.4.14) is not the Boltzmann–Grad limit of the BBGKY hierarchy. The Boltzmann–Grad limit of the BBGKY hierarchy is the stochastic hierarchy (3.5.5), which differs from the ordinary Boltzmann hierarchy (3.4.14) by the boundary conditions for the correlation functions (3.5.6) and the Poisson bracket.

Note that the ordinary Boltzmann hierarchy (3.4.14) has been derived from the BBGKY hierarchy formally in the Boltzmann–Grad limit without taking into account the boundary conditions (3.5.2) because these boundary conditions have been absent in the corresponding erroneous version of the BBGKY hierarchy for hard spheres.

Nevertheless, the ordinary Boltzmann hierarchy (3.4.14) is of great importance because the solutions of the Boltzmann hierarchy (3.4.14) are the Boltzmann–Grad limit of the solutions of the BBGKY hierarchy (3.5.1) for hard spheres in the sense described below.

3.5.2 Solutions of the ordinary Boltzmann hierarchy and the Boltzmann–Grad Limit of solutions of the BBGKY hierarchy

We know that the solutions of the ordinary Boltzmann hierarchy (3.4.14) and BBGKY hierarchy (3.5.1) exist locally in time for initial data from the space $E_{\xi,\beta}$ of sequences of functions bounded with respect to positions and exponentially decreasing with respect to squared momenta. The solutions of both hierarchies exist globally in time for the initial data from the space $\tilde{E}_{\xi,\beta}$ of sequences of functions exponentially decreasing with respect to squared momenta and positions. These solutions can be represented by series of iterations [PGM3, CGP, CIP]. Note that, for some subset from $E_{\xi,\beta}$, namely for local perturbations of the equilibrium states, solutions of the BBGKY hierarchy (3.5.1) exist globally in time. The definition of the spaces $E_{\xi,\beta}$ and $\tilde{E}_{\xi,\beta}$ will be given in the next chapter.

Consider a compact set K_s in the space of positions $(q_1, \ldots, q_s)$ such that $|q_i - q_j| \geq a_0(a)$, $i, j \in \{1, \ldots, s\}$, $s \geq 2$, and $a_0(a) \to 0$ as $a \to 0$ so that

$$\lim_{a \to 0} \frac{a}{a_0(a)} = 0.$$

We also consider a compact set in the momentum space $(p_1, \dots, p_s)$,

$$\sqrt{p_1^2 + \dots + p_s^2} < p_0,$$

and cones V_{ij} with respect to the differences $p_i - p_j$, $i, j \in \{1, \dots, s\}$, $s \geq 2$, with axes parallel to the vectors $q_i - q_j$, the radius of the base equal to $\frac{a}{a_0(a)}$, and height p_0.

Then

$$\lim_{a \to 0} \left(\tilde{F}_s^a(t, x_1, \dots, x_s) - F_s(t, x_1, \dots, x_s) \right) = 0, \quad s \geq 2, \tag{3.5.7}$$

uniformly in $(q_1, \dots, q_s) \subset K_s$ and in the momenta $(p_1, \dots, p_s)$ outside the cones $\bigcup_{i<j=1}^{N} V_{ij}$. We assume that limits (3.5.7) exist for the initial distribution functions $\tilde{F}_s^a(0, x_1, \dots, x_s)$ and $F_s(0, x_1, \dots, x_s)$ for $(x_1, \dots, x_s)$ on arbitrary compact sets of admissible configurations.

In other words, the renormalized solutions of the BBGKY hierarchy (3.5.1) tend in the Boltzmann–Grad limit to the solutions of the ordinary Boltzmann hierarchy (3.4.14) outside the hyperplanes $q_i - q_j = 0$ and the hyperplanes with vectors $p_i - p_j$ parallel to $q_i - q_j$, $i, j \in \{1, \dots, s\}$. Note that all vectors $p_i - p_j$ are parallel to $q_i - q_j = 0$, $s \geq 2$.

For $s = 1$, we have

$$\lim_{a \to 0} (\tilde{F}_1^a(t, x_1) - F_1(t, x_1)) = 0$$

in the entire phase space of one particle.

The rigorous proof of the above-formulated results was presented in papers of the author and Gerasimenko [GeP1, GeP2, GeP3] and summarized in the books [CGM, PMG3].

Thus, the solutions of the ordinary Boltzmann hierarchy are the Boltzmann–Grad limit of the solutions of the BBGKY hierarchy for hard spheres outside the indicated hyperplanes for $s \geq 2$ and in the entire phase space for $s = 1$.

In the next section, we show that the solutions of the stochastic hierarchy (3.5.5) (with boundary condition (3.5.6)) are the Boltzmann–Grad limit of the solutions of the BBGKY hierarchy (3.5.1) *in the entire phase space*. In the previous sections we showed that hierarchy (3.5.5) is associated with the stochastic dynamics, while the BBGKY hierarchy is associated with the Hamiltonian dynamics of hard spheres. For this reason, hierarchy (3.5.5) is called the *stochastic, or proper, Boltzmann hierarchy*, and hierarchy (3.4.14) is called the *ordinary Boltzmann hierarchy*, or the *Boltzmann hierarchy*.

3.5.3 Derivation of the stochastic Boltzmann hierarchy from the evolution operator of the BBGKY hierarchy for hard spheres

Consider the solutions of the BBGKY hierarchy (3.5.1) represented by the group of evolution operators. Denote by $F^a(t)$ the sequence of correlation functions

$$F^a(t) = \left(F_1^a(t, x_1), \ldots, F_s^a(t, (x)_s), \ldots\right), \quad (x)_s = (x_1, \ldots, x_s).$$

The sequence $F^a(t)$ can be represented by using the group of evolution operators $U^a(t)$ as follows:

$$F^a(t) = U^a(t)F(0) = e^b S^a(-t)e^{-b} F(0). \tag{3.5.8}$$

For the meaning of the operators b and $S^a(-t)$, see Chapter 1. Relation (3.5.8) yields

$$F_s^a(t, (x)_s)$$
$$= \sum_{n=0}^{\infty} \sum_{k=0}^{n} \frac{(-1)^k}{k!(n-k)!} \int d(x)_{s+n}^s S_{s+n-k}^a(-t, (x)_{s+n-k}) F_{s+n}^a(0, (x)_{s+n}),$$

$$d(x)_{s+n}^s = dx_{s+1} \ldots dx_{s+n}, \quad (x)_{s+n}^s = (x_{s+1}, \ldots, x_{s+n}), \quad s \geq 1,$$

where $S_{s+n-k}^a(-t, (x)_{s+n-k})$ is the operator of evolution of $s+n-k$ particles with initial phase points $(x)_{s+n-k}$.

The group $U^a(t)$ is a strongly continuous bounded operator in the space L of direct sums of integrable functions. The infinitesimal operator of the group $U^a(t)$ coincides with the operator on the right-hand side of hierarchy (3.5.1) on a certain everywhere-dense set L_0 in L.

The group $U^a(t)$ is also meaningful in the space $E_{\xi,\beta}$ of sequences of functions bounded with respect to positions and exponentially decreasing with respect to squared momenta for a finite interval of time and globally in time for sequences of locally perturbed equilibrium states.

We now use the group property of $U^a(t)$ and represent $F^a(t+\Delta t)$ in terms of $F^a(t)$ as follows:

$$F^a(t + \Delta t) = U^a(\Delta t)U^a(t)F^a(0) = U^a(\Delta t)F^a(t). \tag{3.5.9}$$

We use the following representation of the evolution operator $U^a(t)$ via the series of iterations:

$$U^a(t) = \sum_{n=0}^{\infty} \int_0^t dt_1 \ldots \int_0^{t_{n-1}} dt_n S^a(-t)$$
$$\times S^a(t_1)A^a S^a(-t_1) \ldots S^a(t_n)A^a S^a(-t_n), \tag{3.5.10}$$

where the operator A^a is defined by the second term on the right-hand side of hierarchy (3.5.1), and the operator $S^a(-t)$ is the direct sum of the evolution operators of the s-particle subsystem $S_s^a(-t,(x)_s)$, $s \geq 1$. We restrict ourselves to the case where $t > 0$ and $\Delta t > 0$.

We consider infinitesimal Δt and represent $U^a(\Delta t)$ by the following expression identical to (3.5.10):

$$U^a(\Delta t) = S^a(-\Delta t) + \int_0^{\Delta t} d\tau S^a(\tau - \Delta t) A^a S^a(-\tau). \tag{3.5.11}$$

Substituting (3.5.11) in (3.5.9), we obtain

$$F^a(t+\Delta t) = S^a(-\Delta t)F^a(t) + \int_0^{\Delta t} d\tau S^a(\tau - \Delta t) A^a S^a(-\tau) F^a(t)$$

or, componentwise,

$$\begin{aligned} F^a&(t+\Delta t,(x)_s) \\ &= S_s^a(-\Delta t,(x)_s)F_s^a(t,(x)_s) \\ &\quad + \int_0^{\Delta t} d\tau S_s^a(\tau-\Delta t,(x)_s)\sum_{i=1}^s \int_{S_+^2} d\eta \int dp_{s+1}\eta\cdot(p_i - p_{s+1}) \\ &\quad\quad \times \big[S_{s+1}^a(-\tau,(x^*)_{s+1})F_{s+1}^a(t,(x^*)_{s+1})|_{q_{s+1}=q_i-a\eta} \\ &\quad\quad\quad - S_{s+1}^a(-\tau,(x)_{s+1})F_{s+1}^a(t,(x)_{s+1})|_{q_{s+1}=q_i+a\eta}\big], \end{aligned} \tag{3.5.12}$$

where $(x^*)_{s+1} = (x_1,\ldots,q_i,p_i^*,\ldots,x_s,q_{s+1},p_{s+1}^*)$ in the term with number i.

We now multiply relation (3.5.12) by a^{2s}, use the renormalized functions

$$\tilde{F}_s^a(t,(x)_s) = a^{2s}F_s^a(t,(x)_s), \quad s \geq 1,$$

and let the diameter a tend to zero. This procedure is known as the Boltzmann–Grad limit. Taking into account that

$$\lim_{a\to 0} S_s^a(-\Delta t,(x)_s) = S_s(-\Delta t,(x)_s),$$

where $S_s(-\Delta t,(x)_s)$ is the evolution operator corresponding to the stochastic dynamics [see Chapter 2], and assuming the existence of the Boltzmann–Grad limit

$$\lim_{a\to 0} \tilde{F}_s^a(t,(x)_s) = F_s(t,(x)_s), \quad s \geq 1,$$

we obtain from (3.5.12) the following relation *in the entire phase space* $(x)_s$:

$$
\begin{aligned}
F_s(t+\Delta t,(x)_s)
&= S_s(-\Delta t,(x)_s)F_s(t,(x)_s)\\
&\quad + \int_0^{\Delta t} d\tau S_s(\tau-\Delta t,(x)_s)\sum_{i=1}^{s}\int_{S_+^2} d\eta \int dp_{s+1}\eta\cdot(p_i-p_{s+1})\\
&\qquad \times \big[S_{s+1}(-\tau,(x^*)_{s+1})F_{s+1}(t,(x^*)_{s+1})|_{q_{s+1}=q_i}\\
&\qquad\quad - S_{s+1}(-\tau,(x)_{s+1})F_{s+1}(t,(x)_{s+1})|_{q_{s+1}=q_i}\big]\\
&= S_s(-\Delta t,(x)_s)F_s(t,(x)_s) + \sum_{i=1}^{s}\int dx_{s+1}\int_0^{\Delta t} d\tau \int_{S_+^2} d\eta_{is+1}\\
&\quad \times \eta_{is+1}\cdot(p_i-p_{s+1})|\delta(q_i-p_i\tau-q_{s+1}+p_{s+1}\tau)\\
&\quad \times \big[F_{s+1}(t,q_1-p_1\Delta t,p_1,\dots,q_i-p_i\tau-p_i^*(\Delta t-\tau),p_i^*,\dots,\\
&\qquad\qquad q_{s+1}-p_{s+1}\tau+p_{s+1}^*(\Delta t-\tau),p_{s+1}^*)\\
&\quad - F_{s+1}(t,q_1-p_1\Delta t,p_1,\dots,q_i-p_i\Delta t,p_i,\dots,q_{s+1}-p_{s+1}\Delta t,p_{s+1})\big]\\
&= S_s(-\Delta t,(x)_s)F_s(t,(x)_s) + I_s, \quad s\ge 1. \qquad (3.5.13)
\end{aligned}
$$

Obtaining the final expression for I_s, we replace the operators $S_{s+1}(-\tau,(x^*)_{s+1})$ and $S_{s+1}(-\tau,(x)_{s+1})$ of stochastic evolution by the corresponding operators $S_{s+1}^0(-\tau,(x^*)_{s+1})$ and $S_{s+1}^0(-\tau,(x)_{s+1})$ of free evolution and neglect the terms of higher order with respect to Δt. It is obvious that we can also replace all these operators by the identity operators under the integral sign $\int_0^{\Delta t} d\tau$.

It is known that

$$
\begin{aligned}
&S_s(-\Delta t,(x)_s)F_s(t,(x)_s)\\
&\quad = F_s(t,q_1-p_1\Delta t,p_1,\dots,q_i-p_i\tau-p_i^*(\Delta t-\tau),p_i^*,\dots,\\
&\qquad\qquad q_j-p_j\tau-p_j^*(\Delta t-\tau),p_j^*,\dots,q_s-p_s\Delta t,p_s)
\end{aligned}
$$

if $q_i-q_j=(p_i-p_j)\tau$ for some $0\le\tau\le\Delta t$ and a certain pair $i,j\in\{1,\dots,s\}$ and $\eta_{ij}\in S_+^2$, and

$$
\begin{aligned}
&S_s(-\Delta t,(x)_s)F_s(t,(x)_s)\\
&\quad = F_s(t,q_1-p_1\Delta t,p_1,\dots,q_i-p_i\Delta t,\\
&\qquad\qquad p_i,\dots,q_j-p_j\Delta t,p_j,\dots,q_s-p_s\Delta t,p_s)
\end{aligned}
$$

if $q_i-q_j\ne(p_i-p_j)\tau$ for all pairs $i,j\in\{1,\dots,s\}$ and all $0\le\tau\le\Delta t$, or $q_i-q_j=(p_i-p_j)\tau$ but $\eta_{ij}\in S_-^2$; we assume that Δt is infinitesimal.

Denote by $D_{-\Delta t}$ the set

$$\bigcup_{(i,j)} \bigcup_{0 \le \tau \le \Delta t} (q_i - q_j = (p_i - p_j)\tau).$$

Then the operator $S_s(-\Delta t, (x)_s)$ is equal to $S_s^0(-\Delta t, (x)_s)$ outside the set $D_{-\Delta t}$.

It follows from (3.5.13) that the function $F_s(t + \Delta t, (x)_s)$ depends on the random vectors η_{ij} created on the interval $(t, t + \Delta t)$ for $(x)_s \subset D_{-\Delta t}$ only through the term $S_s(-\Delta t, (x)_s) F_s(t, (x)_s)$; the second term I_s does not depend on these random vectors.

By differentiating relation (3.5.13) with respect to time, using pointwise convergence, and taking into account the definition of the operator $S_s(-\Delta t, (x)_s)$ of evolution of the stochastic dynamics with the infinitesimal operator $\mathcal{H}_s$, one obtains the stochastic Boltzmann hierarchy (3.5.5) with boundary conditions (3.5.6).

Remark 3.3. Note that the second term on the right-hand side of equation (3.5.13) is represented by the integral over the hyperplanes $q_i - p_i\tau = q_{s+1} - p_{s+1}\tau$, $0 \le \tau \le \Delta t$, where the stochastic i-th particles ($i \in \{1, \dots, s\}$) interact with the $(s + 1)$-th particle.

3.5.4 Functional for correlation functions

We now define the functional $(F_s(t + \Delta t), \varphi_s)$ that is the average of the observable $\varphi_s((x)_s)$ over the state $F_s(t + \Delta t, (x)_s)$ with respect to the random vectors η_{is}, $1 \le i < j \le s$, corresponding to the stochastic dynamics described by the operator $S_s(-\Delta t, (x)_s)$ for infinitesimal Δt. In doing this, we take into account the contribution of the set $D_{-\Delta t}$. According to the definition [see Section 3.4], the functional $(F_s(t + \Delta t), \varphi_s)$ consists of two terms. One of them is defined by the average (with respect to the random vectors η_{ij}) of the integral over the set $D_{-\Delta t}$ with the integrand

$$\big[S_s(-\Delta t, (x)_s) - S_s^0(-\Delta t, (x)_s)\big] F_s(t, (x)_s) \varphi_s((x)_s).$$

The second term is defined by the integral over the entire phase space of the s-particle system with the integrand

$$\big[S_s^0(-\Delta t, (x)_s) F_s(t, (x)_s) + I_s\big] \varphi_s((x)_s)$$

because the term I_s does not depend on any random vectors.

Finally, we obtain the functional

$$\int d(x)_s \Big\{ \sum_{i<j=1}^{s} \int_0^{\Delta t} d\tau \int_{S_+^2} d\eta_{ij}\, \eta_{ij} \cdot (p_i - p_j) \delta(q_i - p_i\tau - q_j + p_j\tau)$$

$$\times \big[F_s(t, q_1 - p_1\Delta t, p_1, \dots, q_i - p_i\tau - p_i^*(\Delta t - \tau), p_i^*, \dots,$$

$$q_j - p_j\tau + p_j^*(\Delta t - \tau), p_j^*, \dots, q_s - p_s\Delta t, p_s)$$

$$- F_s(t, q_1 - p_1\Delta t, p_1, \dots, q_i - p_i\Delta t, p_i, \dots,$$

$$q_j - p_j\Delta t, p_j, \dots, q_s - p_s\Delta t, p_s)\big]\varphi_s((x)_s)\big\}$$

$$+ \int d(x)_s S_s^0(-\Delta t, (x)_s) F_s(t, (x)_s)\varphi_s((x)_s)$$

$$+ \int \Big\{\sum_{i=1}^{s} \int d(x)_{s+1} \int_0^{\Delta t} d\tau \int_{S_+^2} d\eta_{is+1}\eta_{is+1} \cdot (p_i - p_{s+1})$$

$$\times \delta(q_i - p_i\tau - q_{s+1} + p_{s+1}\tau)$$

$$\times \big[F_{s+1}(t, q_1 - p_1\Delta t, p_1, \dots, q_i - p_i\tau - p_i^*(\Delta t - \tau), p_i^*, \dots,$$

$$q_s - p_s\Delta t, p_s, q_{s+1} - p_{s+1}\tau + p_{s+1}^*(\Delta t - \tau), p_{s+1}^*)$$

$$- F_{s+1}(t, q_1 - p_1\Delta t, p_1, \dots, q_i - p_i\Delta t, p_i, \dots,$$

$$q_s - p_s\Delta t, p_s, q_{s+1} - p_{s+1}\Delta t, p_{s+1})\big]\varphi_s((x)_s)\Big\}$$

$$= \int d(x)_s \bar{F}_s(t + \Delta t, (x)_s)\varphi_s((x)_s) = (\bar{F}_s(t + \Delta t), \varphi_s). \qquad (3.5.14)$$

Relation (3.5.14) yields

$$\bar{F}_s(t + \Delta t), (x)_s)$$

$$= S_s^0(-\Delta t, (x)_s)\bar{F}_s(t, (x)_s)$$

$$+ \sum_{i<j=1}^{s} \int_0^{\Delta t} d\tau \int_{S_+^2} d\eta_{ij}\eta_{ij} \cdot (p_i - p_j)\delta(q_i - p_i\tau - q_j + p_j\tau)$$

$$\times \big[\bar{F}_s(t, q_1 - p_1\Delta t, p_1, \dots, q_i - p_i\tau - p_i^*(\Delta t - \tau), p_i^*, \dots,$$

$$q_j - p_j\tau - p_j^*(\Delta t - \tau), p_j^*, \dots, q_s - p_s\Delta t, p_s)$$

$$- \bar{F}_s(t, q_1 - p_1\Delta t, p_1, \dots, q_i - p_i\Delta t, p_i, \dots, q_j - p_j\Delta t, \dots,$$

$$q_s - p_s\Delta t, p_s)\big]$$

$$+ \sum_{i=1}^{s} \int dx_{s+1} \int_0^{\Delta t} d\tau \int_{S_+^2} d\eta_{is+1}\eta_{is+1} \cdot (p_i - p_{s+1})$$

$$\times \delta(q_i - p_i\tau - q_{s+1} + p_{s+1}\tau)$$

$$\times \big[\bar{F}_{s+1}(t, q_1 - p_1\Delta t, p_1, \dots, q_i - p_i\tau - p_i^*(\Delta t - \tau), p_i^*, \dots,$$

$$q_s - p_s\Delta t, p_s, q_{s+1} - p_{s+1}\tau + p^*_{s+1}(\Delta t - \tau), p^*_{s+1})$$

$$- \bar{F}_{s+1}(t, q_1 - p_1\Delta t, p_1, \ldots, q_i - p_i\Delta t, p_i, \ldots,$$

$$q_s - p_s\Delta t, p_s, q_{s+1} - p_{s+1}\Delta t, p_{s+1})\big]. \qquad (3.5.15)$$

In the correlation functions that satisfy equation (3.5.15), the averaging procedure is performed with respect to the random vectors η_{ij} at the points $q_i - q_j = \tau(p_i - p_j)$, $0 \leq \tau \leq \Delta t$, $1 \leq i < j \leq s$, where the stochastic particles interact. We use for them the notation $\bar{F}_s(t + \Delta t, (x)_s)$, and for those defined by (3.5.13) we use the notation $F_s(t, (x)_s)$. In expressions (3.5.14)–(3.5.15), the contribution of the hyperplanes $D_{-\Delta t}$ of lower dimension where the stochastic particles interact is taken into account. In statistical mechanics, it is customary to neglect the hyperplanes of lower dimension. In Section 3.6, we will show that the solutions of the Boltzmann equation and hierarchy are expressed in terms of the correlation functions introduced above, which take into account the contribution of the hyperplanes where the stochastic particles interact. For the stochastic hierarchy it this was already shown for infinitesimal time Δt by formula (3.5.13).

Remark 3.4. It should be stressed that the expressions for $F_s(t + \Delta t, (x)_s)$ (3.5.13) and $\bar{F}_s(t + \Delta t, (x)_s)$ (3.5.15) are identical in the following sense: Expression (3.5.13) defines the correlation function $F_s(t+\Delta t, (x)_s)$ numerically as the usual function. Expression (3.5.15) defines the same correlation function as the corresponding generalized function that shows how to calculate the integral of $F_s(t+\Delta t, (x)_s)$ with test function $\varphi_s((x)_s)$ and to average with respect to the random vectors η_{ij}, $i, j \in \{1, \ldots, s\}$. In Subsection 3.3.3 it has been explained how to obtain each function from another one. From (3.5.13) and (3.5.15) it follows that $F_s(t + \Delta t, (x)_s)$ and $\bar{F}_s(t + \Delta t, (x)_s)$ differ by the different representation of the operators $S_s(-\Delta t)$ and $\bar{S}_s(-\Delta t)$. One can also obtain $\tilde{F}_s(t + \Delta t, (x)_s)$ from (3.5.13) or (3.5.15) by replacing $S_s(-\Delta t)$ and $\bar{S}_s(-\Delta t)$ by the operator $\tilde{S}_s(-\Delta t)$.

Note that, analogously to usual statistical mechanics, the average of the observable $\varphi_s((x)_s)$ over the state $\bar{F}_s(t + \Delta t), (x)_s)$ (3.5.15) is equal to the integral

$$(\bar{F}_s(t + \Delta t), \varphi_s) = \int \bar{F}_s(t + \Delta t, (x)_s)\varphi_s((x)_s)d(x)_s.$$

3.5.5 Stochastic Boltzmann hierarchy

It follows directly from (3.5.14) that $\bar{F}_s(t, (x)_s)$, $s \geq 1$, satisfy the following equations in the weak form:

$$\begin{aligned}
&\Big(\frac{\partial \bar{F}_s(t)}{\partial t}, \varphi_s\Big) \\
&= \int d(x)_s\Big[-\sum_{i=1}^{s} p_i \cdot \frac{\partial}{\partial q_i}\bar{F}_s(t,(x)_s)\Big]\varphi_s((x)_s) \\
&+ \int d(x)_s\Big\{\sum_{i<j=1}^{s}\int_{S_+^2} d\eta_{ij}\eta_{ij}\cdot(p_i - p_j)\delta(q_i - q_j) \\
&\qquad \times \big[\bar{F}_s(t, q_1, p_1, \dots, q_i, p_i^*, \dots, q_j, p_j^*, \dots, q_s, p_s) \\
&\qquad - \bar{F}_s(t, q_1, p_1, \dots, q_i, p_i, \dots, q_j, p_j, \dots, q_s, p_s)\big]\Big\}\varphi_s((x)_s) \\
&+ \int\Big\{\sum_{i=1}^{s}\int d(x)_{s+1}\int_{S_+^2} d\eta_{is+1}\eta_{is+1}\cdot(p_i - p_{s+1})\delta(q_i - q_{s+1}) \\
&\qquad \times \big[\bar{F}_{s+1}(t, q_1, p_1, \dots, q_i, p_i^*, \dots, q_s, p_s, q_{s+1}, p_{s+1}^*) \\
&\qquad - \bar{F}_{s+1}(t, q_1, p_1, \dots, q_i, p_i, \dots, q_s, p_s, q_{s+1}, p_{s+1})\big]\Big\}\varphi_s((x)_s), \quad s \geq 1.
\end{aligned} \tag{3.5.16}$$

It follows from (3.5.16) that

$$\begin{aligned}
&\frac{\partial \bar{F}_s(t,(x)_s)}{\partial t} \\
&= -\sum_{i=1}^{s} p_i \cdot \frac{\partial}{\partial q_i}\bar{F}_s(t,(x)_s) + \sum_{i<j=1}^{s}\int_{S_+^2} d\eta_{ij}\eta_{ij}\cdot(p_i - p_j)\delta(q_i - q_j) \\
&\qquad \times \big[\bar{F}_s(t, q_1, p_1, \dots, q_i, p_i^*, \dots, q_j, p_j^*, \dots, q_s, p_s) \\
&\qquad\quad - \bar{F}_s(t, q_1, p_1, \dots, q_i, p_i, \dots, q_j, p_j, \dots, q_s, p_s)\big] \\
&\quad + \sum_{i=1}^{s}\int d(x)_{s+1}\int_{S_+^2} d\eta_{is+1}\eta_{is+1}\cdot(p_i - p_{s+1})\delta(q_i - q_{s+1}) \\
&\qquad \times \big[\bar{F}_{s+1}(t, q_1, p_1, \dots, q_i, p_i^*, \dots, q_s, p_s, q_{s+1}, p_{s+1}^*) \\
&\qquad\quad - \bar{F}_{s+1}(t, q_1, p_1, \dots, q_i, p_i, \dots, q_s, p_s, q_{s+1}, p_{s+1})\big], \quad s \geq 1.
\end{aligned} \tag{3.5.17}$$

The correlation functions (3.5.15) that satisfy (3.5.17) do not depend on any random vectors because the infinitesimal operator in (3.5.17) does not depend on any random vectors (they are averaged with respect to all random vectors).

From (3.5.17), we can derive equations for the distribution functions that depend on the random vectors η_{ij}, $1 \le i, j \le s$. For these functions, we use the notation $\tilde{F}_s(t, (x)_s)$. For arbitrary fixed η_{ij}, we have

$$\begin{aligned}
&\frac{\partial \tilde{F}_s(t,(x)_s)}{\partial t} \\
&\quad = -\sum_{i=1}^{s} p_i \cdot \frac{\partial}{\partial q_i} \tilde{F}_s(t,(x)_s) \\
&\qquad + \sum_{i<j=1}^{s} \Theta(\eta_{ij} \cdot (p_i - p_j))\eta_{ij} \cdot (p_i - p_j)\delta(q_i - q_j) \\
&\qquad\quad \times \big[\tilde{F}_s(t, q_1, p_1, \dots, q_i, p_i^*, \dots, q_j, p_j^*, \dots, q_s, p_s) \\
&\qquad\qquad - \tilde{F}_s(t, q_1, p_1, \dots, q_i, p_i, \dots, q_j, p_j, \dots, q_s, p_s)\big] \\
&\qquad + \sum_{i=1}^{s} \int d(x)_{s+1} \int_{S_+^2} d\eta_{is+1}\eta_{is+1} \cdot (p_i - p_{s+1})\delta(q_i - q_{s+1}) \\
&\qquad\quad \times \big[\tilde{F}_{s+1}(t, q_1, p_1, \dots, q_i, p_i^*, \dots, q_s, p_s, q_{s+1}, p_{s+1}^*) \\
&\qquad\qquad - \tilde{F}_{s+1}(t, q_1, p_1, \dots, q_i, p_i, \dots, q_s, p_s, q_{s+1}, p_{s+1})\big]. \qquad (3.5.18)
\end{aligned}$$

The last term on the right-hand side of (3.5.18) depends on the same random vectors as $\tilde{F}_s(t, (x)_s)$ because the $(s+1)$-th particle can interact with the other s-particles only for momenta p_{s+1} from the set of lower dimension, which does not contribute to the integral with respect to p_{s+1}.

(One can derive hierarchy (3.5.18) by differentiating expression (3.5.13) with respect to time and using representation (3.2.24)–(3.2.25) of the infinitesimal operator $\mathcal{H}_s$. This will be shown in the next subsection.)

Formulas (3.5.13) and (3.5.15) give an exact expression for the semigroup of the evolution operators $U(\Delta t)$ associated with the stochastic Boltzmann hierarchy (3.5.17):

$$\bar{F}(t + \Delta t) = \bar{U}(\Delta t)\bar{F}(t), \qquad (3.5.19)$$

where $\bar{F}(t)$ is the sequence of distribution functions $\bar{F}_s(t, (x)_s)$, $s \ge 1$, and

$$\bar{F}(t + \Delta t, (x)_s) = (\bar{U}(\Delta t)\bar{F}(t))_s((x)_s).$$

The semigroup $\bar{U}(t)$ with arbitrary t is formally defined as follows:

$$\bar{U}(t) = \lim_{n\to\infty} \prod_{i=1}^{n} \bar{U}(\Delta t_i), \quad \sum_{i=1}^{n} \Delta t_i = t. \qquad (3.5.20)$$

The corresponding semigroup $U(\Delta t)$ for hierarchy (3.5.5) is defined by (3.5.13). For this semigroup, formulas (3.5.19) and (3.5.20) are also true. If one substitutes the operator $\tilde{S}_s(-\Delta t, (x)_s)$ for $S_s(-\Delta t, (x)_s)$ in (3.5.13), one obtains $\tilde{U}(\Delta t)$ and $\tilde{U}(t)$ that are evolution operators for hierarchy (3.5.18).

By using the evolution operators $U(t)$, $\tilde{U}(t)$, and $\bar{U}(t)$ one formally obtains the Boltzmann–Grad limit of $F_s^a(t, (x)_s)$ in the entire phase space.

3.5.6 Different representations of the infinitesimal operator

We have obtained the stochastic Boltzmann hierarchy (3.5.17) in the weak sense from hierarchy (3.5.17). We now derive it by differentiating relation (3.5.13) with the use of pointwise convergence and show that the stochastic Boltzmann hierarchy (3.5.18) is equivalent to hierarchy (3.5.5) with boundary conditions for the Poisson bracket and functions $F_s(t, (x)_s)$ (3.5.6).

We have

$$\begin{aligned}\frac{\partial F_s(t,(x)_s)}{\partial t} &= \frac{\partial S_s(-\Delta t,(x)_s)}{\partial(\Delta t)}\Big|_{\Delta t=0} F_s(t,(x)_s)\\ &\quad + \sum_{i=1}^{s}\int dx_{s+1}\int_{S_+^2} d\eta_{is+1}\eta_{is+1}\cdot(p_i - p_{s+1})\delta(q_i - q_{s+1})\\ &\qquad\times\big[F_{s+1}(t,q_1,p_1,\dots,q_i,p_i^*,\dots,q_{s+1},p_{s+1}^*)\\ &\qquad\quad - F_{s+1}(t,q_1,p_1,\dots,q_i,p_i,\dots,q_{s+1},p_{s+1})\big]. \quad (3.5.21)\end{aligned}$$

For the infinitesimal generator of the semigroup $S_s(-t, (x)_s)$, we have the following expression (see Chapter 2): We repeat the corresponding calculation:

$$\begin{aligned}&\frac{\partial S_s(-\Delta t,(x)_s)}{\partial \Delta t} F_s(t,(x)_s)\\ &= -\sum_{i=1}^{s} p_i\cdot\frac{\partial}{\partial q_i}F_s(t,(x)_s)\\ &\quad + \sum_{i<j=1}^{s}\Theta(\eta_{ij}\cdot(p_i - p_j))\delta(\Delta t - \tau_{ij})\big[F_s(t,(x)_s^*) - F_s(t,(x)_s)\big]\big|_{q_i=q_j}, \quad (3.5.22)\\ &\qquad (x)_s^* = (x_1,\dots,q_i,p_i^*,\dots,q_j,p_j^*,\dots,x_s).\end{aligned}$$

Consider the function $\delta(\Delta t - \tau_{ij})|_{\Delta t=0}$. In the coordinate system where the first component of the vector (q_i, p_i, q_j, p_j) is directed along the vector η_{ij}, the time of collision τ_{ij} is defined as follows:

$$\tau_{ij} = \frac{q_i^1 - q_j^1}{p_i^1 - p_j^1}.$$

The (i, j)-th term in (3.5.22) can be expressed as follows:

$$\Theta(p_i^1 - p_j^1)\delta(q_i^1 - q_j^1) \cdot (p_i^1 - p_j^1)\big[F_s(t, (x)_s^*) - F_s(t, (x)_s)\big]\big|_{q_i^2 - q_j^2 = 0,\, q_i^3 - q_j^3 = 0},$$

$$\eta_{ij} \cdot (p_i - p_j) = (p_i^1 - p_j^1). \tag{3.5.23}$$

This term is different from zero on the first axis $q_i^1 - q_j^1$ (with respect to the vector $q_i - q_j$, i.e., for $q_i^2 - q_j^2 = 0$ and $q_i^3 - q_j^3 = 0$) and, regarded as a generalized function in the three-dimensional space, is equal to

$$\begin{aligned}\Theta(p_i^1 - p_j^1)\delta(q_i^1 - q_j^1)\delta(q_i^2 - q_j^2)\delta(q_i^3 - q_j^3) \cdot (p_i^1 - p_j^1)\big[F_s(t, (x)_s^*) - F_s(t, (x)_s)\big] \\ = \Theta(\eta_{ij} \cdot (p_i - p_j))\delta(q_i - q_j)\eta_{ij} \cdot (p_i - p_j)\big[F_s(t, (x)_s^*) - F_s(t, (x)_s)\big].\end{aligned} \tag{3.5.24}$$

This means that expression (3.5.23) is considered as a certain generalized function in the phase space, and the integral of expression (3.5.23) with test function $\varphi_s((x)_s)$ should be calculated using expression (3.5.24). (For analogous calculations, see [GGV]). The expression obtained does not depend on the choice of a coordinate system because $\delta(q_i - q_j)$ and $\eta_{ij} \cdot (p_i - p_j)$ are invariant under rotation.

Substituting expression (3.5.21) in (3.5.23), we finally obtain

$$\begin{aligned}&\frac{\partial S_s(-\Delta t, (x)_s)}{\partial \Delta t}\bigg|_{\Delta t = 0} F_s(t, (x)_s) \\ &\quad = -\sum_{i=1}^{s} p_i \cdot \frac{\partial}{\partial q_i} F_s(t, (x)_s) \\ &\qquad + \sum_{i<j=1}^{s} \Theta(\eta_{ij} \cdot (p_i - p_j))\delta(q_i - q_j)\big[F_s(t, (x)_s^*) - F_s(t, (x)_s)\big].\end{aligned} \tag{3.5.25}$$

Substituting the last expression in hierarchy (3.5.21), we obtain the stochastic Boltzmann hierarchy (3.5.18).

We stress that expressions (3.5.22) or (3.5.23) and (3.5.24) are identical in the following sense: Numerically,

$$\frac{\partial S_s(-\Delta t, (x)_s)}{\partial \Delta t}\bigg|_{\Delta t = 0} F_s(t, (x)_s)$$

is given by expressions (3.5.22) and (3.5.23) and, for the calculation of functionals (averages), one should use expressions (3.5.24)–(3.5.25).

In expressions (3.5.22) and (3.5.23), we have the one-dimensional δ-function $\delta(\tau_{ij}) = \delta(q_i^1 - q_j^1)(p_i^1 - p_j^1)$. As is known, a one-dimensional δ-function is equivalent to the following boundary conditions for $q_i = q_j$ and $\eta_{ij} \cdot (p_i - p_j) \geq 0$:

$$F_s(t + 0, x_1, \ldots, x_i, \ldots, x_j, \ldots, x_s) = F_s(t, x_1, \ldots, q_i, p_i^*, \ldots, q_j, p_j^*, \ldots, x_s),$$

$$F_s(t-0, x_1, \dots, x_i, \dots, x_j, \dots, x_s) = F_s(t, x_1, \dots, x_i, \dots, x_j, \dots, x_s). \quad (3.5.26)$$

For $q_i = q_j$ and $\eta_{ij} \cdot (p_i - p_j) \leq 0$, we have

$$F_s(t-0, (x)_s) = F_s(t+0, (x)_s).$$

This means that, in formulas (3.5.22), (3.5.25), one can omit the term with δ-function but take into account the boundary condition (3.5.26) and the boundary condition in the Poisson bracket, according to which for $q_i = q_j$, $\eta_{ij} \cdot (p_i - p_j) \geq 0$, the momenta p_i, p_j should be replaced by p_i^*, p_j^* in $-\sum_{i=1}^{s} p_i \cdot \frac{\partial}{\partial q_i} F_s(t, x_1, \dots, x_s)$.

3.5.7 Different equivalent forms of the stochastic Boltzmann hierarchy

Thus, the stochastic Boltzmann hierarchies (3.5.17)–(3.5.18) are equivalent to the stochastic Boltzmann hierarchy (3.5.5′) without δ-function, namely,

$$\begin{aligned}
&\frac{\partial F_s(t,(x)_s)}{\partial t} \\
&\quad = -\sum_{i=1}^{s} p_i \cdot \frac{\partial}{\partial q_i} F_s(t,(x)_s) \\
&\qquad + \sum_{i=1}^{s} \int dx_{s+1} \int_{S_+^2} d\eta_{is+1} \eta_{is+1} \cdot (p_i - p_{s+1}) \delta(q_i - q_{s+1}) \\
&\qquad\quad \times \big[F_{s+1}(t, x_1, \dots, q_i, p_i^*, \dots, q_{s+1}, p_{s+1}^*) \\
&\qquad\qquad - F_{s+1}(t, x_1, \dots, q_i, p_i, \dots, q_{s+1}, p_{s+1})\big], \quad s \geq 1, \qquad (3.5.5')
\end{aligned}$$

but with the boundary conditions (3.5.26) and boundary conditions for the Poisson bracket, according to which, for $q_i = q_j$ and $\eta_{ij} \cdot (p_i - p_j) \geq 0$, the momenta (p_i, p_j) in it should be replaced by (p_i^*, p_j^*).

The stochastic Boltzmann hierarchies with the three-dimensional δ-function (3.5.17)–(3.5.18) have the same structure as the BBGKY hierarchy for smooth potentials because the integral term is obtained from the infinitesimal operator with δ-functions, describing the interaction of stochastic particles, by integration with respect to x_{s+1} and averaging over η_{is+1}. Of course, we have boundary conditions for the Poisson bracket and correlation functions in (3.5.5′). In both forms (3.5.5′) and (3.5.17)–(3.5.18) of the stochastic Boltzmann hierarchy, we use different equivalent representations (3.5.22)–(3.2.24) and (3.5.25) of the infinitesimal operator $\mathcal{H}_s$ of the stochastic evolution operator $S_s(-t, (x)_s)$.

Note that, in (3.5.22)–(3.5.25), the term with δ-function can be omitted and, therefore, in what follows we will use the stochastic hierarchy (3.5.5′) where the infinitesimal operator $\mathcal{H}_s$ is calculated in the sense of pointwise differentiation and is represented in the form

$$\mathcal{H}_s f_s(x_1, \dots, x_s) = -\sum_{i=1}^{s} p_i \cdot \frac{\partial}{\partial q_i} f_s(x_1, \dots, x_s)$$

with the known boundary condition.

Thus, we derive the hierarchy for the correlation functions $F_s(t,(x)_s)$, $\tilde{F}_s(t,(x)_s)$ and $\bar{F}_s(t,(x)_s)$, $s \geq 1$, from the BBGKY hierarchy in the Boltzmann–Grad limit.

3.6 Boltzmann equation and its solutions in terms of stochastic dynamics

3.6.1 Iterations of the Boltzmann equation

In this section, we show that the Boltzmann equation and its solutions can be represented in terms of the stochastic dynamics and the corresponding functional.

It is obvious that the Boltzmann equation (3.4.11) with initial conditions (3.4.12) can be represented as the following integral equation:

$$\begin{aligned} F_1(t,x_1) &= F_1(0, q_1 - p_1 t, p_1) \\ &\quad + e^{p_1 \nabla_{q_1}(\tau - t)} \int_0^t d\tau \int_{S_+^2} d\eta \int dp_2 \eta \cdot (p_1 - p_2) \\ &\quad \times \big[F_1(\tau, q_1, p_1^*) F_1(\tau, q_1, p_2^*) - F_1(\tau, q_1, p_1) F_1(\tau, q_1, p_2)\big], \quad (3.6.1) \end{aligned}$$

$$\nabla_{q_1} = \frac{\partial}{\partial q_1}.$$

In the first approximation, the solution of (3.6.1) has the form

$$F_1^{(1)}(t,x_1) = F_1(0, q_1 - p_1 t, p_1). \quad (3.6.2)$$

Substituting (3.6.2) in (3.6.1), we obtain the second approximation of solutions of (3.6.1):

$$\begin{aligned} F_1^{(2)}(t,x_1) &= F_1(0, q_1 - p_1 t, p_1) + \int_0^t d\tau \int_{S_+^2} d\eta \int dp_2 \eta \cdot (p_1 - p_2) e^{p_1 \nabla_{q_1}(\tau - t)} \\ &\qquad \times \big[F_1(0, q_1 - p_1^* \tau, p_1^*) F_1(0, q_1 - p_2^* \tau, p_2^*) \\ &\qquad\quad - F_1(0, q_1 - p_1 \tau, p_1) F_1(0, q_1 - p_2 \tau, p_2)\big] \\ &= F_1(0, q_1 - p_1 t, p_1) + \int_0^t d\tau \int_{S_+^2} d\eta \int dp_2 \eta \cdot (p_1 - p_2) \end{aligned}$$

$$\times \big[F_1(0, q_1 - p_1(t-\tau) - p_1^*\tau, p_1^*)F_1(0, q_1 - p_1(t-\tau) - p_2^*\tau, p_2^*)$$
$$- F_1(0, q_1 - p_1\tau, p_1)F_1(0, q_1 - p_2\tau - p_1(t-\tau), p_2)\big]$$
$$= F_1(0, q_1 - p_1 t, p_1) + \int_0^t d\tau \int_{S_+^2} d\eta \int dp_2 \eta \cdot (p_1 - p_2)$$
$$\times \big[F_1(0, q_1 - p_1\tau - p_1^*(t-\tau), p_1^*)F_1(0, q_1 - p_1\tau - p_2^*(t-\tau), p_2^*)$$
$$- F_1(0, q_1 - p_1 t, p_1)F_1(0, q_1 - p_1\tau - p_2(t-\tau), p_2)\big]$$
$$= F_1(0, q_1 - p_1 t, p_1) + \int_0^t d\tau \int_{S_+^2} d\eta \int dp_2 dq_2 \eta \cdot (p_1 - p_2)$$
$$\times \delta(q_1 - p_1\tau - q_2 + p_2\tau)$$
$$\times \big[F_1(0, q_1 - p_1\tau - p_1^*(t-\tau), p_1^*)F_1(0, q_2 - p_2\tau - p_2^*(t-\tau), p_2^*)$$
$$- F_1(0, q_1 - p_1 t, p_1)F_1(0, q_2 - p_2 t, p_2)\big]. \tag{3.6.3}$$

We now consider the two-particle stochastic system with the initial distribution function

$$F_2(0, x_1, x_2) = F_1(0, x_1)F_1(0, x_2), \qquad \int dx F_1(0, x) = 1. \tag{3.6.4}$$

It is easy to see that the second approximation (3.6.3) can be identically represented as follows:

$$F_1^{(2)}(t, x_1) = F_1(t, x_1)$$
$$= \int S_2^0(-t, x_1, x_2)F_2(0, x_1, x_2)dx_2$$
$$+ \int_0^t d\tau \int_{S_+^2} d\eta \int dp_2 dq_2 \eta \cdot (p_1 - p_2)\delta(q_1 - p_1\tau - q_2 + p_2\tau)$$
$$\times \big[S_2(-t, x_1, x_2)F_2(0, x_1, x_2)$$
$$- S_2^0(-t, x_1, x_2)F_2(0, x_1, x_2)\big], \tag{3.6.5}$$

where $F_1(t, x_1)$ coincides with the one-particle distribution function of the given two-particle stochastic system.

Multiplying both sides of (3.6.5) by a test function $\varphi_1(x_1)$ and integrating with respect to the variable x_1, we get

$$\begin{aligned}
\int & F_1(t,x_1)\varphi_1(x_1)dx_1 \\
&= \int S_2^0(-t,x_1,x_2)F_2(0,x_1,x_2)\times\varphi_1(x_1)dx_1dx_2 \\
&\quad + \int_0^t d\tau\int_{S_+^2}d\eta\int dx_1dx_2\eta\cdot(p_1-p_2) \\
&\qquad \times\delta(q_1-p_1\tau-q_2+p_2\tau)\big[S_2(-t,x_1,x_2)F_2(0,x_1,x_2) \\
&\qquad\quad - S_2^0(-t,x_1,x_2)F_2(0,x_1,x_2)\big]\varphi_1(x_1) \\
&= \frac{1}{2}\Big(S_2(-t)F_2(0),\varphi_2\Big),
\end{aligned} \tag{3.6.6}$$

where $(S_2(-t)F_2(0),\varphi_2)$ is functional (3.2.4), which represents the average of the one-particle observable

$$\varphi_2(x_1,x_2)=\varphi_1(x_1)+\varphi_1(x_2)$$

over the state $S_2(-t,x_1,x_2)F_2(0,x_1,x_2)$.

Thus, we have shown that the second approximation of the solution of the Boltzmann equation coincides with the one-particle correlation function of the two-particle stochastic system with the chaotic initial correlation function (3.6.4).

Note that, in the definition of one-particle correlation function, the contribution of the two-particle correlation function on the hyperplanes of lower dimension $q_1-p_1\tau-q_2+p_2\tau=0$ is taken into account. In statistical mechanics, the sets of lower dimension are usually neglected.

We now proceed to the general case.

We introduce the function

$$\begin{aligned}
\tilde{F}_1(\tau,q_1-p_1\tau,p_1) &= F_1(\tau,q_1,p_1), \\
F_1(\tau,q_1,p_1) &= e^{+p_1\nabla_{q_1}\tau}\tilde{F}_1(\tau,q_1,p_1), \\
\tilde{F}_1(\tau,q_1,p_1) &= e^{+p_1\nabla_{q_1}\tau}F_1(\tau,q_1,p_1)=F_1(\tau,q_1+p_1\tau,p_1)
\end{aligned} \tag{3.6.7}$$

and represent equation (3.6.1) as follows:

$$\begin{aligned}
\tilde{F}_1(t,q_1-p_1t,p_1) &= F_1(t,x_1) \\
&= F_1(0,q_1-p_1t,p_1) \\
&\quad + \int_0^t d\tau\int_{S_+^2}d\eta\eta\cdot(p_1-p_2)\big[\tilde{F}_1(\tau,q_1-p_1(t-\tau)-p_1^*\tau,p_1^*) \\
&\qquad \times\tilde{F}_1(\tau,q_1-p_1(t-\tau)-p_2^*\tau,p_2^*)
\end{aligned}$$

$$
\begin{aligned}
&- \tilde{F}_1(\tau, q_1 - p_1 t, p_1)\tilde{F}_1(\tau, q_1 - p_2\tau - p_1(t-\tau) - p_2)\big] \\
&= F_1(0, q_1 - p_1 t, p_1) + \int_0^t d\tau \int_{S_+^2} d\eta\eta \cdot (p_1 - p_2) \\
&\quad \times \big[\tilde{F}_1(t-\tau, q_1 - p_1\tau - p_1^*(t-\tau), p_1^*) \\
&\qquad \times \tilde{F}_1(t-\tau, q_1 - p_1\tau - p_2^*(t-\tau), p_2^*) \\
&\qquad - \tilde{F}_1(t-\tau, q_1 - p_1 t, p_1)\tilde{F}_1(t-\tau, q_1 - p_2(t-\tau) - p_1\tau, p_2)\big] \\
&= F_1(0, q_1 - p_1 t, p_1) + \int_0^t d\tau \int_{S_+^2} d\eta \int dp_2 dq_2 \eta \cdot (p_1 - p_2) \\
&\quad \times \delta(q_1 - p_1\tau - q_2 + p_2\tau)\big[\tilde{F}_1(t-\tau, q_1 - p_1\tau - p_1^*(t-\tau), p_1^*) \\
&\qquad \times \tilde{F}_1(t-\tau, q_2 - p_2\tau - p_2^*(t-\tau), p_2^*) \\
&\qquad - \tilde{F}_1(t-\tau, q_1 - p_1 t, p_1)\tilde{F}_1(t-\tau, q_2 - p_2 t, p_2)\big] \\
&= F_1(0, q_1 - p_1 t, p_1) + \int_0^t d\tau \int_{S_+^2} d\eta \int dp_2 dq_2 \eta \cdot (p_1 - p_2) \\
&\quad \times \delta(q_1 - p_1\tau - q_2 + p_2\tau)\Big[S_2(-t, x_1, x_2) \\
&\qquad \times \big\{\tilde{F}_1(t-\tau, q_1, p_1)\tilde{F}_1(t-\tau, q_2, p_2)\big\} \\
&\qquad - S_2^0(-t, x_1, x_2)\big\{\tilde{F}_1(t-\tau, q_1, p_1)\tilde{F}_1(t-\tau, q_2, p_2)\big\}\Big].
\end{aligned}
\tag{3.6.8}
$$

We represent the Boltzmann equation in terms of the operator of evolution of two stochastic particles. In the weak form, we have

$$
\begin{aligned}
&\int F_1(t, q_1 - p_1 t, p_1)\varphi_1(q_1, p_1)dq_1 dp_1 \\
&\quad = \int \{S_2^0(-t, x_1, x_2)F_1(0, q_1, p_1)F_1(0, q_2, p_2)\}\varphi_1(x_1)dx_1 dx_2 \\
&\qquad + \int_0^t d\tau \int_{S_+^2} d\eta\eta \cdot (p_1 - p_2)\delta(q_1 - p_1\tau - q_2 + p_2\tau) \\
&\qquad \times \Big[S_2(-t, x_1, x_2)\big\{\tilde{F}_1(t-\tau, x_1)F_1(t-\tau, x_2)\big\} \\
&\qquad - S_2^0(-t, x_1, x_2)\big\{\tilde{F}_1(t-\tau, x_1)F_1(t-\tau, x_2)\big\}\Big]\varphi_1(x_1)dx_1 dx_2
\end{aligned}
\tag{3.6.9}
$$

Note that in (3.6.8) and (3.6.9), the functions $S_2(-t, x_1, x_2)\tilde{F}_1(t-\tau, q_1, p_1)\tilde{F}_1(t-\tau, q_2, p_0)$ and $S_2^0(-t, x_1, x_2)\tilde{F}_1(t-\tau, q_1, p_1)F_1(t-\tau, q_1, p_1)$ are integrated over the hyperplane $q_1 - p_1\tau - q_2 + p_2\tau = 0$, $0 \le \tau \le t$. By using (3.6.8), we obtain the representation of the n-th approximation of the solution $F_1^{(n)}(t, x_1)$ in terms of the $(n-1)$-th approximation:

$$\begin{aligned}
F_1^{(n)}(t, x_1) &= \tilde{F}_1^{(n)}(t, q_1 - p_1 t, p_1) \\
&= F_1(0, q_1 - p_1 t, p_1) \\
&\quad + \int_0^t d\tau \int_{S_+^2} d\eta \int dp_2 dq_2 \eta \cdot (p_1 - p_2)\delta(q_1 - p_1\tau - q_2 + p_2\tau) \\
&\qquad \times \big[\tilde{F}_1^{(n-1)}(t-\tau, q_1 - p_1\tau - p_1^*(t-\tau), p_1^*) \\
&\qquad \times \tilde{F}_1^{(n-1)}(t-\tau, q_2 - p_2\tau - p_2^*(t-\tau), p_2^*) \\
&\qquad - \tilde{F}_1^{(n-1)}(t-\tau, q_1 - p_1 t, p_1)\tilde{F}_1^{(n-1)}(t-\tau, q_2 - p_2 t, p_2)\big] \\
&= F_1(0, q_1 - p_1 t, p_1) + \int_0^t d\tau \int_{S_+^2} d\eta \int dp_2 dq_2 \eta \cdot (p_1 - p_2) \\
&\quad \times \delta(q_1 - p_1\tau - q_2 + p_2\tau) \\
&\qquad \times \Big[S_2(-t, x_1, x_2)\{\tilde{F}_1^{(n-1)}(t-\tau, q_1, p_1)\tilde{F}_1^{(n-1)}(t-\tau, q_2, p_2)\} \\
&\qquad - S_2^0(-t, x_1, x_2)\{\tilde{F}_1^{(n-1)}(t-\tau, q_1, p_1)\tilde{F}_1^{(n-1)}(t-\tau, q_2, p_2)\}\Big].
\end{aligned} \tag{3.6.10}$$

It follows from (3.6.10) that each n-th approximation takes into account one new collision corresponding to the δ-function $\delta(q_1 - p_1\tau - q_2 + p_2\tau)$. The previous collisions were taken into account by $\tilde{F}^{(n-1)}$.

Thus, arbitrary n-th iterations (3.6.10) of the Boltzmann equation are represented in terms of averages of the stochastic dynamics by a formula completely analogous to (3.6.6) through $n-1$ iterations $\tilde{F}^{(n-1)}(t, x_1)$. Solutions of the Boltzmann equation are represented through integrals over the hypersurfaces of lower dimension where the stochastic point particles interact.

3.6.2 Iterations of the Boltzmann hierarchies

The solutions of the ordinary Boltzmann hierarchy (3.4.14) can be represented as the series of iterations

$$F(t) = \sum_{n=0}^{\infty} \int_0^t dt_1 \ldots \int_0^{t_{n-1}} dt_n S^0(-t)$$
$$\times S^0(t_1) A S^0(-t_1) \ldots S^0(t_n) A S^0(-t_1) F(0), \tag{3.6.11}$$

where the operator A is defined by the second term on the right-hand side of (3.4.14) and $S^0(t)$ is the direct sum of the evolution operators $S_s^0(t, (x)_s)$ of free systems. We consider the initial data $F(0)$ for which series (3.6.11) is convergent (see Chapter 4).

The projection of (3.6.11) onto the s-particle space is equal to the s-particle correlation function

$$F_s(t, (x)_s) = \sum_{n=0}^{\infty} \int_0^t dt_1 \ldots \int_0^{t_{n-1}} dt_n S_s^0(-t, (x)_s) S_s^0(t_1, (x)_s)$$
$$\times \sum_{i=1}^{s} \int dx_{s+1} \delta(q_i - q_{s+1}) + \int_{S_+^2} d\eta_{is+1} \eta_{is+1} \cdot (p_1 - p_{s+1})$$
$$\times [S_{s+1}^0(-t_1, (x)_{s+1}^*) - S_{s+1}^0(-t, (x)_{s+1})] \ldots$$
$$\times S_{s+n+1}^0(t_{n-1}, (x)_{s+n-1}) \sum_{i=1}^{s+n-1} \int dx_{s+n} \delta(q_i - q_{s+n})$$
$$+ \int_{S_+^2} d\eta_{is+1} \eta_{is+n} \cdot (p_i - p_{s+n}) \big[S_{s+n}^0(-t_n, (x)_{s+n}^*)$$
$$- S_{s+n}^0(-t_n, (x)_{s+n}) \big] F_{s+n}(0, (x)_{s+n}). \tag{3.6.12}$$

It follows from representation (3.6.12) that the integral with respect to $x_{s+1}, \ldots, x_{s+n}$ is taken over hypersurfaces whose dimension is lower than the dimension of the phase space $(x_{s+1}, \ldots, x_{s+n})$. Indeed, the initial positions of the i-th and $(s+j)$-th particles $(1 \le j \le n, 1 \le i \le s+j-1)$ coincide $(q_i = q_{s+j})$, and, after the action of the operators $S_{s+j}^0(-t_j, (x)_{s+j}^*)$, $S_{s+j}^0(-t_j, (x)_{s+j})$, these particles are located on the hyperplanes of lower dimension on which the vector of difference of their positions is parallel to the vector of difference of their momenta.

An analogous result is also true for the stochastic Boltzmann hierarchy (3.5.25). Its solutions are also represented by series (3.6.11) and (3.6.12), but, instead of the operator $S^0(t)$, one should take the operator $S(t)$ of stochastic evolution. The results of the action of the operators of stochastic evolution $S_{s+j}(-t_j, (x)_{s+j}^*)$ and $S_{s+j}(-t_j, (x)_{s+j})$ differ from the results of the action of the operators of free evolution $S_{s+j}^0(-t_j, (x)_{s+j}^*)$ and $S_{s+j}^0(-t_j, (x)_{s+j})$ only on hyperplanes of lower dimension with respect to p_{s+j}, and we neglect this difference in integrals with respect to p_{s+j} (the integration with respect to q_{s+j} is performed by using the δ-function).

Outside these hyperplanes with respect to phase points $(x)_s$, the solutions of the stochastic Boltzmann hierarchy coincide with the solution of the ordinary Boltzmann hierarchy (3.6.12), and one-particle distribution functions of both hierarchies coincide completely. This means that the one-particle distribution function of the stochastic hierarchy coincides with the solution of the Boltzmann equation because the one-particle distribution function of the Boltzmann hierarchy coincides with the solution of the Boltzmann equation.

The detailed proof of these assertions will be presented in Chapter 4.

Chapter 4

Solutions of the stochastic Boltzmann hierarchy

4.1 Introduction

In this chapter, we consider the problem of the existence of solutions of the stochastic hierarchy in two functional spaces, namely, in the space E_ξ of sequences of functions bounded with respect to positions and exponentially decreasing with respect to squared momenta and in the space $\tilde{E}_{\xi,\beta}$ of sequences of functions exponentially decreasing with respect to squared positions and momenta.

The stochastic hierarchy can be considered in the above-mentioned spaces as an abstract evolution equation with an unbounded operator on the right-hand side. Therefore, the problem of the existence of the solutions is very complicated.

We consider a solution represented by a series of iterations and prove that the series converges uniformly on a compact set in the phase space for a finite interval of time for initial data from the space E_ξ and globally in time for initial data from the space $\tilde{E}_{\xi,\beta}$.

We prove that solutions of the stochastic Boltzmann hierarchy constructed by the series of iterations coincide with the corresponding solutions of the ordinary Boltzmann hierarchy outside the hyperplanes where the stochastic point particles interact, i.e., on the hyperplanes where the vectors of difference of positions of two particles are parallel to the vectors of difference of momenta of the corresponding particles.

As is known, solutions of the ordinary Boltzmann hierarchy have the chaos property, provided that the initial data have, these properties and the one-particle correlation function satisfies the Boltzmann equation. This means that solutions of the stochastic Boltzmann hierarchy also possess the chaos property outside the hyperplanes where the vectors of difference of positions are parallel to the vectors of difference of momenta.

4.2 Solutions of the stochastic hierarchy in the space of bounded functions

4.2.1 Abstract form of the stochastic hierarchy

Denote by $\mathcal{H}$ the direct sum of the infinitesimal operators $\mathcal{H}_s$ (3.5.29)–(3.5.32):

$$\mathcal{H} = \sum_{s=1}^{\infty} \oplus \mathcal{H}_s. \tag{4.2.1}$$

If f is a sequence of functions $f_s(x_1, \dots, x_s)$ defined on the s-particle phase space, then

$$(\mathcal{H} f)_s(x_1, \dots, x_s) = \mathcal{H}_s f_s(x_1, \dots, x_s) = -\sum_{i=1}^{s} p_i \cdot \frac{\partial}{\partial q_i} f_s(x_1, \dots, x_s) \tag{4.2.2}$$

with the boundary condition according to which, for $q_i = q_j$, $i, j \in \{1, \dots, s\}$, $\eta_{ij} \cdot (p_i - p_j) \geq 0$, the momenta p_i, p_j should be replaced by p_i^*, p_j^*. If $\eta_{ij} \cdot (p_i - p_j) \leq 0$, then the momenta p_i, p_j do not change. Denote by $\mathcal{A}$ the operator that acts on the sequence f as follows:

$$\begin{aligned}
&(\mathcal{A} f)_s(x_1, \dots, x_s) \\
&= \sum_{i=1}^{s} \int dx_{s+1} \int_{S_+^2} d\eta_{is+1} \eta_{is+1} \cdot (p_i - p_{s+1}) \delta(q_i - q_{s+1}) \\
&\quad \times \big[f_{s+1}(x_1, \dots, x_i^*, \dots, x_s, x_{s+1}^*) \\
&\quad - f_{s+1}(x_1, \dots, x_i, \dots, x_s, x_{s+1}) \big], \quad s \geq 1.
\end{aligned} \tag{4.2.3}$$

Then hierarchy (3.5.28) can be represented in the abstract form

$$\frac{dF(t)}{dt} = \mathcal{H} F(t) + \mathcal{A} F(t) \tag{4.2.4}$$

with initial data

$$F(t)|_{t=0} = F(0) = F. \tag{4.2.5}$$

Denote by $S(\pm t)$ the direct sum of the operators $S_s(\pm t)$:

$$\begin{aligned}
S(\pm t) &= \sum_{s=1}^{\infty} \oplus S_s(\pm t), \\
(S(\pm t) f)_s(x_1, \dots, x_s) &= S_s(\pm t) f_s(x_1, \dots, x_s).
\end{aligned} \tag{4.2.6}$$

Then mild solutions of hierarchy (4.2.4) with initial data (4.2.5) can be represented by the series of iterations

$$F(t) = \sum_{n=0}^{\infty} \int_0^t dt_1 \ldots \int_0^{t_{n-1}} dt_n S(-t)S(t_1)$$
$$\times \mathcal{A}S(-t_1)\ldots S(t_n)\mathcal{A}S(-t_n)F(0), \tag{4.2.7}$$

or componentwise

$$F_s(t,(x)_s) = \sum_{n=0}^{\infty} \int_0^t dt_1 \ldots \int_0^{t_{n-1}} dt_n S_s(-t,(x)_s)S_s(t_1,(x)_s)$$
$$\times \sum_{i=1}^{s} \int dx_{s+1}\delta(q_i - q_{s+1}) \int_{S_+^2} d\eta_{is+1}\eta_{is+1}\cdot(p_i - p_{s+1})$$
$$\times \left[S_{s+1}(-t_1,(x)_{s+1}^*) - S_{s+1}(-t_1,(x)_{s+1})\right]\ldots$$
$$\times S_{s+n-1}(t_{n-1},(x)_{s+n-1}) \sum_{i=1}^{s+n-1} \int dx_{s+n}\delta(q_i - q_{s+n})$$
$$\times \int_{S_+^2} d\eta_{is+n}\eta_{is+n}\cdot(p_i - p_{s+n})$$
$$\times \left[S_{s+n}(-t_n,(x)_{s+n}^*) - S_{s+n}(-t_n,(x)_{s+n})\right]$$
$$\times F_{s+n}(0,(x)_{s+n}), \tag{4.2.8}$$

where $(x)_{s+n}^* = (x_1,\ldots,x_i^*,\ldots,x_s,x_{s+n}^*)$ in the i-th term and $S_s(\pm t,(x)_s)$ is the operator of shift along the trajectory $X(\pm t,(x)_s)$ of s particles with initial data $(x)_s$ at $t = 0$.

We consider a solution of hierarchy (4.2.4) in the functional space $E_{\xi,\beta}$ that consists of sequences f of continuous functions $f_s(x_1,\ldots,x_s)$ defined on the s-particle phase space, i.e.,

$$f = \big(f_1(x_1),\ldots,f_s(x_1,\ldots,x_s),\ldots\big), \tag{4.2.9}$$

with componentwise linear operations and the following norm:

$$\|f\| = \sup_{s\ge 1}\frac{1}{\xi^s}\sup_{(x)_s}|f_s(x_1,\ldots,x_s)|e^{\beta\sum_{i=1}^{s}\frac{p_i^2}{2}}, \tag{4.2.10}$$

where $\xi > 0$ and $\beta > 0$ are fixed numbers, and mass $m = 1$. Note that equilibrium correlation functions with stable regular potential belong to the space $E_{\xi,\beta}$ with inverse temperature β [PGM3, Ru5].

It follows from (4.2.10) that

$$|f_s(x_1,\dots,x_s)| < \xi^s e^{-\frac{\beta}{2}\sum_{i=1}^{s} p_i^2}\|f\|. \tag{4.2.11}$$

We suppose that initial data $F(0) \in E_{\xi,\beta}$ and prove that series (4.2.8) is convergent uniformly with respect to $(x)_s$ on compact sets and with respect to time on the finite interval $0 \le t \le t_0$. The number t_0 will be determined later.

4.2.2 Convergence of series (4.2.8) in the space $E_{\xi,\beta}$

Consider an arbitrary n-th term in (4.2.8). Using δ-functions, we perform integration with respect to $q_{s+1},\dots,q_{s+n}$. As a result, we obtain

$$\begin{aligned}
F_s^{(n)}(t,(x)_s) &= \int_0^t dt_1 \dots \int_0^{t_{n-1}} dt_n S_s(-t,(x)_s)S_s(t_1,(x)_s)\\
&\quad\times \sum_{i=1}^{s}\int dp_{s+1}\int_{S_+^2} d\eta_{is+1}\eta_{is+1}\cdot(p_i - p_{s+1})\\
&\quad\times \big[S_{s+1}(-t_1,(x)^*_{s+1}) - S_{s+1}(-t_1,(x)_{s+1})\big]\big|_{q_{s+1}=q_i}\cdots\\
&\quad\times S_{s+n-1}(t_{n-1},(x)_{s+n-1})\sum_{i=1}^{s+n-1}\int dp_{s+n}\int_{S_+^2} d\eta_{is+n}\\
&\quad\times \eta_{is+n}\cdot(p_i - p_{s+n})\big[S_{s+n}(-t_n,(x)^*_{s+n}) - S_{s+n}(-t_n,(x)_{s+n})\big]\\
&\quad\times F_{s+n}(0,(x)_{s+n})\big|_{q_{s+n}=q_i}.
\end{aligned} \tag{4.2.12}$$

Note that $q_{s+1},\dots,q_{s+n}$ are expressed through $q_1,\dots,q_s$ according to δ-functions in (4.2.8).

Let us estimate the expression

$$\begin{aligned}
I_{s+n} = \Bigg|&\sum_{i=1}^{s+n-1}\int dp_{s+n}\int_{S_+^2} d\eta_{is+n}\eta_{is+n}\cdot(p_i - p_{s+n})\\
&\times\big[S_{s+n}(-t_n,(x)^*_{s+n}) - S_{s+n}(-t_n,(x)_{s+n})\big]F_{s+n}(0,(x)_{s+n})\Bigg|.
\end{aligned}$$

Using estimate (4.2.11) with $s+n$ instead of s, we get

$$I_{s+n} \le 4\pi\xi^{s+n}\int dp_{s+n}\sum_{i=1}^{s+n-1}(|p_i| + p_{s+n})e^{-\frac{\beta}{2}\sum_{i=1}^{s+n} p_i^2}\|f\|,$$

where we have used the fact that, according to the law of conservation of energy,

$$\left| \left[S_{s+n}(-t,(x)^*) - S_{s+n}(-t,(x)_{s+n}) \right] e^{-\frac{\beta}{2}\sum_{i=1}^{s+n} p_i^2} \right| \leq 2e^{-\frac{\beta}{2}\sum_{i=1}^{s+n} p_i^2}.$$

We use the equality

$$e^{-\frac{\beta}{2}\sum_{i=1}^{s+n} p_i^2} = e^{-\frac{\beta}{4}p_{s+n}^2} e^{-\frac{\beta}{4}p_{s+n}^2} e^{-\frac{\beta'}{2}\sum_{i=1}^{s+n-1} p_i^2} e^{-\frac{n(\beta-\beta')}{2n}\sum_{i=1}^{s+n-1} p_i^2}$$

with some fixed $\beta' < \beta$.

Using these inequalities and the last equality, one obtains

$$\begin{aligned} I_{s+n} \leq 4\pi \xi^{s+n} \sum_{i=1}^{s+n-1} \int dp_{s+n} \left(\sum_{i=1}^{s+n} |p_i| + (s+n-1)|p_{s+n}| \right) \\ \times e^{-\frac{\beta}{4}p_{s+n}^2} e^{-\frac{\beta}{4}p_{s+n}^2} e^{-\frac{\beta'}{2}\sum_{i=1}^{s+n-1} p_i^2} e^{-\frac{(\beta-\beta')}{2n}\sum_{i=1}^{s+n-1} p_i^2} \\ \times e^{-(n-1)\frac{(\beta-\beta')}{2n}\sum_{i=1}^{s+n-1} p_i^2} \|f\|. \end{aligned} \tag{4.2.13}$$

We now use the elementary inequalities

$$e^{-\frac{\beta}{4}p_{s+n}^2}(s+n-1)|p_{s+n}| \leq (s+n-1)C_1(\beta)^{-\frac{1}{2}}, \quad C_1 = 2^{\frac{1}{2}}e^{-\frac{1}{2}},$$

$$\sum_{i=1}^{s+n-1} |p_i| e^{-\frac{(\beta-\beta')}{n}\sum_{i=1}^{s+n-1} p_i^2} \leq C_2(s+n-1)^{\frac{1}{2}}(\beta-\beta')^{-\frac{1}{2}}n^{\frac{1}{2}}, \quad C_2 = (2e)^{-\frac{1}{2}},$$

$$\int e^{-\frac{\beta}{4}p_{s+n}^2} dp_{s+n} = (4\pi)^{\frac{3}{2}}(\beta)^{-\frac{3}{2}} \leq (4\pi)^{\frac{3}{2}}(\beta')^{-\frac{3}{2}}.$$

Note that, in the second inequality, we have used the fact that the maximum of the left-hand side is attained at $|p_1| = |p_2| = \ldots = |p_{s+n-1}|$.

Finally, one obtains

$$\begin{aligned} I_{s+n} &\leq 4\pi \xi^{s+n}(s+n-1)(4\pi)^{\frac{3}{2}}(\beta')^{-\frac{3}{2}}\left[C_1(\beta')^{-\frac{1}{2}} \right. \\ &\quad \left. + C_2(s+n-1)^{-\frac{1}{2}}n^{\frac{1}{2}}(\beta-\beta')^{-\frac{1}{2}}\right] \\ &\quad \times e^{-\frac{\beta'}{2}\sum_{i=1}^{s+n-1} p_i^2} e^{-(n-1)\frac{(\beta-\beta')}{2n}\sum_{i=1}^{s+n-1} p_i^2} \|f\| \\ &\leq \xi^{s+n}(s+n)A(\beta,\beta')e^{-\frac{\beta'}{2}\sum_{i=1}^{s+n-1} p_i^2} e^{-(n-1)\frac{(\beta-\beta')}{2n}\sum_{i=1}^{s+n-1} p_i^2} \|f\|, \end{aligned} \tag{4.2.14}$$

$$A(\beta,\beta') = (4\pi)^{\frac{5}{2}}(\beta')^{-\frac{3}{2}}\left[C_1(\beta')^{-\frac{1}{2}} + C_2(\beta-\beta')^{-\frac{1}{2}}\right].$$

Using analogous estimates $n-1$ times with $n-i$ instead of $n-1$ and $\frac{\beta'}{4}$ instead of $\frac{\beta}{4}$, $1 \le i \le n-1$, in (4.2.13) and integrating with respect to p_{s+n-i}, one obtains the following estimate for $F_s^{(n)}(t,(x)_s)$:

$$|F_s^{(n)}(t,(x)_s)| \le \xi^s \frac{t^n}{n!}(A(\beta,\beta')\xi)^n \prod_{i=1}^{n}(s+n-i)e^{-\frac{\beta'}{2}\sum_{i=1}^{s} p_i^2}\|f\|. \qquad (4.2.15)$$

In the proof of (4.2.15), we have used the identity

$$\int_0^t dt_1 \int_0^{t_1} dt_2 \ldots \int_0^{t_{n-1}} dt_n = \frac{t^n}{n!}.$$

From (4.2.15), we obtain the following estimate for series (4.2.8):

$$|F_s(t,(x)_s)| \le e^{-\frac{\beta'}{2}\sum_{i=1}^{s} p_i^2}\xi^s \sum_{n=0}^{\infty}\frac{(s+n)^n}{n!}t^n A(\beta,\beta')^n \xi^n \|f\|$$

$$\le \xi^s e^{-\frac{\beta'}{2}\sum_{i=1}^{s} p_i^2}\sum_{n=0}^{\infty} n^n e^s (2\pi)^{-\frac{1}{2}} e^n n^{-n-\frac{1}{2}} t^n A(\beta,\beta')^n \xi^n \|f\|; \quad (4.2.16)$$

this series is convergent for $t < t_0$, where

$$t_0 = e^{-1}\xi^{-1}A(\beta,\beta')^{-1}.$$

We summarize the results obtained above in the following theorem:

Theorem 4.1. *For initial correlation functions $F(0) \in E_{\xi,\beta}$, series (4.2.8) are uniformly convergent with respect to $(x)_s$, $s \ge 1$, on compact sets and time $t < t_0$. These series are a mild solution of the stochastic Boltzmann hierarchy.*

Note that we have used some modification of the proof of the existence of a the mild solution of the BBGKY hierarchy for a system of hard spheres proposed by Lanford [Lan3] and Cercignani, Illner and Pulvirenti [CIP].

4.2.3 One auxiliary lemma

Lemma 4.1. *The following equality holds:*

$$\begin{aligned} S_s(-t)\left[q_1^2 + \ldots + q_s^2\right] &= S_s(-t,x_1,\ldots,x_s)\left[q_1^2 + \ldots + q_s^2\right] \\ &= Q_1^2(-t,x_1,\ldots,x_s) + \ldots + Q_s^2(-t,x_1,\ldots,x_s) \\ &= (q_1 - p_1 t)^2 + \ldots + (q_s - p_s t)^2, \quad s \ge 2, \quad t > 0. \end{aligned} \qquad (4.2.17)$$

Proof. According to the stochastic dynamics, particles move freely until their positions coincide. Then they elastically collide and continue to move freely. Only pair collisions are considered. Therefore, it is sufficient to prove the equality for two particles.

Let the initial positions and momenta be (q_1, p_1), (q_2, p_2) and let $-t_1$ be the time of collision, i.e., $q_1 - p_1 t_1 = q_2 - p_2 t_1$. Then, for $t > t_1$, the positions and momenta are

$$(q_1 - p_1 t_1 - p_1^*(t - t_1), p_1^*), \quad (q_2 - p_2 t_1 - p_2^*(t - t_1), p_2^*)$$

$$p_1^* = p_1 - \eta\eta\cdot(p_1 - p_2), \quad p_2^* = p_2 + \eta\eta\cdot(p_1 - p_2).$$

Consider the expression

$$\begin{aligned}
&Q_1^2(-t)^2 + Q_2^2(-t)\\
&= \big(q_1 - p_1 t_1 - p_1^*(t - t_1)\big)^2 + \big(q_2 - p_2 t_1 - p_2^*(t - t_1)\big)^2\\
&= \big[q_1 - p_1 t + \eta\eta\cdot(p_1 - p_2)(t - t_1)\big]^2 + \big[q_2 - p_2 t - \eta\eta\cdot(p_1 - p_2)(t - t_1)\big]^2\\
&= (q_1 - p_1 t)^2 + (q_2 - p_2 t)^2 + \Big\{2\big[\eta\cdot(p_1 - p_2)(t - t_1)\big]^2\\
&\qquad + 2\big[(q_1 - p_1 t - q_2 + p_2 t)\cdot\eta\eta\cdot(p_1 - p_2)(t - t_1)\big]\Big\}.
\end{aligned}$$

To prove equality (4.2.17), we show that

$$2\big[\eta\cdot(p_1 - p_2)(t - t_1)\big]^2 + 2\big[(q_1 - p_1 t - q_2 + p_2 t)\cdot\eta\eta\cdot(p_1 - p_2)(t - t_1)\big] = 0.$$

Indeed, we have

$$\begin{aligned}
&2\big[\eta\cdot(p_1 - p_2)(t - t_1)\big]^2 + 2\big[(q_1 - p_1 t - q_2 + p_2 t)\cdot\eta\eta\cdot(p_1 - p_2)(t - t_1)\big]\\
&\quad = 2\big[\eta\cdot(p_1 - p_2)(t - t_1)\big]^2 + 2\big[(q_1 - p_1 t_1 - p_1(t - t_1) - q_2\\
&\qquad + p_2 t_1 + p_2(t - t_1))\big]\cdot\eta\eta\cdot(p_1 - p_2)(t - t_1)\\
&\quad = 2\big[\eta\cdot(p_1 - p_2)(t - t_1)\big]^2 - 2\big[\eta\cdot(p_1 - p_2)(t - t_1)\big]^2 = 0.
\end{aligned}$$

We have used the fact that $q_1 - p_1 t_1 = q_2 - p_2 t_1$.

Now consider the general case of s-particles, denote the times of collision by $t_1, \ldots, t_n$, and represent

$$\begin{aligned}
&S_s(-t)[q_1^2 + \ldots + q_s^2]\\
&\quad = S_s(-t + t_n)S_s(-t_n + t_{n_1})S_s(-t_2 + t_1)S_s(-t_1)[q_1^2 + \ldots + q_s^2],
\end{aligned}$$

using the group property of the operator $S_s(-t)$. Since we consider only pair collisions, for every $S_s(-t_i + t_{i-1})$ equality (4.2.17) holds and equality (4.2.17) is also true for $S_s(-t)$.

4.2.4 Convergence of series (4.2.8) in the space $\tilde{E}_{\xi,\beta}$

Consider series (4.2.8) for initial data $F(0)$ that belong to the space $\tilde{E}_{\xi,\beta}$ that consists of sequences $f = (f_1(x_1), \ldots, f_s(x_1, \ldots, x_s), \ldots)$ with the norm

$$\|f\| = \sup_{s\geq 1} \frac{1}{\xi^s} \sup_{(x)_s} |f_s(x_1, \ldots, x_s)| e^{\frac{\beta}{2}\left(\sum_{i=1}^{s}(q_i^2+p_i^2)\right)}, \quad \beta > 0. \tag{4.2.18}$$

This means that the space $\tilde{E}_{\xi,\beta}$ consists of sequences of functions exponentially decreasing with respect to the squared position and momenta [IlP1, IlP2].

In this subsection, it would be useful to represent series (4.2.7)–(4.2.8) in the following identical form:

$$F(t) = \sum_{n=0}^{\infty} \int_0^t dt_1 \ldots \int_0^{t_{n-1}} dt_n S(-t + t_1) A S(-t_1 + t_2) \ldots$$
$$\times S(-t_{n-1} + t_n) A S(-t_n) F(0), \tag{4.2.7$'$}$$

$$F_s(t, (x)_s) = \sum_{n=0}^{\infty} \int_0^t dt_1 \ldots \int_0^{t_{n-1}} dt_n S_s(-t + t_1, (x)_s) \sum_{i=1}^{s} \int dp_{s+1}$$
$$\times \int_{S^2} d\eta_{is+1} \eta_{is+1} \cdot (p_i - p_{s+1}) S_{s+1}(-t_1 + t_2, (x)_{s+1})|_{q_{s+1}=q_i} \cdots$$
$$\times S_{s+n-1}(-t_{n-1} + t_n, (x)_{s+n-1}) \sum_{i=1}^{s+n-1} \int dp_{s+n} \int_{S^2} d\eta_{is+n}$$
$$\times \eta_{is+n} \cdot (p_i - p_{s+n}) \big[S_{s+n}(-t_n, (x)_{s+n}) F_{s+n}(0, (x)_{s+n})\big]|_{q_{s+n}=q_i}$$
$$= \sum_{n=0}^{\infty} F_s^{(n)}(t, (x)_s). \tag{4.2.8$'$}$$

Note that, in (4.2.8), we integrate with respect to η over the entire sphere S^2, and it is assumed that, for $\eta \in S_+^2$, one should replace $(x)_{s+1}, \ldots, (x)_{s+n}$ by the corresponding $(x)_{s+1}^*, \ldots, (x)_{s+n}^*$; for $\eta \in S_-^2$ one has $(x)_{s+1}, \ldots, (x)_{s+n}$.

We begin with the estimation of the expression

$$I_{s+n} = \Bigg| \sum_{i=1}^{s+n-1} \int dp_{s+n} \int_{S^2} d\eta_{is+n} \eta_{is+n} \cdot (p_i - p_{s+n})$$
$$\times \big[S_{s+n}(-t_n, (x)_{s+n}) F_{s+n}(0, (x)_{s+n})\big]|_{q_{s+n}=q_i} \Bigg|$$

$$\leq 4\pi\xi^{s+n}\int dp_{s+n}\sum_{i=1}^{s+n-1}\left(|p_i|+|p_{s+n}|\right)$$
$$\times e^{-\frac{\beta}{2}\sum\limits_{i=1}^{s+n-1}(q_i-p_it_n)^2-\frac{\beta}{2}\sum\limits_{i=1}^{s+n-1}p_i^2-\frac{\beta}{2}(q_i-p_{s+n}t_n)^2-\frac{\beta}{2}p_{s+n}^2}\|f\|. \quad (4.2.19)$$

To obtain estimate (4.2.19), we have used Lemma 4.1 and the law of conservation of kinetic energy.

In (4.2.19), we use equalities

$$e^{-\frac{\beta}{2}\sum\limits_{i=1}^{s+n-1}p_i^2}=e^{-\frac{\beta'}{2}\sum\limits_{i=1}^{s+n-1}p_i^2}e^{-\frac{\beta-\beta'}{2n}n\sum\limits_{i=1}^{s+n-1}p_i^2},$$
$$e^{-\frac{\beta}{2}p_{s+n}^2}=e^{-\frac{\beta}{4}p_{s+n}^2}e^{-\frac{\beta}{4}p_{s+n}^2}\leq e^{-\frac{\beta'}{4}p_{s+n}^2}e^{-\frac{\beta'}{4}p_{s+n}^2} \quad (4.2.20)$$

with some fixed $\beta'<\beta$.

Using estimate (4.2.14) from Subsection 4.2.2, equality (4.2.20), and the estimate

$$\sup_{q_i}\int e^{-\frac{\beta'}{4}p_{s+n}^2-\frac{\beta}{2}(q_i-p_{s+n}t_n)^2}dp_{s+n}$$
$$\leq\sup_{q_i}\int e^{-\frac{\beta'}{4}p_{s+n}^2-\frac{\beta'}{4}(q_i-p_{s+n}t_n)^2}dp_{s+n}\leq\frac{C}{(1+t_n^2)^{\frac{3}{2}}}, \quad (4.2.21)$$

which will be proved later, one obtains the following relation from (4.2.19):

$$I_{s+n}\leq\xi^{s+n}e^{-\frac{\beta}{2}\sum\limits_{i=1}^{s+n-1}(q_i-p_it_n)^2-\frac{\beta'}{2}-\frac{(\beta-\beta')}{2n}(n-1)\sum\limits_{i=1}^{s+n-1}p_i^2}$$
$$\times\sum_{i=1}^{s+n-1}p_i^2\frac{C}{(1+t_n^2)^{\frac{3}{2}}}(s+n)A(\beta,\beta')\|f\|. \quad (4.2.22)$$

Using Lemma 4.1 again, one obtains the following relation from (4.2.22):

$$S_{s+n-1}(-t_{n-1}+t_n,(x)_{s+n-1})I_{s+n}$$
$$=4\pi\xi^{s+n}e^{-\frac{\beta}{2}\sum\limits_{i=1}^{s+n-1}(q_i-p_it_{n-1})^2-\frac{\beta'}{2}\sum\limits_{i=1}^{s+n-1}p_i^2-\frac{(\beta-\beta')}{2n}(n-1)\sum\limits_{i=1}^{s+n-1}p_i^2}$$
$$\times\frac{C}{(1+t_n^2)^{\frac{3}{2}}}(s+n)A(\beta,\beta')\|f\|. \quad (4.2.23)$$

Using analogous estimates $n-1$ times with $n-i$ instead of $n-1$, $1\leq i\leq n-1$, in (4.2.23) and integrating with respect to p_{s+n-i}, one obtains the following estimate for $F_s^{(n)}(t,(x)_s)$ (4.2.12):

$$
\begin{aligned}
|F_s^{(n)}(t,(x)_s)| \\
&\le 4\pi\xi^s\big[A(\beta,\beta')\xi\big]^n (s+n)^n e^{-\frac{\beta'}{2}\sum\limits_{i=1}^{s} p_i^2} \\
&\quad \times e^{-\frac{\beta}{2}\sum\limits_{i=1}^{s}(q_i-p_it)^2} \int_0^t dt_1 \dots \int_0^{t_{n-1}} dt_n \prod_{i=1}^{n} \frac{C}{(1+t_i^2)^{\frac{3}{2}}} \|f\| \\
&\le \frac{1}{n!}\xi^s\big[A(\beta,\beta')\xi\big]^n (s+n)^n C_1^n e^{-\frac{\beta'}{2}\sum\limits_{i=1}^{s} p_i^2} e^{-\frac{\beta}{2}\sum\limits_{i=1}^{s}(q_i-p_it)^2} \|f\|, \qquad (4.2.24)
\end{aligned}
$$

where

$$C_1 = C\int_0^\infty \frac{1}{(1+t^3)}dt.$$

Substituting (4.2.24) in (4.2.8′), we get

$$
\begin{aligned}
|F_s(t,(x)_s)| \\
&\le \sum_{n=0}^{\infty} \frac{1}{n!}\xi^s\big[A(\beta,\beta')\xi C_1\big]^n (s+n)^n e^{-\frac{\beta'}{2}\sum\limits_{i=1}^{s} p_i^2} e^{-\frac{\beta}{2}\sum\limits_{i=1}^{s}(q_i-p_it)^2} \|f\|. \qquad (4.2.25)
\end{aligned}
$$

Series (4.2.25) with arbitrary s is convergent for sufficiently small ξ because, for large n, we have

$$n! \simeq \sqrt{2\pi n}\Big(\frac{n}{e}\Big)^n e^{-\frac{\Theta(n)}{12}}, \qquad 0 < \Theta(n) < 1$$

$$(s+n)^n = n^n\Big(1+\frac{s}{n}\Big)^n \le n^n e^s.$$

Series (4.2.25) do not depend on t, and, therefore, they are uniformly convergent with respect to $(x)_s$ on compact sets for arbitrary $t > 0$.

We summarize the results obtained above in the following theorem:

Theorem 4.2. *For initial correlation functions $F(0) \in \tilde{E}_{\xi,\beta}$ and sufficiently small ξ, series (4.2.8)–(4.2.8′) are uniformly convergent on compact sets with respect to $(x)_s$ for arbitrary $t > 0$.*

As is known, series (4.2.8)–(4.2.8′) are a mild solution of the stochastic hierarchy.

Remark 4.1. Let us prove estimate (4.2.21). We have

$$
\begin{aligned}
\sup_q \int e^{(-\frac{\beta'}{4}p^2 - \frac{\beta'}{4}(q-pt)^2)}dp \\
&= \sup_q \int e^{(-\frac{\beta'}{4}(1+t^2)p^2 + \frac{\beta'}{4}2tqp - \frac{\beta'}{4}q^2)}dp
\end{aligned}
$$

$$= \sup_q \int e^{(-\frac{\beta'}{4}(1+t^2)p^2 - \frac{\beta'}{4(1+t^2)}q^2)} dp$$

$$\leq \int e^{-\frac{\beta'}{4}(1+t^2)p^2} dp = \Big(\frac{4\pi}{\beta'(1+t^2)}\Big)^{\frac{3}{2}} = C\frac{1}{(1+t^2)^{\frac{3}{2}}}.$$

Note that our proof of Theorem 4.2 is a modification of the proof of a theorem of Illner and Pulvirenti [IlP1, IlP2] for a system of hard spheres.

4.3 Chaos property of solutions of the stochastic hierarchy

4.3.1 New representation of the series of iterations

Consider again the series of iterations (4.2.8) and perform integration with respect to $q_{s+1}, \dots, q_{s+n}$, using δ-functions. We get

$$\begin{aligned} F_s(t,(x)_s) &= \sum_{n=0}^{\infty} \int_0^t dt_1 \dots \int_0^{t_{n-1}} dt_n S_s(-t,(x)_s) S_s(t_1,(x)_s) \\ &\times \sum_{i=1}^{s} \int dp_{s+1} \int_{S_+^2} d\eta_{is+1} \eta_{is+1} \cdot (p_i - p_{s+1}) \\ &\times \big[S_{s+1}(-t_1,(x)_{s+1}^*) - S_{s+1}(-t_1,(x)_{s+1})\big]|_{q_{s+1}=q_i} \cdots \\ &\times S_{s+n-1}(t_{n-1},(x)_{s+n-1}) \sum_{i=1}^{s+n-1} \int dp_{s+n} \int_{S_+^2} d\eta_{is+n} \\ &\times \eta_{is+n} \cdot (p_i - p_{s+n}) \big[S_{s+n}(-t_n,(x)_{s+n}^*) - S_{s+n}(-t_n,(x)_{s+n})\big] \\ &\times F_{s+n}(0,(x)_{s+n})|_{q_{s+n}=q_i}. \end{aligned} \tag{4.3.1}$$

Recall that the operators of evolution of the stochastic dynamics differ from those of free dynamics only on the hypersurfaces V_{ij}, $i < j$, where the vectors of difference of positions are parallel to the vectors of difference of momenta, i.e., $q_i - q_j = \tau(p_i - p_j)$, $0 \leq \tau \leq t$.

Consider first the phase points $(x)_{s+n}$. Then, according to (4.3.1), each of $q_{s+1}, \dots, q_{s+n}$ is equal to one of $q_1, \dots, q_s$. The corresponding hypersurfaces V_{is+j} with at least one momenta from the set $p_{s+1}, \dots, p_{s+n}$ and fixed $q_1, \dots, q_s$ are hypersurfaces of lower dimension with respect to p_{s+j}, $1 \leq j \leq n$. Integrals with respect to $p_{s+1}, \dots, p_{s+n}$ are Lebesgue ones, and, therefore, all hypersurfaces V_{is+j} can be neglected. This means that we can consider the evolution of all particles with number $s+1, \dots, s+n$ as free one, i.e.,

$$S_{s+1}(-t_1, x_1, \dots, x_s, x_{s+1}) = S_{s+1}(-t, x_1, \dots, x_s) S_1^0(-t, x_{s+1}), \dots,$$

$$\begin{aligned} &S_{s+n}(-t_n, x_1, \dots, x_s, x_{s+1}, \dots, x_{s+n}) \\ &\quad = S_{s+n}(-t_n, x_1, \dots, x_s) S_s^0(-t_n, x_{s+1}, \dots, x_{s+n}), \qquad (4.3.2) \\ &S_{s+1}(-t_1, x_1, \dots, x_i^*, \dots, x_s, x_{s+1}^*) \\ &\quad = S_s(-t_1, x_1, \dots, x_i^*, \dots, x_s) S_1^0(-t_1, x_{s+1}^*), \dots, \\ &S_{s+n}(-t_n, x_1, \dots, x_i^*, \dots, x_{s+n-1}, x_{s+n}^*) \\ &\quad = S_s(-t_n, x_1, \dots, x_i^*, \dots, x_s) S_n^0(-t_n, x_{s+1}, \dots, x_{s+n}^*). \end{aligned}$$

Here $S_i^0(-t)$ means the operator of evolution of free point particles. The same formula (4.3.2) holds for $S_i(t)$, with arbitrary i.

Taking (4.3.2) into account, we represent (4.3.1) identically as follows:

$$\begin{aligned} F_s(t, (x)_s) = &\sum_{n=0}^{\infty} \int_0^t dt_1 \dots \int_0^{t_{n-1}} dt_n S_s(-t, (x)_s) S_s(t_1, (x)_s) \\ &\times \sum_{i=1}^{s} \int dp_{s+1} \int_{S_+^2} d\eta_{is+1} \eta_{is+1} \cdot (p_i - p_{s+1}) [S_s(-t_1, (x)_s^*) \\ &\times S_1^0(-t_1, x_{s+1}^*) - S_s(-t_1, (x)_s) S_1^0(-t_1, x_{s+1})]|_{q_{s+1}=q_i} \cdots \\ &\times S_s(t_{n-1}, (x)_s) S_{n-1}^0(t_{n-1}, (x)_{s+n-1}^s) \\ &\times \sum_{i=1}^{s+n-1} \int dp_{s+n} \int_{S_+^2} d\eta_{is+n} \eta_{is+n} \cdot (p_i - p_{s+n}) \\ &\times \big[S_s(-t_n, (x)_s^*) S_n^0(-t_n, (x)_{s+n}^{*n}) - S_s(-t_n, (x)_s) S_n^0(-t_n, (x)_{s+n}^n)\big] \\ &\times F_{s+n}(0, (x)_{s+n})|_{q_{s+n}=q_i}, \quad (x)_{s+i}^i = (x_{s+1}, \dots, s_{s+i}). \qquad (4.3.3) \end{aligned}$$

It follows from (4.3.3) that $F_s(t, (x)_s)$ depends only on random vectors η of the operators $S_s(\pm t, (x)_s)$. Note that $S_1(\pm t) = S_1^0(\pm t)$, and, therefore, for the one-particle correlation function one has

$$\begin{aligned} F_1(t, x_1) = &\sum_{n=0}^{\infty} \int_0^t dt_1 \dots \int_0^{t_{n-1}} dt_n S_1^0(-t, x_1) S_1^0(t_1, x_1) \\ &\times \int dp_2 \int_{S_+^2} d\eta_{12} \eta_{12} \cdot (p_1 - p_2) \\ &\times \big[S_2^0(-t_1, (x)_2^*) - S_2^0(-t_1, (x)_2)\big]|_{q_1=q_2} \dots S_n^0(t_{n-1}, (x)_n) \end{aligned}$$

$$\times \sum_{i=1}^{n} \int dp_{1+n} \int_{S_+^2} d\eta_{in+1} \eta_{in+1} \cdot (p_i - p_{n+1}) \big[S_{n+1}^0(-t_n, (x)_{n+1}^*)$$

$$- S_{n+1}^0(-t_n, (x)_{n+1}) \big] F_{n+1}(0, (x)_{n+1}) |_{q_{s+n}=q_i}. \tag{4.3.4}$$

4.3.2 Chaos property

Suppose that the initial correlation functions possess the chaos property

$$F_s(0, x_1, \ldots, x_s) = F_1(0, x_1) \ldots F_1(0, x_s), \tag{4.3.5}$$

i.e., s-particle correlation functions are equal to the product of one-particle correlation functions. Consider the ordinary Boltzmann hierarchy

$$\frac{\partial F_s(t, (x)_s)}{\partial t} = - \sum_{i=1}^{s} p_i \cdot \frac{\partial}{\partial q_i} F_s(t, (x)_s) + \sum_{i=1}^{s} \int dp_{s+1} \int_{S_+^2} d\eta$$

$$\times \eta \cdot (p_i - p_{s+1}) \big[F_{s+1}(t, (x)_{s+1}^*) - F_{s+1}(t, (x)_{s+1}) \big] |_{q_{s+1}=q_i} \tag{4.3.6}$$

with initial data $F_s(0, (x)_s)$, $s \geq 1$.

We stress that, in the ordinary Boltzmann hierarchy, the boundary conditions of the stochastic Boltzmann hierarchy are absent and the operator $-\sum_{i=1}^{s} p_i \cdot \frac{\partial}{\partial q_i}$ is the infinitesimal operator of the evolution operator of free dynamics, i.e., $S_s^0(-t, (x)_s)$.

Solutions of hierarchy (4.3.6) can be represented by the following series of iterations:

$$F_s(t, (x)_s) = \sum_{n=0}^{\infty} \int_0^t dt_1 \ldots \int_0^{t_{n-1}} dt_n S_s^0(-t, (x)_s) S_s^0(t_1, (x)_s)$$

$$\times \sum_{i=1}^{s} \int dp_{s+1} \int_{S_+^2} d\eta_{is+1} \eta_{is+1} \cdot (p_i - p_{s+1})$$

$$\times \big[S_{s+1}^0(-t_1, (x)_{s+1}^*) - S_{s+1}^0(-t_1, (x)_{s+1}) \big] |_{q_{s+1}=q_i}$$

$$\times S_{s+n-1}^0(t_{n-1}, (x)_{s+n-1}) \sum_{i=1}^{s+n-1} \int dp_{s+n} \int_{S_+^2} d\eta_{is+n}$$

$$\times \eta_{is+n} \cdot (p_i - p_{s+n}) \big[S_{s+n}^0(-t_n, (x)_{s+n}^*) - S_{s+n}^0(-t_n, (x)_{s+n}) \big]$$

$$\times F_{s+n}(0, (x)_{s+n}) |_{q_{s+n}=q_i}, \quad s \geq 1. \tag{4.3.7}$$

It is obvious that the theorems from the previous section concerning the convergence of series of iterations of the stochastic Boltzmann hierarchy are also true for the series of iterations (4.3.7) of the ordinary Boltzmann hierarchy.

Comparing (4.3.7) with (4.3.4), one concludes that the one-particle correlation function (4.3.7) of the ordinary Boltzmann hierarchy (4.3.6) coincides with the one-particle correlation function (4.3.4) of the stochastic Boltzmann hierarchy (4.2.4).

The s-particle correlation functions of the stochastic Boltzmann hierarchy (4.3.3) with $s > 1$ do not coincide with those of the ordinary Boltzmann hierarchy (4.3.7) because the operators of evolution of the stochastic dynamics $S_s(\pm t, (x)_s)$ do not coincide with the evolution operators of free dynamics $S_s^0(-t, (x)_s)$.

The operators $S_s(\pm t, (x)_s)$ and $S_s^0(\pm t, (x)_s)$ differ on the hypersurfaces V_{ij}, $i, j \in \{1, \dots, s\}$, where the stochastic particles interact and completely coincide outside all these hypersurfaces V_{ij}. This means that solutions of the stochastic Boltzmann hierarchy coincide with solutions of the ordinary Boltzmann hierarchy outside all hypersurfaces V_{ij}, $i, j \in \{1, \dots, s\}$, $s \geq 2$. The one-particle correlation functions of both hierarchies coincide in the entire phase space.

As is known, solutions of the ordinary Boltzmann hierarchy possess the chaos property if initial correlation functions have the chaos property (4.3.5). Therefore, solutions (4.3.7) of the ordinary Boltzmann hierarchy (4.3.6) are equal to the product of one-particle correlation functions

$$F_s(t, x_1, \dots, x_s) = F_1(t, x_1) \dots F_1(t, x_s), \tag{4.3.8}$$

where $F_1(t, x_1)$ is a solution of the nonlinear Boltzmann equation

$$\begin{aligned} \frac{\partial F_1(t, x_1)}{\partial t} &= -p_1 \frac{\partial}{\partial q_1} F_1(t, x_1) + \int dp_2 \int_{S_+^2} d\eta \eta \cdot (p_1 - p_2) \\ &\quad \times \left[F_1(t, x_1^*) F_1(t, x_2^*) - F_1(t, x_1) F_1(t, x_2) \right] \big|_{q_2 = q_1}. \end{aligned} \tag{4.3.9}$$

Solutions of the stochastic Boltzmann hierarchy $F_s(t, (x)_s)$ coincide outside all hypersurfaces V_{ij}, $i, j \in \{1, \dots, s\}$, with the solutions of the ordinary Boltzmann hierarchy and have there the chaos property (4.3.8).

4.3.3 Justification of the thermodynamic limit

Consider the stochastic Boltzmann hierarchy for N particles (see Chapter 3, hierarchy (3.3.7))

$$\begin{aligned} &\frac{\partial F_s^{(N)}(t, x_1, \dots, x_s)}{\partial t} \\ &\quad = -\sum_{i=1}^{s} p_i \cdot \frac{\partial}{\partial q_i} F_s^{(N)}(t, x_1, \dots, x_s) \\ &\qquad + \sum_{i=1}^{s} \int dp_{s+1} \int_{S_+^2} d\eta_{is+1} \eta_{is+1} \cdot (p_i - p_{s+1}) \delta(q_i - q_{s+1}) \end{aligned}$$

$$\times \big[F_{s+1}^{(N)}(t, x_1, \ldots, x_i^*, \ldots, x_s, x_{s+1}^*)$$
$$- F_{s+1}^{(N)}(t, x_1, \ldots, x_i, \ldots, x_s, x_{s+1})\big], \quad 1 \le s \le N, \qquad (4.3.10)$$

with initial data $F_s^{(N)}(t, x_1, \ldots, x_s)|_{t=0} = F_s^{(N)}(0, x_1, \ldots, x_s)$ and a known boundary condition. Suppose that the sequence $F_s^{(N)}(0, (x)_s)$, $1 \le s \le N$, belongs to the space $E_{\xi,\beta}$ or $\tilde{E}_{\xi,\beta}$.

The solution of the Cauchy problem (4.3.10) can be represented by the series of iterations (4.2.8) with $F_{s+n}^{(N)}(0, (x)_s)$, $s + n \le N$, instead of $F_{s+n}(0, (x)_s)$, and this series is uniformly convergent with respect to $(x)_s$ on compact sets on a finite time interval for $E_{\xi,\beta}$ and globally in time for $\tilde{E}_{\xi,\beta}$. Note that the series is also uniformly convergent with respect to N.

We consider series (4.2.8) for $F_s(t, (x)_s)$ with initial data such that the first N initial correlation functions coincide with $F_s^{(N)}(0, (x)_s)$, i.e.,

$$F_s(0, (x)_s) = F_s^{(N)}(0, (x)_s), \quad 1 \le s \le N.$$

Then it is obvious that

$$\lim_{N\to\infty} F_s^{(N)}(t, (x)_s) = F_s(t, (x)_s) \qquad (4.3.11)$$

uniformly with respect to $(x)_s$ on compact sets for both $E_{\xi,\beta}$ and $\tilde{E}_{\xi,\beta}$. In this sense, one obtains systems with infinite average number of particles from a system with finite number of particles.

4.3.4 Connection between the correlation functions

We have solutions of the stochastic Boltzmann hierarchy (4.2.4) for the correlation functions $F_s(t, (x)_s)$, $s \ge 1$, considered as usual functions determined numerically by series (4.2.7)–(4.2.8) or (4.3.3). In hierarchy (4.2.4) we have the infinitesimal operator

$$\mathcal{H}_s = -\sum_{i=1}^{s} p_i \cdot \frac{\partial}{\partial q_i}$$

with known boundary condition at $q_i = q_j$, $i, j \in \{1, \ldots, s\}$.

We have also derived a hierarchy for the sequences $\tilde{F}(t) = (\tilde{F}_s(t, (x)_s))_{s=1}^{\infty}$, $\bar{F}(t) = (\bar{F}_s(t, (x)_s))_{s=1}^{\infty}$ of correlation functions $\tilde{F}_s(t, (x)_s)$, $\bar{F}_s(t, (x)_s)$, $s \ge 1$, which can be derived from hierarchy (4.2.4) by replacing the operators

$$\mathcal{H} = \sum_{s=1}^{\infty} \oplus \mathcal{H}_s$$

by

$$\tilde{\mathcal{H}} = \sum_{s=1}^{\infty} \oplus \tilde{\mathcal{H}}_s, \quad \bar{\mathcal{H}} = \sum_{s=1}^{\infty} \oplus \bar{\mathcal{H}}_s,$$

where

$$\tilde{\mathcal{H}}_s f_s(x_1,\ldots,x_s)$$
$$= -\sum_{i=1}^{s} p_i \cdot \frac{\partial}{\partial q_i} f_s(x_1,\ldots,x_s)$$
$$+ \sum_{i<j=1}^{s} \Theta(\eta_{ij}\cdot(p_i-p_j))\eta_{ij}\cdot(p_i-p_j)\delta(q_i-q_j)$$
$$\times \big[f_s(x_1,\ldots,x_i^*,\ldots,x_j^*,\ldots,x_s) - f_s(x_1,\ldots,x_i,\ldots,x_j,\ldots,x_s)\big], \quad (4.3.12)$$

$$\bar{\mathcal{H}}_s f_s(x_1,\ldots,x_s)$$
$$= -\sum_{i=1}^{s} p_i \cdot \frac{\partial}{\partial q_i} f_s(x_1,\ldots,x_s)$$
$$+ \sum_{i<j=1}^{s} \int_{S_+^2} d\eta_{ij}\, \eta_{ij}\cdot(p_i-p_j))\delta(q_i-q_j)$$
$$\times \big[f_s(x_1,\ldots,x_i^*,\ldots,x_j^*,\ldots,x_s) - f_s(x_1,\ldots,x_i,\ldots,x_j,\ldots,x_s)\big],$$

but without the boundary condition.

We are not able to rigorously construct the operators $\tilde{S}_s(-t,(x)_s)$, $\bar{S}_s(-t,(x)_s)$ that correspond to $\tilde{\mathcal{H}}_s$, $\bar{\mathcal{H}}_s$ and to prove directly the existence of solutions of the stochastic hierarchies for $\tilde{F}(t)$, $\bar{F}(t)$ because we have only a formal expression for the operators

$$\bar{S}_s(-t,(x)_s) = \lim_{n\to\infty} \prod_{i=1}^{n} \bar{S}_s(-\Delta t_i,(x)_s),$$

$$\tilde{S}_s(-t,(x)_s) = \lim_{n\to\infty} \prod_{i=1}^{n} \tilde{S}_s(-\Delta t_i,(x)_s), \qquad \sum_{i=1}^{n} \Delta t_i = t.$$

But from the definitions of $\tilde{F}_s(t,(x)_s)$, $\bar{F}_s(t,(x)_s)$ (3.3.14), (3.3.16) we know that, in order to obtain them from $F_s(t,(x)_s)$, it is sufficient to replace the operator $S_s(-t,(x)_s)$ in $F_s(t,(x)_s)$ by the operators $\tilde{S}_s(-t,(x)_s)$ and $\bar{S}_s(-t,(x)_s)$, respectively. All correlation functions $\tilde{F}_s(t,(x)_s)$, $\bar{F}_s(t,(x)_s)$ and $F_s(t,(x)_s)$ coincide outside all hyperplanes $q_i - p_i\tau = q_j - p_j\tau$, $0 \le \tau \le t$, $i,j \in \{1,\ldots,s\}$. This means that all correlation functions have the chaos property, provided that initial correlation functions have this property.

According to the duality principle, $F_s(t,(x)_s)$ given by series (4.2.8) or (4.3.3) determines the s-particles correlation function as a usual function, i.e., numerically. The correlation functions $\tilde{F}_s(t,(x)_s)$, $\bar{F}_s(t,(x)_s)$ determine the corresponding generalized functions by series (4.3.3) in which the operators $S_s(-t,(x)_s)$ are replaced by the operators $\bar{S}_s(-t,(x)_s)$, $\tilde{S}_s(-t,(x)_s)$, respectively. The correlation functions $\tilde{F}_s(t,(x)_s)$, $\bar{F}_s(t,(x)_s)$ are used in integrals with test functions $\varphi_s((x)_s)$ (observables).

Chapter 5

Spatially homogeneous Boltzmann hierarchy

5.1 Introduction

Equations of classical statistical mechanics are derived from equations of classical mechanics. For example, the BBGKY hierarchy is derived from the Hamilton equations via the Liouville equation for the distribution function on the phase space. The Liouville equation is obtained for the distribution functions, at given time t, that are the result of the action of the operator of evolution (the operator of shift along the trajectory) on the initial distribution functions.

The BBGKY hierarchy and the Boltzmann equation, which are a basis of nonequilibrium classical statistical mechanics, are obtained by the following method:

1) The Hamilton equation $\rightarrow$ 2) the operator of evolution (the operator of shift along the trajectory) on the initial distribution functions $\rightarrow$ 3) the Liouville equation for the distribution functions (state) $\rightarrow$ 4) the BBGKY hierarchy for (reduced) correlation functions [PGM3, CGP, CIP] $\rightarrow$ 5) the Boltzmann equation as a certain limit of the BBGKY hierarchy.

The Boltzmann equation is proved to be of great importance in classical statistical mechanics. As far as we know, Kac [Kac1, Kac2] was the first who made an attempt to modify the above-described method for the derivation of the BBGKY hierarchy and Boltzmann equation in the spatially homogeneous case where the correlation functions depend only on time and momenta and do not depend on the positions of particles. He proposed the following method for the derivation of the spatially homogeneous Boltzmann equation:

1) A certain stochastic Markov process in the momentum space $\rightarrow$ 2) the Kolmogorov–Itô equation for the distribution functions $\rightarrow$ 3) the hierarchy for (reduced) correlation functions in the mean-field approximation. (As pointed out in the previous chapters, Leontovich [Leo] made analogous attempts in the spatially inhomogeneous case.)

It was shown that, in the thermodynamic limit, the phenomenon of propagation of chaos takes place, and all many-particle correlation functions are the products of one-particle correlation functions that satisfy the Boltzmann equation in the spatially homogeneous case [Kac1, Kac2, Grn, Szn1, Szn2, Tan, Wag].

This result was a great achievement of nonequilibrium statistical mechanics, but it also raised a series of questions. The first question is related to the stochastic dynamics in the momentum space because only dynamics in the phase space, where states of real particles are determined by their momenta and positions, has a physical meaning. As shown in the previous chapters, the spatially inhomogeneous Boltzmann equation is associated with certain stochastic dynamics in the phase space, and it is natural to suppose that it reduces to the Kac dynamics in the momentum space in the spatially homogeneous case.

The second question is related to the mean-field approximation, which is not completely justified from the physical point of view because it uses an assumption that does not follow from the postulates of physics. The last remark becomes extremely important in connection with the last achievements concerning the Boltzmann–Grad limit for systems of hard spheres, which were exposed in the previous chapter.

Namely, it was shown that, in the Boltzmann–Grad limit, the BBGKY hierarchy for a system of hard spheres is reduced to the corresponding stochastic Boltzmann hierarchy, and, for initial data that satisfy the condition of chaos (i.e., initial many-particle correlation functions are products of one-particle correlation functions), solutions of the stochastic Boltzmann hierarchy also satisfy the chaos condition in the following sense: Many-particle correlation functions at arbitrary time (from the interval where solutions exist) and outside certain hyperplanes of lower dimension in the phase space are equal to products of one-particle correlation functions, and the latter are solutions of the Boltzmann equation. No mean-field approximations have been made.

It was established that the Hamilton dynamics of a system of hard spheres in the Boltzmann–Grad limit degenerates into certain stochastic dynamics of point particles. The stochastic dynamics consists of the following: point particles move freely until the positions of a certain pair of them coincide, then this pair elastically collides, but the vector that determines the elastic collision is random and uniformly distributed on the unit sphere. After collision, particles move freely until the next collision. Point particles interact on the time interval $[0, t]$ only if their phase points belong to the hyperplane V_{ij} where the vector of difference of their positions is parallel to the vector of their momenta, i.e., $q_i - q_j = \tau(p_i - p_j)$, $0 \leq \tau \leq t$.

In connection with this, there arises the problem of how to define correlation functions and averages of observables because this stochastic dynamics differs from the free one only on the hyperplanes V_{ij} where particles interact. In standard statistical mechanics, in which particles interact through a short-range potential, the sets of lower dimension are neglected because correlation functions and averages of observables are determined by Lebesgue integrals.

As shown in the previous chapters, it turns out that, in this case, it is necessary to take into account the contributions of the hyperplanes V_{ij} where point particles interact. It was a big surprise that, in the solutions of the Boltzmann equation and the hierarchy represented by series of iterations, the contributions of the hyperplanes V_{ij} were taken into account, and in a like manner as we determined the correlation functions. Thus, a new concept of correlation functions that take into account the contributions of the hyperplanes where point particles interact was proposed.

For these correlation functions, a hierarchy was derived, which was named the stochastic Boltzmann hierarchy and which differs from the ordinary Boltzmann hierarchy by certain terms with δ-functions in the weak sense, or by the boundary condition. Solutions of the stochastic Boltzmann hierarchy satisfy the chaos condition (or the condition of propagation of chaos), i.e., for all correlation functions we have $F_s(t, x_1, \dots, x_s) = F_1(t, x_1) \dots F_1(t, x_s)$, outside all hyperplanes V_{ij}, $1 \leq i < j \leq s$, if the initial correlation functions satisfy the chaos condition. The one-particle correlation function is a solution of the Boltzmann equation.

In the present chapter, we establish that the Kac results concerning the spatially homogeneous case of systems of hard spheres follow directly from our results described above [PeP1, PeP2, PeP3, LaPe1, LaPe2, LaPe3, LaPe4].

Namely, it is established that the evolution operator in the spatially homogeneous case can be obtained from the evolution operator of the stochastic dynamics in the phase space by means of specific averaging over the space of positions. It is shown that, within the framework of the stochastic dynamics, the functional average of spatially homogeneous observables over distribution functions with spatially homogeneous initial functions diverges as the volume of the system tends to infinity. After some specific averaging over the space of positions, we obtain the functional average of spatially homogeneous observables over spatially homogeneous distribution functions. This functional defines the operator of evolution of spatially homogeneous distribution functions, and the infinitesimal generator of the evolution operator obtained coincides with that proposed by Kac.

We show that this infinitesimal generator of the operator of evolution in the spatially homogeneous case can also be obtained from the infinitesimal generator (in the phase space) of the evolution operator of the stochastic dynamics by the same specific averaging over the space of positions. An equation for the spatially homogeneous distribution function is derived. It is shown that the Kac results in the mean-field approximation can be obtained from our functional average by a simple modification. By using the equation for spatially homogeneous distribution functions, the hierarchy for spatially homogeneous correlation functions is derived. In the mean-field approximation, the hierarchy obtained coincides with that derived by Kac. The operator that determines the spatially homogeneous hierarchy can be obtained from the corresponding operator of the stochastic spatially inhomogeneous Boltzmann hierarchy by averaging over the space of positions.

We also have the chaos property or propagation of chaos in the following sense: If one considers solutions of the stochastic spatially inhomogeneous Boltzmann hierarchy with spatially homogeneous initial correlation functions that satisfy the chaos condition $F_s(0, x_1, \dots, x_s) = F_s(0, p_1, \dots, p_s) = F_1(0, p_1) \dots F_1(0, p_s)$, then, outside all hyperplanes V_{ij}, $1 \leq i < j \leq s$, the correlation functions do not depend on positions, $F_s(t, x_1, \dots, x_s) = F_s(t, p_1, \dots, p_s)$, and satisfy the condition of chaos (or propagation of chaos), i.e.,

$$F_s(t, p_1, \dots, p_s) = F_1(t, p_1) \dots F_s(t, p_s).$$

The one-particle correlation function $F_s(t, p_1, \dots, p_s)$ satisfies the spatially homogeneous Boltzmann equation.

In this sense, the results completely equivalent to the mean-field approximation are obtained by neglecting the fact that the correlation functions $F_s(t, x_1, \dots, x_s)$, for $F_s(0, x_1, \dots, x_s) = F_s(0, p_1, \dots, p_s) = F_1(0, p_1) \dots F_1(0, p_s)$, depend on positions only on the hyperplanes $V_{ij}(q_i - q_j = \tau(p_i - p_j), 0 \le \tau \le t)$, $1 \le i < j \le s$. For arbitrary fixed $(p_1, \dots, p_s)$, we identify them on the hyperplanes V_{ij} with their values outside the hyperplanes V_{ij}, where $F_s(t, x_1, \dots, x_s)$, $s \ge 1$, do not depend on positions, i.e., $F_s(t, x_1, \dots, x_s) = F_s(t, p_1, \dots, p_s)$, and coincide with solutions of the spatially homogeneous hierarchy in the mean-field approximation.

5.2 Stochastic dynamics for a spatially homogeneous stochastic Boltzmann hierarchy

5.2.1 System of *N* particles

Consider N particles with unit mass in the three-dimensional space $\mathbb{R}^3$ and denote by $x_1 = (q_1, p_1), \dots, x_N = (q_N, p_N)$ their phase points. The stochastic dynamics of this system for negative time is defined as follows: particles move as free ones until the positions of two arbitrary particles with numbers i and j coincide at time $-\tau$: $q_i - p_i\tau = q_j - p_j\tau$, $\tau > 0$. Then these two particles collide, their momenta become

$$p_i^* = p_i - \eta_{ij}\, \eta_{ij} \cdot (p_i - p_j), \quad p_j^* = p_j + \eta_{ij}\, \eta_{ij} \cdot (p_i - p_j),$$

$$|\eta_{ij}| = 1, \quad \eta_{ij} \cdot (p_i - p_j) \ge 0, \quad S_+^2(\eta_{ij} | \eta_{ij} \cdot (p_i - p_j) \ge 0),$$

at time $-t$ their phase points are $x_i(-t) = (q_i - p_i\tau - p_i^*(t - \tau), p_i^*)$, $x_j(-t) = (q_j - p_j\tau - p_j^*(t - \tau), p_j^*)$, and they move freely until the next collision. The vectors η_{ij} are random and uniformly distributed on the sphere $|\eta_{ij}| = 1$. If $\eta_{ij} \in S_2^-$, $\eta_{ij} \cdot (p_i - p_j) \le 0$, then the particles continue moving freely. We neglect the case where three or more particles collide at the same point.

We consider infinitesimal time $-\Delta t$ and introduce, as in the previous chapters, the functional

$$\begin{aligned}
&(S_N(-\Delta t) f_N, \varphi_N) \\
&= \int dx_1 \dots dx_N\, f_N(q_1 - p_1\Delta t, p_1, \dots, q_N - p_N\Delta t, p_N)\varphi_N(q_1, p_1, \dots, q_N, p_N) \\
&\quad + \sum_{i<j=1}^{N} \int dx_1 \dots dx_N \int_0^{\Delta t} d\tau \int_{S_+^2} d\eta_{ij}\, \eta_{ij} \cdot (p_i - p_j)\delta(q_i - p_i\tau - q_j + p_j\tau)
\end{aligned}$$

$$\times \Big[f_N(q_1 - p_1\Delta t, p_1, \dots, q_i - p_i\tau - p_i^*(\Delta t - \tau), p_i^*, \dots,$$
$$q_j - p_j\tau - p_j^*(\Delta t - \tau), p_j^*, \dots, q_N - p_N\Delta t, p_N)$$
$$- f_N(q_1 - p_1\Delta t, p_1, \dots, q_i - p_i\Delta t, p_i \dots, q_j - p_j\Delta t,$$
$$p_j, \dots, q_N - p_N\Delta t, p_N)\Big] \varphi_N(q_1, p_1, \dots, q_N, p_N), \tag{5.2.1}$$

which is equal to the average of the observable $\varphi_N(x_1, \dots, x_N)$ over the state

$$S_N(-\Delta t) f_N(x_1, \dots, x_N) = f_N(x_1(-\Delta t), \dots, x_N(-\Delta t)).$$

As usual, it is assumed that f_N is a real, symmetric, differentiable, normalized function and φ_N is a real symmetric test function.

Note that, in the average (functional) (5.2.1), the contributions of the hyperplanes $q_i - q_j = \tau(p_i - p_j)$, $0 \le \tau \le \Delta t$, $1 \le i < j \le N$, of lower dimension, where the stochastic particles interact, are taken into account. These contributions are equal to the second term on the right-hand side of (5.2.1).

Now consider the case where the functions f_N and ϕ_N do not depend on positions:

$$f_N(q_1, p_1, \dots, q_N, p_N) = f_N(p_1, \dots, p_N),$$
$$\phi_N(q_1, p_1, \dots, q_N, p_N) = \varphi_N(p_1, \dots, p_N),$$
$$f_N(q_1 - p_1\Delta t, p_1, \dots, q_N - p_N\Delta t, p_N) = f_N(p_1, \dots, p_N),$$
$$f_N(q_1 - p_1\Delta t, p_1, \dots, q_i - p_i\tau - p_i^*(\Delta t - \tau),$$
$$p_i^*, \dots, q_j - p_j\tau - p_j^*(\Delta t - \tau), p_j^*, \dots, q_N - p_N\Delta t, p_N)$$
$$= f_N(p_1, \dots, p_i^*, \dots, p_j^*, \dots, p_N),$$

$$q_i - p_i\tau = q_j - p_j\tau.$$

In this case, functional (5.2.1) is divergent, and the first and the second terms are proportional to V^N and V^{N-1}, respectively (V is the volume of $\mathbb{R}^3$). Instead of functional (5.2.1), we introduce the following functional:

$$\big(\bar{S}_N(-\Delta t) f_N, \phi_N\big)$$
$$= \lim_{V\to\infty} \frac{1}{V^N} \int_\Lambda dq_1 \int dp_1 \dots \int_\Lambda dq_N \int dp_N\, f_N(p_1, \dots, p_N)\varphi_N(p_1, \dots, p_N)$$
$$+ \lim_{V\to\infty} \frac{1}{V^{N-1}} \sum_{i<j=1}^{N} \int_\Lambda dq_1 \int dp_1 \dots \int_{\Lambda_i} dq_i \int dp_i \dots \int_{\Lambda_j} dq_j$$

$$\times \int dp_j \ldots \int_\Lambda dq_N \int dp_N \int_0^{\Delta t} d\tau$$

$$\times \int_{S_+^2} d\eta_{ij}\, \eta_{ij} \cdot (p_i - p_j)\, \delta(q_i - p_i\tau - q_j + p_j\tau)$$

$$\times \Big[f_N(p_1, \ldots, p_i^*, \ldots, p_j^*, \ldots, p_N)$$

$$- f_N(p_1, \ldots, p_i, \ldots, p_j, \ldots, p_N) \Big] \varphi_N(p_1, \ldots, p_N)$$

$$= \int dp_1 \ldots dp_N\, f_N(p_1, \ldots, p_N) \varphi_N(p_1, \ldots, p_N)$$

$$+ \Delta t \sum_{i<j=1}^{N} \int dp_1 \ldots dp_N \int_{S_+^2} d\eta_{ij}\, \eta_{ij} \cdot (p_i - p_j)$$

$$\times \Big[f_N(p_1, \ldots, p_i^*, \ldots p_j^*, \ldots, p_N)$$

$$- f_N(p_1, \ldots, p_i, \ldots, p_j, \ldots, p_N) \Big] \varphi_N(p_1, \ldots, p_N). \tag{5.2.2}$$

We have denoted by Λ_i and Λ_j the spheres centered at the points $-p_i\tau$ and $-p_j\tau$, respectively, and having volumes $V(\Lambda_i) = V(\Lambda_j) = V$. The sphere Λ is centered at the origin, $V(\Lambda) = V$. Functional (5.2.2) was obtained from functional (5.2.1) by averaging over the space of positions (configurational space) and is the average of the state $\bar{S}_N(-\Delta t) f_N(p_1, \ldots, p_N)$ over the observables $\varphi_N(p_1, \ldots, p_N)$. Formulas (5.2.2) defines the operator of evolution $\bar{S}_N(-\Delta t)$ of states $f_N(p_1, \ldots, p_N)$ in the spatially homogeneous case. The second term is associated with the contribution of the hypersurface $q_i - q_j = \tau(p_i - p_j)$, $1 \le i < j \le N$, where the stochastic particles interact. Formulas (5.2.2) holds for arbitrary test functions $\varphi_N(p_1, \ldots, p_N)$. Therefore, it follows from (5.2.2) that the operator of evolution $\bar{S}_N(-\Delta t)$ of the state of an N-particle system in the spatially homogeneous case is defined by the following formula for infinitesimal time $-\Delta t$:

$$\bar{S}_N(-\Delta t) f_N(p_1, \ldots, p_N) = \bar{f}_N(\Delta t, p_1, \ldots, p_N)$$

$$= f_N(p_1, \ldots, p_N) + \Delta t \sum_{i<j=1}^{N} \int_{S_+^2} d\eta_{ij}\, \eta_{ij} \cdot (p_i - p_j)$$

$$\times \Big[f_N(p_1, \ldots, p_i^*, \ldots, p_j^*, \ldots, p_N)$$

$$- f_N(p_1, \ldots, p_i, \ldots, p_j, \ldots, p_N) \Big]. \tag{5.2.3}$$

In (5.2.3), the state $\bar{S}_N(-\Delta t)f_N(p_1,\ldots,p_N)$ is averaged over random vectors η_{ij}, $i \le j \le N$. For fixed random vectors η_{ij}, we have

$$\tilde{S}_N(-\Delta t)f_N(p_1,\ldots,p_N) = \tilde{f}_N(\Delta t, p_1,\ldots,p_N)$$

$$= f_N(p_1,\ldots,p_N) + \Delta t \sum_{i<j=1}^{N} \eta_{ij}\cdot(p_i-p_j)\Theta\left((\eta_{ij}\cdot(p_i-p_j)\right)$$

$$\times \Big[f_N(p_1,\ldots,p_i^*,\ldots,p_j^*,\ldots,p_N)$$

$$- f_N(p_1,\ldots,p_i,\ldots,p_j,\ldots,p_N)\Big], \tag{5.2.4}$$

where $\Theta(\alpha) = 1$, $\alpha > 0$, and $\Theta(\alpha) = 0$, $\alpha < 0$. One obtains (5.2.3) by averaging the second term on the right-hand side of (5.2.4) with respect to η_{ij}.

We do not know how to construct the functional $(\bar{S}_N(-t)f_N, \varphi_N)$ directly for arbitrary time t and $N \ge 3$. Therefore, we define this functional using the following procedure: Assume that the function $\bar{f}_N(t,p_1,\ldots,p_N) = \bar{S}_N(-t)f_N(p_1,\ldots,p_N)$ is already defined. Also assume that the operator $\bar{S}_N(-t)$ possesses the group property and is formally defined as follows:

$$\bar{S}_N(-t) = \lim_{n\to\infty}\prod_{i=1}^{n}\bar{S}_N(-\Delta t_i), \quad \sum_{i=1}^{n}\Delta t_i = t,$$

where $\bar{S}_N(-\Delta t_i)$ for infinitesimal Δt_i is already defined by (5.2.3). By using the group property of the operator of evolution

$$\bar{S}_N(-t-\Delta t) = \bar{S}_N(-\Delta t)\bar{S}_N(-t) = \bar{S}_N(-t)\bar{S}_N(-\Delta t),$$

one can define the function

$$\bar{f}_N(t+\Delta t, p_1,\ldots,p_N) = \bar{S}_N(-\Delta t)\bar{f}_N(t,p_1,\ldots,p_N)$$

and obtain the functional equal to the average of the state $\bar{S}_N(-\Delta t) \times \bar{f}_N(t,p_1,\ldots,p_N)$ over the observable $\varphi_N(p_1,\ldots,p_N)$ with respect to the random vectors η_{ij} that appear on the interval $(-t,-t-\Delta t)$ with infinitesimal Δt.

The average $\left(\bar{S}_N(-\Delta t)f_N(t),\varphi_N\right)$ of the state $\bar{S}_N(-\Delta t)f_N(t,p_1,\ldots,p_N)$ over the observable $\varphi_N(p_1,\ldots,p_N)$ is determined by formula (5.2.2) if one substitutes the function $f_N(t,p_1,\ldots,p_N)$ for $f_N(p_1,\ldots,p_N)$:

$$\left(\bar{S}_N(-\Delta t)\bar{f}_N(t),\varphi_N\right)$$

$$= \int dp_1\ldots,dp_N\,\bar{f}_N(t,p_1,\ldots,p_N)\varphi_N(p_1,\ldots,p_N)$$

$$+\,\Delta t \sum_{i<j=1}^{N} \int dp_1 \dots dp_N \int_{S_+^2} d\eta_{ij}\, \eta_{ij} \cdot (p_i - p_j)$$
$$\times \Big[\bar{f}_N(t, p_1, \dots p_i^*, \dots, p_j^*, \dots, p_N)$$
$$- \bar{f}_N(t, p_1, \dots, p_i, \dots, p_j, \dots, p_N) \Big] \varphi_N(p_1, \dots, p_N). \quad (5.2.5)$$

It follows from (5.2.5) that

$$\bar{S}_N(-\Delta t) \bar{f}_N(t, p_1, \dots, p_N)$$
$$= \bar{S}_N(-\Delta t) \bar{S}_N(-t) f_N(p_1, \dots, p_N) = \bar{f}_N(t + \Delta t, p_1, \dots, p_N)$$
$$= \bar{f}_N(t, p_1, \dots, p_N) + \Delta t \sum_{i<j=1}^{N} \int_{S_+^2} d\eta_{ij}\, \eta_{ij} \cdot (p_i - p_j)$$
$$\times \Big[\bar{f}_N(t, p_1, \dots, p_i^*, \dots, p_j^*, \dots, p_N)$$
$$- \bar{f}_N(t, p_1, \dots, p_i, \dots, p_j, \dots, p_N) \Big]. \quad (5.2.6)$$

5.2.2 Equation for spatially homogeneous distribution functions

We obtain from (5.2.6) the following equation for the function $\bar{f}_N(t, p_1, \dots, p_N)$:

$$\frac{\partial \bar{f}_N(t, p_1, \dots, p_N)}{\partial t} = \sum_{i<j=1}^{N} \int_{S_+^2} d\eta_{ij}\, \eta_{ij} \cdot (p_i - p_j)$$
$$\times \Big[\bar{f}_N(t, p_1, \dots, p_i^*, \dots, p_j^*, \dots, p_N)$$
$$- \bar{f}_N(t, p_1, \dots, p_i, \dots, p_j, \dots, p_N) \Big],$$

$$\bar{f}_N(t, p_1, \dots, p_N) \Big|_{t=0} = f_N(p_1, \dots, p_N). \quad (5.2.7)$$

In equation (5.2.7), the right-hand side is averaged with respect to random unit vectors η_{ij}, $1 \le i < j \le N$.

Equation (5.2.6)–(5.2.7) yields the following equation for $f_N(t, p_1, \dots, p_N)$ with fixed random vectors η_{ij}:

$$\tilde{S}_N(-\Delta t)\tilde{f}_N(t,p_1,\ldots,p_N) = \tilde{f}_N(t+\Delta t,p_1,\ldots,p_N)$$
$$= \tilde{f}_N(t,p_1,\ldots,p_N) + \Delta t \sum_{i<j=1}^{N} \eta_{ij}\cdot(p_i-p_j)\Theta(\eta_{ij}\cdot(p_i-p_j))$$
$$\times\Big[\tilde{f}_N(t,p_1,\ldots,p_i^*,\ldots,p_j^*,\ldots,p_N)$$
$$-\tilde{f}_N(t,p_1,\ldots,p_i,\ldots,p_j,\ldots,p_N)\Big], \tag{5.2.8}$$

$$\frac{\partial \tilde{f}_N(t,p_1,\ldots p_N)}{\partial t} = \sum_{i<j=1}^{N} \eta_{ij}\cdot(p_i-p_j)\,\Theta\left(\eta_{ij}\cdot(p_i-p_j)\right)$$
$$\times\Big[\tilde{f}_N(t,p_1,\ldots,p_i^*,\ldots,p_j^*,\ldots,p_N)$$
$$-\tilde{f}_N(t,p_1,\ldots,p_i,\ldots,p_j,\ldots,p_N)\Big],$$
$$\tilde{f}_N(t,p_1,\ldots,p_N)|_{t=0} = f_N(p_1,\ldots,p_N).$$

It can be obtained if the integration with respect to the random vectors η_{ij} is omitted in functionals (5.2.1)–(5.2.2) and (5.2.5), and, consequently, in (5.2.6)–(5.2.7).

The function $\bar{f}_N(t,p_1,\ldots,p_N)$ from equation (5.2.7) does not depend on any random vectors if the initial function $f_N(p_1,\ldots,p_N)$ does not depend on them, whereas the function $\tilde{f}_N(t,p_1,\ldots,p_N)$ from equation (5.2.8) depends on the random vectors η_{ij}, $1\le i<j\le N$.

Thus, we have obtained equations (5.2.7), (5.2.8) for the evolution of a state that does not depend on position (in the case of a spatially homogeneous state) beginning with the stochastic dynamics in the phase space and averaging functional (5.2.1) over the configurational space.

Kac [Kac1, Kac2] postulated an analogous equation for the explanation of the phenomenon of propagation of chaos, but with mean-field multipliers $1/N$ on the right-hand side. His equation

$$\frac{\partial \bar{f}_N(t,p_1,\ldots,p_N)}{\partial t} = \frac{1}{N}\sum_{i<j=1}^{N}\int_{S_+^2} d\eta_{ij}\,\eta_{ij}\cdot(p_i-p_j)$$
$$\times\Big[\bar{f}_N(t,p_1,\ldots,p_i^*,\ldots,p_j^*,\ldots,p_N)$$
$$-\bar{f}_N(t,p_1,\ldots,p_i,\ldots,p_j,\ldots,p_N)\Big],$$
$$\bar{f}_N(t,p_1,\ldots,p_N)\big|_{t=0} = f_N(p_1,\ldots,p_N), \tag{5.2.9}$$

can also be obtained from the stochastic dynamics if one introduces the multiplier $1/N$ in the second term of functionals (5.2.1)–(5.2.2) and (5.2.5).

We also need the operators $\bar{S}_N(t)$ and equation (5.2.7) for positive t. To obtain them, it suffices to replace S_+^2 in (5.2.6) and (5.2.7) by $S_-^2\ \big(\eta_{ij}\,|\eta_{ij}\cdot(p_i - p_j) \leq 0\big)$ (for details concerning the stochastic dynamics and evolution operators $\bar{S}_N(t)$ for $t > 0$, see the previous chapters).

5.3 Derivation of the spatially homogeneous hierarchy

5.3.1 Spatially homogeneous hierarchy within the framework of canonical and grand canonical ensemble

We consider an N-particle system with normalized state $\bar{f}_N(t, p_1, \ldots, p_N)$ that satisfies equation (5.2.7) and introduce the following sequence of reduced correlation functions:

$$\bar{F}_s^{(N)}(t,(p)_s) = \bar{F}_s^{(N)}(t, p_1, \ldots, p_s)$$

$$= \frac{N!}{(N-s)!}\int dp_{s+1}\ldots dp_N\, \bar{f}_N(t, p_1, \ldots, p_s, p_{s+1}, \ldots, p_N), \qquad (5.3.1)$$

$$1 \leq s \leq N-1,$$

$$\bar{F}_N^{(N)}(t,(p)_N) = \bar{F}_N^{(N)}(t, p_1, \ldots, p_N) = \bar{f}_N^{(N)}(t, p_1, \ldots, p_N),$$

$$(p)_s = (p_1, \ldots, p_s).$$

Integrating both sides of (5.2.7) over momenta $p_{s+1}, \ldots, p_N$ and taking into account that $f_N(t, p_1, \ldots, p_N)$ is symmetric with respect to $p_1, \ldots, p_N$ and the Jacobian of the transformation $(p_i, p_j) \to (p_i^*, p_j^*)$ is equal to one, we obtain the following hierarchy (see the derivation of the spatially inhomogeneous hierarchy in Chapter 3):

$$\frac{\partial \bar{F}_s^{(N)}(t, p_1, \ldots, p_s)}{\partial t}$$

$$= \sum_{i<j=1}^{s} \int_{S_+^2} d\eta_{ij}\, \eta_{ij}\cdot(p_i - p_j)$$

$$\times \Big[\bar{F}_s^{(N)}(t, p_1, \ldots, p_i^*, \ldots, p_j^*, \ldots, p_s)$$

$$- \bar{F}_s^{(N)}(t, p_1, \ldots, p_i \ldots, p_j, \ldots, p_s)\Big]$$

$$+ \sum_{i=1}^{s} \int dp_{s+1} \int_{S_+^2} d\eta_{is+1}\, \eta_{is+1}\cdot(p_i - p_{s+1})$$

$$\times \Big[\bar{F}_{s+1}^{(N)}(t, p_1, \ldots, p_i^*, \ldots, p_s, \ldots, p_{s+1}^*)$$

$$- \bar{F}_{s+1}^{(N)}(t, p_1, \ldots, p_i, \ldots, p_s, p_{s+1}) \Big], \quad 1 \leq s \leq N-1, \tag{5.3.2}$$

$$\bar{F}_s^{(N)}(t, p_1, \ldots, p_s)\big|_{t=0} = F_s^{(N)}(p_1, \ldots, p_s).$$

Note that $\bar{F}_N^{(N)}(t, p_1, \ldots, p_N)$ satisfies equation (5.2.7).

As is known, in the grand canonical ensemble, the system can be in states $\bar{f}_N(t, p_1, \ldots, p_N)$ with arbitrary $N = 0, 1 \ldots,$ with certain probability. The functions $\bar{f}_N(t, p_1, \ldots, p_N)$ are not normalized in this case.

The infinite sequence of reduced correlation functions is defined as follows:

$$\bar{F}_s(t, (p)_s) = \frac{1}{\Xi} \sum_{n=0}^{\infty} \frac{1}{n!} \int dp_{s+1} \ldots dp_{s+n}$$

$$\times \bar{f}_{s+n}(t, p_1, \ldots, p_s, p_{s+1}, \ldots, p_{s+n}), \quad s \geq 1, \tag{5.3.3}$$

where the grand partition function Ξ is equal to

$$\Xi = \sum_{n=0}^{\infty} \frac{1}{n!} \int dp_1 \ldots dp_n \bar{f}_n(t, p_1, \ldots, p_n)$$

$$= \sum_{n=0}^{\infty} \frac{1}{n!} \int dp_1 \ldots dp_n \, \bar{f}_n(p_1, \ldots, p_n),$$

$$f_0 = 1.$$

In the last equality, we have used the fact that

$$\int dp_1 \ldots dp_n \, \bar{f}_n(t, p_1, \ldots, p_n) = \int dp_1 \ldots dp_n \, f_n(p_1, \ldots, p_n),$$

which follows directly from (5.2.7). It is easy to check that sequence (5.3.3) satisfies hierarchy (5.3.2) with $1 \leq s < \infty$ (see analogous calculations for the spatially inhomogeneous case in Chapter 3).

In order to derive the corresponding hierarchy in the "mean-field" case, one should use equation (5.2.9) and the functions $F_s(t, p_1, \ldots, p_s)$ defined according to (5.3.1) but without the factors $\frac{N!}{(N-s)!}$. The corresponding hierarchy has the form

$$\frac{\partial \bar{F}_s^{(N)}(t, p_1, \dots, p_s)}{\partial t}$$

$$= \frac{1}{N} \sum_{i<j=1}^{s} \int_{S_+^2} d\eta_{ij}\, \eta_{ij} \cdot (p_i - p_j)$$

$$\times \Big[\bar{F}_s^{(N)}(t, p_1, \dots, p_i^*, \dots, p_j^*, \dots, p_s)$$

$$- \bar{F}_s^{(N)}(t, p_1, \dots, p_i, \dots, p_j, \dots, p_s) \Big]$$

$$+ \frac{N-s}{N} \sum_{i=1}^{s} \int dp_{s+1} \int_{S_+^2} d\eta_{is+1}\, \eta_{is+1} \cdot (p_i - p_{s+1})$$

$$\times \Big[\bar{F}_{s+1}^{(N)}(t, p_1, \dots, p_i^*, \dots, p_s, \dots, p_{s+1}^*)$$

$$- \bar{F}_{s+1}^{(N)}(t, p_1, \dots, p_i, \dots, p_s, p_{s+1}) \Big], \quad 1 \le s \le N, \tag{5.3.4}$$

$$\bar{F}_s^{(N)}(t, p_1, \dots, p_s)\big|_{t=0} = F_s^{(N)}(p_1, \dots, p_s).$$

Passing formally to the thermodynamical limit as $N \to \infty$ in (5.3.4) and taking into account that the first term on the right-hand side of (5.3.4) tends to zero as $N \to \infty$, one gets the limit hierarchy

$$\frac{\partial \bar{F}_s(t, p_1, \dots, p_s)}{\partial t}$$

$$= \sum_{i=1}^{s} \int dp_{s+1} \int_{S_+^2} d\eta_{is+1} \eta_{is+1} \cdot (p_i - p_{s+1})$$

$$\times \Big[\bar{F}_{s+1}(t, p_1, \dots, p_i^*, \dots, p_s, \dots, p_{s+1}^*)$$

$$- \bar{F}_{s+1}(t, p_1, \dots, p_i, \dots, p_s, p_{s+1}) \Big], \quad s \ge 1, \tag{5.3.5}$$

$$\bar{F}_s(t, p_1, \dots, p_s)\big|_{t=0} = F_s(p_1, \dots, p_s).$$

The corresponding limit hierarchy for (5.3.2) is the same as (5.3.2), but $s \ge 1$ and, instead of $\bar{F}_s^{(N)}(t, p_1, \dots, p_s)$, one should take

$$\bar{F}_s(t, p_1, \dots, p_N) = \lim_{N\to\infty} \bar{F}_s^{(N)}(t, p_1, \dots, p_s).$$

(For justification, see [Grn, Szn1, Szn2, Tan, Wag].)

The "mean-field" hierarchy (2.3.5) has the following characteristic property: it preserves the chaos property, i.e., if the initial functions

$$F_s(p_1, \dots, p_s) = F_1(p_1) \dots F_1(p_s) \tag{5.3.6}$$

have the chaos property, then the functions $\bar{F}_s(t, p_1, \dots, p_s)$ also have the chaos property,

$$\bar{F}_s(t, p_1, \dots, p_s) = \bar{F}_1(t, p_1) \dots \bar{F}_1(t, p_s), \tag{5.3.7}$$

and the function $\bar{F}_1(t, p_1)$ satisfies the nonlinear Boltzmann equation

$$\begin{aligned} \frac{\partial \bar{F}_1(t, p_1)}{\partial t} = \int dp_2 \int_{S_+^2} d\eta\, \eta \cdot (p_1 - p_2) \\ \times \Big[\bar{F}_1(t, p_1^*) \bar{F}_1(t, p_2^*) - \bar{F}_1(t, p_1) \bar{F}_1(t, p_2) \Big]. \end{aligned} \tag{5.3.8}$$

This property follows directly from hierarchy (5.3.5) because it permits the separation of variables if the initial data satisfy (5.3.6). It is also assumed that a solution of equation (5.3.8) exists.

5.3.2 Hierarchy with fixed random vectors

In hierarchy (5.3.2), the functions $F_s^{(N)}(p_1, \dots, p_N)$ do not depend on any random vectors. This hierarchy is derived from equation (5.2.7). We also need the hierarchy with fixed random vectors. It can be derived from the state $\tilde{f}_N(t, p_1, \dots, p_N)$ that satisfies equation (5.2.8) and for the sequence of reduced correlation functions defined as follows:

$$\begin{aligned} \tilde{F}_s^{(N)}(t, (p)_s) &= \tilde{F}_s^{(N)}(t, p_1, \dots, p_s) \\ &= \frac{N!}{(N-s)!} \int dp_{s+1} \dots dp_N \prod_{i<j=1}^{N}{}^{(s)} \int_{S_+^2} d\eta_{ij}\, \tilde{f}_N(t, p_1, \dots, p_s, p_{s+1}, \dots, p_N), \end{aligned} \tag{5.3.9}$$

$$\tilde{F}_N^{(N)}(t, (p)_N) = \tilde{F}_N^{(N)}(t, p_1, \dots, p_N) = \tilde{f}_N(t, p_1, \dots, p_N),$$

where $\prod_{i<j=1}^{N}{}^{(s)}$ means that the pairs (i, j) with $1 \le i \le s$ and $1 \le j \le s$ are excluded. Note that only the random vectors η_{ij} that appear on the interval $[t, t + \Delta t]$ and are present in equation (5.2.8) are considered.

It follows directly from equation (5.2.8) that sequence (5.3.9) satisfies the following hierarchy:

$$\frac{\partial \tilde{F}_s^{(N)}(t, p_1, \dots, p_s)}{\partial t}$$
$$= \sum_{i<j=1}^{s} \eta_{ij} \cdot (p_i - p_j)\, \Theta(\eta_{ij} \cdot (p_i - p_j))$$
$$\times \Big[\tilde{F}_s^{(N)}(t, p_1, \dots, p_i^*, \dots, p_j^*, \dots, p_s)$$
$$- \tilde{F}_s^{(N)}(t, p_1, \dots, p_i, \dots, p_j, \dots, p_s) \Big]$$
$$+ \sum_{i=1}^{s} \int dp_{s+1} \int_{S_+^2} d\eta_{is+1}\, \eta_{is+1} \cdot (p_i - p_{s+1})$$
$$\times \Big[\tilde{F}_{s+1}^{(N)}(t, p_1, \dots, p_i^*, \dots, p_s, p_{s+1}^*)$$
$$- \tilde{F}_{s+1}^{(N)}(t, p_1, \dots, p_i, \dots, p_s, p_{s+1}) \Big], \quad 1 \le s < N-1, \tag{5.3.10}$$

$$\tilde{F}_s^{(N)}(t, p_1, \dots, p_s)\big|_{t=0} = F_s^{(N)}(p_1, \dots, p_s).$$

It is obvious that the function $\tilde{F}_N^{(N)}(t, p_1, \dots, p_N)$ satisfies equation (5.2.8).

We have not indicated that functions (5.3.9) depend on random vectors.

The "mean-field" version of hierarchy (5.3.10) can easily be obtained if one omits the sign of integration with respect to $\eta_{i,j}$ in the first term on the right-hand side of (5.3.4). The limit "mean-field" hierarchy is the same as (5.3.5).

5.4 Representation of solutions of the spatially homogeneous hierarchy

5.4.1 Representation of solutions of the spatially homogeneous hierarchy through series of iterations

We represent hierarchy (5.3.2) with $1 \le s < \infty$ and (N) omitted in an abstract form. Denote by $\bar{F}(t)$ the sequence of the functions $\bar{F}_s(t, p_1, \dots, p_s)$, i.e.,

$$\bar{F}(t) = \Big(\bar{F}_1(t, p_1), \dots, \bar{F}_s(t, p_1, \dots, p_s), \dots \Big), \tag{5.4.1}$$

and let $\bar{\mathcal{H}}$ and $\bar{\mathcal{A}}$ denote the operators defined, respectively, by the first and the second terms on the right-hand side of (5.3.2), i.e.,

$$\begin{aligned}\left(\bar{\mathcal{H}}\bar{F}(t)\right)_s(p_1,\dots,p_s) &= \bar{\mathcal{H}}_s\bar{F}_s(t,p_1,\dots,p_s)\\ &= \sum_{i<j=1}^{s}\int_{S_+^2} d\eta_{ij}\,\eta_{ij}\cdot(p_i-p_j)\\ &\quad\times\Big[\bar{F}_s(t,p_1,\dots,p_i^*,\dots,p_j^*,\dots,p_s)\\ &\qquad-\bar{F}_s(t,p_1,\dots,p_i,\dots,p_j,\dots,p_s)\Big],\\ \left(\bar{\mathcal{A}}\bar{F}(t)\right)_s(p_1,\dots,p_s) &= (\bar{\mathcal{A}}_s\bar{F}(t))_s(p_1,\dots,p_s)\\ &= \sum_{i=1}^{s}\int dp_{s+1}\int_{S_+^2} d\eta_{is+1}\,\eta_{is+1}\cdot(p_i-p_{s+1})\\ &\quad\times\Big[\bar{F}_{s+1}(t,p_1,\dots,p_i^*,\dots,p_s,p_{s+1}^*)\\ &\qquad-\bar{F}_{s+1}(t,p_1,\dots,p_i,\dots,p_s,p_{s+1})\Big].\end{aligned} \tag{5.4.2}$$

Then hierarchy (5.3.2) has the following abstract form:

$$\frac{d\bar{F}(t)}{dt} = \bar{\mathcal{H}}\bar{F}(t) + \bar{\mathcal{A}}\bar{F}(t) = \bar{\mathcal{L}}\bar{F}(t), \quad \bar{F}(t)|_{t=0} = F(0). \tag{5.4.3}$$

Recall that the operator $\bar{\mathcal{H}}_s$ is the infinitesimal generator of the operator $\bar{S}_s(-t)$, which, for an arbitrary t, is formally defined as follows:

$$\bar{S}_s(-t) = \prod_{i=1}^{n}\bar{S}_s(-\Delta t_i), \quad \sum_{i=1}^{n}\Delta t_i = t;$$

for infinitesimal Δt_i, $\bar{S}_s(-\Delta t_i)$ is defined by (5.2.3).

As is known, a solution of (5.4.3) can be represented by the following series of iterations:

$$\begin{aligned}\bar{F}(t) = \sum_{n=0}^{\infty}\int_0^t dt_1\dots\int_0^{t_{n-1}} dt_n \bar{S}(-t)\bar{S}(t_1)\bar{\mathcal{A}}\bar{S}(-t_1)\dots\\ \times\bar{S}(t_n)\bar{\mathcal{A}}\bar{S}(-t_n)F(0),\end{aligned} \tag{5.4.4}$$

where $\bar{S}(-t)$ is the direct sum of the operators $\bar{S}_s(-t)$, i.e.,

$$\bar{S}(-t) = \sum_{s=1}^{\infty} \oplus \bar{S}_s(-t).$$

From (5.4.4), we get

$$\begin{aligned}
&\bar{F}_s(t,(p)_s) \\
&= \sum_{n=0}^{\infty} \int_0^t dt_1 \ldots \int_0^{t_{n-1}} dt_n \bar{S}_s(-t,(p)_s)\bar{S}_s(t_1,(p)_s) \\
&\times \sum_{i=1}^{s} \int dp_{s+1} \int_{S_+^2} d\eta_{is+1}\, \eta_{is+1}\cdot(p_i - p_{s+1})\big[\bar{S}_{s+1}(-t_1,(p)^*_{s+1}) \\
&\qquad - \bar{S}_{s+1}(-t_1,(p)_{s+1})\big]\ldots \bar{S}_{s+n-1}(t_{n-1},(p)_{s+n-1}) \\
&\times \sum_{i=1}^{s+n-1} \int dp_{s+n} \int_{S_+^2} d\eta_{i,s+n}\eta_{i,s+n}\cdot(p_i - p_{s+n}) \\
&\times \Big[\bar{S}_{s+n}(-t_n,(p)^*_{s+n}) - \bar{S}_{s+n}(-t_n,(p)_{s+n})\Big] F_{s+n}(0,(p)_{s+n})
\end{aligned} \tag{5.4.5}$$

where, for the sake of simplicity, we have used the notation

$$\begin{aligned}
&\Big[\bar{S}_{s+n}(-t_n,(p)^*_{s+n}) - \bar{S}_{s+n}(-t_n,(p)_{s+n})\Big] F_{s+n}(0,(p)_{s+n}) \\
&\quad \equiv \bar{S}_{s+n}(-t_n,(p)^*_{s+n})F_{s+n}(0,(p)^*_{s+n}) - \bar{S}_{s+n}(-t_n,(p)_{s+n})F_{s+n}(0,(p)_{s+n}),
\end{aligned}$$

and $(p)^*_{s+n} = (p_1,\ldots,p_i^*,\ldots,p_{s+n-1},p^*_{s+n})$ in terms with number i.

We have defined the operator $\bar{S}_N(-\Delta t)$ for infinitesimal $\Delta t > 0$ by formulas (5.2.4). For arbitrary $t > 0$, the operator $\bar{S}_N(-t)$ is formally defined by using the group property, namely,

$$\bar{S}_N(-t) = \lim_{n\to\infty} \prod_{i=1}^{n} \bar{S}_N(-\Delta t_i), \quad \sum_{i=1}^{n} \Delta t_i = t, \tag{5.4.6}$$

but we do not prove the existence of this limit.

The operator $\bar{S}_N(-t)$ can be defined as the group of operators with infinitesimal generator $\bar{\mathcal{H}}_N$ determined by (5.2.8), but again only formally because we do not give a rigorous meaning to $\bar{\mathcal{H}}_N$ as an operator in some functional space.

The operator $\bar{S}_N(-t)$ can also be determined by the corresponding stochastic process $p_1(-t),\ldots,p_N(-t)$ according to the standard formula for the operator of shift along the "trajectory":

$$\bar{S}_N(-t) f_N(p_1, \dots, p_N) = f_N(p_1(-t), \dots, p_N(-t)),$$

$$p_1(t)|_{t=0} = p_1, \ \dots \ , \ p_N(-t)|_{t=0} = p_N$$

where the stochastic process is such that the function

$$f_N(t, p_1, \dots, p_N) = \bar{S}_N(-t) f_N(p_1, \dots, p_N)$$

satisfies equation (5.2.8). However, this approach is not elaborated yet (for details, see Subsection 5.2.3).

To define the operator $\bar{S}_N(\Delta t)$ with $\Delta t > 0$, it suffices to replace S_+^2 by S_-^2 in formulas (5.2.8) and, for arbitrary $t > 0$, to use (5.4.6) with $\bar{S}_N(\Delta t_i)$.

We cannot rigorously prove the existence of solutions of the spatially homogeneous hierarchy (5.4.3) because we have not yet necessary information concerning the operators $\bar{\mathcal{H}}_s$ and $\bar{\mathcal{A}}_s$. These operators are obviously unbounded because they contain operators of multiplication by p_i^1, p_i^2, and p_i^3, $i \in \{1, \dots, s+1\}$ and sums that grow together with s as s^2 and s. But we have detailed information concerning solutions of the inhomogeneous stochastic Boltzmann hierarchy. Namely, in Chapter 4 we have proved the existence of mild solutions that are local in time for initial data in the space $E_{\xi,\beta}$ and global in time for initial data in the space $\tilde{E}_{\xi,\beta}$. One can use, for spatially homogeneous hierarchy, only results obtained in $E_{\xi,\beta}$ because spatially homogeneous initial data $F(0)$ do not belong to the space $\tilde{E}_{\xi,\beta}$.

It will be shown that the spatially homogeneous Boltzmann equation can be derived from the stochastic spatially inhomogeneous Boltzmann hierarchy without using the mean-field approximation.

We now obtain a representation for $\bar{F}_s(t, (p)_s)$ and deduce the Boltzmann equation for $\bar{F}_1(t, p_1)$ by using solutions of the spatially inhomogeneous hierarchy. It can also be represented in the following abstract form:

$$\frac{d\bar{F}(t)}{dt} = \bar{\mathcal{H}}\bar{F}(t) + \bar{\mathcal{A}}\bar{F}(t),$$

$$\bar{F}(t) = \Big(\bar{F}_1(t, x_1), \dots, \bar{F}_s\big(t, (x)_s\big), \dots\Big),$$

$$\bar{F}(t)|_{t=0} = F(0), \tag{5.4.7}$$

or componentwise (see Chapters 3 and 4)

$$\begin{aligned}
&\frac{\partial \bar{F}_s(t,x_1,\dots,x_s)}{\partial t}\\
&\quad= -\sum_{i=1}^{s} p_i \cdot \frac{\partial}{\partial q_i}\bar{F}_s(t,x_1,\dots,x_s)\\
&\qquad+\sum_{i<j=1}^{s}\int_{S_+^2} d\eta_{ij}\delta(q_i-q_j)\eta_{ij}\cdot(p_i-p_j)\\
&\qquad\quad\times\Big[\bar{F}_s(t,x_1,\dots,x_i^*,\dots,x_j^*,\dots,x_s)\\
&\qquad\qquad-\bar{F}_s(t,x_1,\dots,x_i,\dots,x_j,\dots,x_s)\Big]\\
&\qquad+\sum_{i=1}^{s}\int dq_{s+1}\delta(q_i-q_{s+1})\int dp_{s+1}\int_{S_+^2} d\eta_{is+1}\,\eta_{is+1}\cdot(p_i-p_{s+1})\\
&\qquad\quad\times\Big[\bar{F}_{s+1}(t,x_1,\dots,x_i^*,\dots,x_s,x_{s+1}^*)\\
&\qquad\qquad-\bar{F}_{s+1}(t,x_1,\dots,x_i,\dots,x_s,x_{s+1})\Big],
\end{aligned}\tag{5.4.8}$$

$$\bar{F}_s(t,x_1,\dots,x_s)|_{t=0} = F_s(x_1,\dots,x_s), \quad s\ge 1.$$

It is easy to see that the operators $\bar{\mathcal{H}}_s$ and $\bar{\mathcal{A}}_s$ of hierarchy (5.4.3) can be obtained by the spatial averaging of the operators $\bar{\mathcal{H}}_s$ and $\bar{\mathcal{A}}_s$ of the spatially inhomogeneous hierarchy (5.4.8) with spatially homogeneous functions $f_s(p_1,\dots,p_s)$ and $f_{s+1}(p_1,\dots,p_s,p_{s+1})$ by analogy with (5.2.2). Namely, one should use the multiplier $\frac{1}{V^{s-1}}$ with $\bar{\mathcal{H}}_s$ and $\frac{1}{V^s}$ with $\bar{\mathcal{A}}_s$.

Note that one can also derive hierarchy (5.3.10) for $\tilde{F}_s(t,p_1,\dots,p_s)$ from the hierarchy for $\bar{F}_s(t,x_1,\dots,x_s)$.

As is known, solutions of hierarchy (5.4.8) coincide with solutions of the hierarchy for correlation functions $F_s(t,x_1,\dots,x_s)$ (which satisfy hierarchy (5.4.8) with omitted second term on the right-hand side, but with the known boundary conditions at $q_i = q_j$, $i,j\in\{1,\dots,s\}$) outside all hypersurfaces V_{ij} ($q_i - p_i\tau = q_j - p_j\tau$, $0\le\tau\le t$).

The solution of (5.4.8) with omitted second term but with boundary condition can be represented as the following series of iterations:

$$\begin{aligned}
F(t) = \sum_{n=0}^{\infty}\int_0^t dt_1\dots\int_0^{t_{n-1}} dt_n S(-t)S(t_1)\mathcal{A}S(-t_1)\dots\\
\times S(t_n)\mathcal{A}S(-t_n)F(0),
\end{aligned}\tag{5.4.9}$$

or componentwise

$$
\begin{aligned}
&F_s(t,(x)_s)\\
&\quad=\sum_{n=0}^{\infty}\int_0^t dt_1\dots\int_0^{t_{n-1}} dt_n\, S_s(-t,(x)_s)S_s(t_1,(x)_s)\\
&\quad\times\sum_{i=1}^{s}\int dx_{s+1}\delta(q_i-q_{s+1})\int_{S_+^2} d\eta_{is+1}\,\eta_{is+1}\cdot(p_i-p_{s+1})\\
&\quad\times\Big[S_{s+1}(-t_1,(x)_{s+1}^*)-S_{s+1}(-t_1,(x)_{s+1})\Big]\dots\\
&\quad\times S_{s+n-1}(t_{n-1},(x)_{s+n-1})\sum_{i=1}^{s+n-1}\int dx_{s+n}\delta(q_i-q_{s+n})\\
&\quad\times\int_{S_+^2} d\eta_{i,s+n}\,\eta_{i,s+n}\cdot(p_i-p_{s+n})\\
&\quad\times\Big[S_{s+n}(-t_n,(x)_{s+n}^*)-S_{s+n}(-t_n,(x)_{s+n})\Big]F_{s+n}(0,(x)_{s+n}). \qquad (5.4.10)
\end{aligned}
$$

Recall that $S_s(-t)$ is the operator of evolution of s stochastic particles, $\mathcal{H}_s$ is its infinitesimal generator, and $S(-t)$ is the direct sum of $S_s(-t)$. In (5.4.8), we have used $\mathcal{H}_s$ in a weak form. Recall that $\mathcal{H}_s$ can also be represented numerically as follows:

$$
\mathcal{H}_s f_s(x_1,\dots,x_s)=-\sum_{i=1}^{s} p_i\cdot\frac{\partial}{\partial q_i} f_s(x_1,\dots,x_s)
$$

with known boundary condition. The operators $S_s(-t)$ corresponding to $\mathcal{H}_s$ are rigorously defined as the operators of shift along trajectories in the phase space (see Chapter 2).

As shown in Chapter 4, series (5.4.10) is uniformly convergent with respect to $(x)_s$ on a finite time interval $[0,t_0]$ if the sequence of initial functions $F(0)$ belongs to the space $E_{\xi,\beta}$ with the norm

$$
\|F(0)\|=\sup_{s\ge 1}\frac{1}{\xi^s}e^{\beta\sum_{i=1}^{s}\frac{p_i^2}{2}}\sup_{(x)_s}\big|F_s(0,(x)_s)\big|,
$$

where $\xi>0$ and $\beta>0$ are fixed numbers and t_0 is a certain constant dependent on (ξ,β).

It is an open question whether series (5.4.4)–(5.4.5) are also uniformly convergent with respect to $(p)_s$ on the time interval $[0,t_0]$ if the sequence of initial functions $F_s\big(0,(p)_s\big)$, $s\ge 1$, belongs to $E_{\xi,\beta}$. In this case, series (5.4.4)–(5.4.5) would represent a *mild solution* of the spatially homogeneous hierarchy (5.4.3).

5.4.2 One-particle distribution function as a solution of the Boltzmann equation

Consider series (5.4.10) again and denote by V_{ij} the set (hyperplanes) $q_i - q_j = \tau(p_i - p_j)$ with all $0 \le \tau \le t$. It is known from Chapter 4 that if phase points $x_1, \dots, x_s$ are outside all V_{ij}, $1 \le i < j \le s$, then all operators $S_{s+i}(\pm t)$ can be replaced by the operators of evolution of the free systems $S^0_{s+i}(\pm t)$, and representation (5.4.10) reduces to the representation of solutions of the Boltzmann hierarchy:

$$
\begin{aligned}
&F_s(t,(x)_s) \\
&= \sum_{n=0}^{\infty} \int_0^t dt_1 \dots \int_0^{t_{n-1}} dt_n S^0_s(-t,(x)_s) S^0_s(t_1,(x)_s) \\
&\quad \times \sum_{i=1}^{s} \int dx_{s+1} \delta(q_i - q_{s+1}) \int_{S^2_+} d\eta_{is+1}\, \eta_{is+1} \cdot (p_i - p_{s+1}) \\
&\quad \times \left[S^0_{s+1}(-t_1,(x)^*_{s+1}) - S^0_{s+1}(-t_1,(x)_{s+1}) \right] \dots \\
&\quad \times S^0_{s+n-1}(t_{n-1},(x)_{s+n-1}) \sum_{i=1}^{s+n-1} \int dx_{s+n} \delta(q_i - q_{s+n}) \\
&\quad \times \int_{S^2_+} d\eta_{i,s+n}\, \eta_{i,s+n} \cdot (p_i - p_{s+n}) \\
&\quad \times \left[S^0_{s+n}(-t_n,(x)^*_{s+n}) - S^0_{s+n}(-t_n,(x)_{s+n}) \right] F_{s+n}(0,(x)_{s+n}). \qquad (5.4.11)
\end{aligned}
$$

It was shown in Chapter 4 that, for $s = 1$, series (5.4.11) represents $F_1(t,x_1)$ in the entire phase space of one particle and coincides with a solution of the Boltzmann equation

$$
\begin{aligned}
\frac{\partial F_1(t,x_1)}{\partial t} &= -p_1 \frac{\partial}{\partial q_1} F_1(t,x_1) + \int dp_2 \int_{S^2_+} d\eta_{12}\, \eta_{12} \cdot (p_1 - p_2) \\
&\quad \times \left[F_1(t,x_1^*) F_1(t,x_2^*) - F_1(t,x_1) F_1(t,x_2) \right] \qquad (5.4.12)
\end{aligned}
$$

if the initial functions $F_s(0,x_1,\dots,x_s)$ possess the chaos property

$$F_s(0,x_1,\dots,x_s) = F_1(0,x_1)\dots F_1(0,x_s). \qquad (5.4.13)$$

The functions $F_s(t,(x)_s)$ determined by series (5.4.11) coincide with the corresponding correlation functions that satisfy the ordinary Boltzmann hierarchy (4.8.6) and also possess the chaos property

$$F_s(t,x_1,\dots,x_s) = F_1(t,x_1)\dots F_1(t,x_s) \qquad (5.4.14)$$

for time on the interval $[0, t_0]$ and $F(0) \in E_{\xi,\beta}$. This fact is regarded as a rigorous derivation of the Boltzmann equation.

Note that the functions $F_1(t, x_1) \dots F_1(t, x_s)$ do not coincide with the solutions $F_s(t, (x)_s)$ of the stochastic hierarchy (5.4.7)–(5.4.8) given by (5.4.10) in the entire phase space of s particles, i.e.,

$$F_s(t, x_1, \dots, x_s) \neq F_1(t, x_1) \dots F_1(t, x_s), \quad s > 1,$$

if $q_i - q_j = \tau(p_i - p_j)$ for some $0 \leq \tau \leq t$ for at least one pair (i, j). This can be proved as follows: In the entire phase space of s particles, representation (5.4.10) reduces to the following one:

$$\begin{aligned} F_s(t, (x)_s) & \\ = \sum_{n=0}^{\infty} \int_0^t dt_1 \dots & \int_0^{t_{n-1}} dt_n S_s\big(-t, (x)_s\big) S_s\big(t_1, (x)_s\big) \\ \times \sum_{i=1}^{s} \int dx_{s+1} & \delta(q_i - q_{s+1}) \int_{S_+^2} d\eta_{is+1}\, \eta_{is+1} \cdot (p_i - p_{s+1}) \\ \times \Big[(S_s(-t_1) & S_1^0(-t_1))((x)_{s+1}^*) - S_s(-t_1) S_1^0(-t_1))((x)_{s+1}) \Big] \dots \\ \times \Big(S_s(t_{n-1}) & S_{n-1}^0(t_{n-1}) \Big) \big((x)_{s+n-1}\big) \\ \times \sum_{i=1}^{s+n-1} \int & dx_{s+n} \delta(q_i - q_{s+n}) \int_{S_+^2} d\eta_{i,s+n}\, \eta_{i,s+n} \cdot (p_i - p_{s+n}) \\ \times \Big[\Big(S_s(-t_n) & S_n^0(-t_n) \Big) \big((x)_{s+n}^* \big) - \big(S_s(-t_n) S_n^0(-t_n) \big) \big((x)_{s+n} \big) \Big] \\ \times F_{s+n} & \big(0, (x)_{s+n}\big), \end{aligned} \tag{5.4.15}$$

where the notation $\big(S_s(-t_i) S_i^0(-t_i)\big) \big((x)_{s+i}^*\big)$ means that the operator $S_s(-t_i)$ acts on the first s phase points of the set $((x)_{s+i}^*)$ and the operator $S_i^0(-t_n)$ acts on the other i phase points of the set $\big((x)_{s+i}^*\big)$. The same notation is used for $\big(S_s(-t_i) S_i^0(-t_i)\big) \big((x)_{s+i}\big)$. Recall that $S_s(\pm t_i)$ is the operator of evolution of stochastic particles and $S_i^0(\pm t_i)$ is the operator of evolution of free particles. Representation (5.4.15) follows from the fact that one has the Lebesgue integral with respect to momenta $p_{s+1}, \dots, p_{s+n}$ and can neglect the hyperplanes V_{ij} where particles with numbers $(s+1, \dots, s+n)$ interact with each other and with those with numbers $(1, \dots, s)$.

On the hyperplanes V_{ij}, the operators $S_s(\pm t_i)$ do not coincide with the operators $S_s^0(\pm t_i)$, and, thus,

$$F_s(t, x_1, \dots, x_s) \neq F_1(t, x_1) \dots F_1(t, x_s), \qquad s > 1,$$

even if the initial functions $F_s(0, x_1, \dots, x_s)$ satisfy (5.4.13).

Assume that all initial functions $F_{s+n}\big(0, (x)_{s+n}\big)$ are spatially homogeneous, i.e.,

$$F_{s+n}\big(0, (x)_{s+n}\big) = F_{s+n}\big(0, (p)_{s+n}\big).$$

Then it follows from (5.4.15) that the functions $F_s\big(t, (x)_s\big)$ depend on the position $(q)_s$ only on the hyperplanes V_{ij} because, outside all hyperplanes V_{ij}, $1 \le i < j \le s$, the operators of stochastic and free evolution coincide $S_s(\pm t_i) = S_s^0(\pm t_i)$, the action of the operators $S_s^0(\pm t_i)$ on spatially homogeneous functions again gives spatially homogeneous functions, $S_s^0(\pm t) = I$, and $S_i^0(\pm t) = I$.

If we have $S_s(\pm t_i) S_i^0(\pm t_i)\big((x)_{s+i}^*\big)$ with some x_k^*, $1 \le k \le s$, then the corresponding V_{kl}, $1 \le l \ne k \le s$, should be excluded from consideration because p_k^* depends on certain momenta from the set $(p_{s+1}, \dots, p_{s+i})$ and one can neglect V_{kl} as hyperplanes of lower dimension in the Lebesgue integrals with respect to $(p_{s+1}, \dots, p_{s+i})$.

Thus, the functions $F_s\big(t, (x)_s\big)$ depend only on momenta $(p)_s$ almost everywhere with respect to positions $(q)_s$, i.e., outside all V_{ij}, $1 \le i < j \le s$, if the initial functions do not depend on positions. The function $F_1(t, x_1)$ does not depend on q_1, i.e., $F_1(t, x_1) = F_1(t, p_1)$.

Now consider (5.4.15) with $(x)_s$ outside all V_{ij}, $1 \le i < j \le s$. Then, in the spatially homogeneous case, representation (5.4.15) is reduced to the following one:

$$\begin{aligned} F_s(t, (p)_s) &\\ = \sum_{n=0}^{\infty} \int_0^t dt_1 \dots \int_0^{t_{n-1}} dt_n &\\ \times \sum_{i=1}^{s} \int dp_{s+1} \int_{S_+^2} d\eta_{is+1} \eta_{is+1} \cdot (p_i - p_{s+1}) \Big[I((p)_{s+1}^*) - I((p)_{s+1}) \Big] \dots &\\ \times \sum_{i=1}^{s+n-1} \int dp_{s+n} \int_{S_+^2} d\eta_{is+n}\, \eta_{is+n} \cdot (p_i - p_{s+n}) &\\ \times \Big[I((p)_{s+n}^*) - I((p)_{s+n}) \Big] F_{s+n}(0, (p)_{s+n}). & \qquad (5.4.16) \end{aligned}$$

For $F_1(t, p_1)$, we get the following representation:

$$\begin{aligned} F_1(t, p_1) = \sum_{n=0}^{\infty} \int_0^t dt_1 \dots \int_0^{t_{n-1}} dt_n \sum_{i=1}^{s} \int dp_2 \int_{S_+^2} d\eta_{i2} &\\ \times \eta_{i2} \cdot (p_i - p_{1+1}) \Big[I\big((p)_{1+1}^*\big) - I(p)_{1+1}) \Big] &\\ \times \sum_{i=1}^{1+n-1} \int dp_{1+n} \int_{S_+^2} d\eta_{i1+n}\, \eta_{i1+n} \cdot (p_i - p_{1+n}) & \end{aligned}$$

$$\times \Big[I((p)^*_{1+n}) - I((p)_{1+n}) \Big] F_{1+n}\big(0, (p)_{1+n}\big), \tag{5.4.17}$$

where

$$\Big[I((p)^*_{s+i}) - I((p)_{s+i}) \Big] F_{s+i}(0, (p)_{s+i}) = F_{s+i}(0, (p)^*_{s+i}) - F_{s+i}(0, (p)_{s+i}).$$

Representation (5.4.16)–(5.4.17) follows from (5.4.15) because $S_s(\pm t) = S_s^0(\pm t)$ outside V_{ij}, $i, j \in \{1, \dots, s\}$, and $S^0_{s+i}(\pm t) = I$ on $F_{s+n}(0, (x)_{s+n}) = F_{s+n}(0, (p)_{s+n})$. Note that formulas (5.4.16)–(5.4.17) hold in the entire momentum space. We can now explain why one cannot conclude that $F_s(t, (p)_s)$ (5.4.16) do not coincide with solution (5.4.5) of hierarchy (5.4.3). The reason is that, obtaining (5.4.16), we have neglected contributions of the hyperplanes V_{kl} $(1 \le k < l \le s)$. To obtain $\tilde{F}_s(t, (p)_s)$ (5.4.5) that satisfy (5.4.3), one should perform the averaging procedure on these hyperplanes (see (5.2.2)).

If the initial functions possess the chaos property (5.4.13), then series (5.4.17) represents a solution of the spatially homogeneous Boltzmann equation

$$\frac{\partial F_1(t, p_1)}{\partial t} = \int dp_2 \int_{S^2_+} d\eta_{12}\, \eta_{12} \cdot (p_1 - p_2) \times \Big[F_1(t, p_1^*) F_1(t, p_2^*) - F_1(t, p_1^*) F_1(t, p_2^*) \Big], \tag{5.4.18}$$

which follows from (5.4.12).

Note that functions (5.4.16) also possess the chaos properties because functions (5.4.11) possess these properties outside all V_{ij}, $1 \le i < j \le s$.

Thus, we get the chaos property, or propagation of chaos, without using the mean-field approach if we consider the functions $F_s(t, (x)_s)$ outside the hyperplanes V_{ij}, $1 \le i < j \le s$, because, for spatially homogeneous initial functions $F_s(0, (x)_s)$, series (5.4.16) possess this property.

Note that series (5.4.16) and (5.4.17) are uniformly convergent with respect to $(p)_s$ and time $0 \le t < t_0$ if the initial functions $F_s(0, (p)_s) = F_1(0, p_1) \dots F_1(0, p_s)$ belong to the space $E_{\xi,\beta}$ because series (5.4.10) are convergent in this case (see Chapter 4).

There is another method for obtaining the propagation of chaos without using the mean-field approach. Namely, consider the following average of the functions $F_s(t, (x)_s)$ (5.4.15) over the space of positions:

$$\lim_{V \to \infty} \frac{1}{V^s} \int_\Lambda dq_1 \dots \int_\Lambda dq_s F_s(t, q_1, p_1, \dots, q_s, p_s) = F_s(t, (p)_s), \tag{5.4.19}$$

where the Lebesgue integral is used and the behavior of functions $F_s(t, (x)_s)$ (5.4.15) on the hyperplanes V_{ij}, $1 \le i < j \le s$, is neglected. It is obvious that the obtained functions $F_s(t, (p)_s)$ (5.4.19) coincide with those defined by (5.4.16), and if the initial functions $F_s(0, (x)_s)$ possess the chaos property (5.4.19) and do not depend on $(q)_s$, then functions (5.4.19) also possess the chaos property and $F_1(t, p_1)$ satisfies the spatially homogeneous Boltzmann equation (5.4.18).

Chapter 6

Stochastic dynamics for the Boltzmann equation with arbitrary differential scattering cross section

6.1 Introduction

In the previous chapters, we have introduced the stochastic dynamics of point particles, which is obtained from the Hamilton dynamics of a system of hard spheres in the Boltzmann–Grad limit. According to this stochastic dynamics, point particles move as free ones until their positions coincide, and then they undergo elastic scattering. The unit vector that determines elastic scattering is a random vector uniformly distributed on the unit sphere. Then particles move as free ones until the next collision.

The present chapter is a generalization of the results obtained in the previous chapters to the case of stochastic dynamics in which the unit vector that determines elastic scattering of point particles is distributed on the unit sphere with distribution density corresponding to an arbitrary differential scattering cross section. As is customary in classical statistical mechanics [PGM3], the initial state of a system is defined by a distribution function on the phase space. The state of the system at arbitrary time is defined as the result of the action of an evolution operator, i.e., the operator of transition along the trajectory, on the initial distribution function. The distribution function thus defined differs from the distribution function of the free system of particles at arbitrary time only on the hyperplanes of lower dimension where the point particles interact.

From the viewpoint of traditional classical statistical mechanics, the system of point particles moving according to the stochastic dynamics should be regarded as a free system. Indeed, in traditional statistical mechanics, averages are calculated via the Lebesgue integral, and the behavior of distribution functions on hyperplanes of lower dimension is not taken into account. In this connection, by analogy with the previous chapters, a new concept of averages of observables over distribution functions was introduced; this concept takes into account, in a special way, the contribution of the hyperplanes where the particles interact. With the use of averages thus introduced, correlation functions are defined that also take into account, in a special way, the contribution of the hyperplanes where the particles interact.

The hierarchy of equations derived for the sequence of correlation functions has the form of the ordinary Boltzmann hierarchy [CIP] but takes into account the boundary conditions corresponding to the case where the positions of particles coincide. If the hierarchy is considered in the weak sense, then it contains δ-functions that differ from zero when the positions of particles coincide. The hierarchy obtained was called the stochastic Boltzmann hierarchy.

We construct solutions of the stochastic Boltzmann hierarchy on a finite time interval for a sequence of correlation functions that belong to the space of functions bounded with respect to coordinates and exponentially decreasing with respect to momenta. We also construct solutions of the hierarchy on an arbitrary time interval for initial correlation functions that belong to the space of functions exponentially decreasing with respect to coordinates and momenta. These solutions are equal to the sum of certain contributions of the hyperplanes where point particles interact.

It should be noted that the solutions of the Boltzmann equation and the ordinary Boltzmann hierarchy are also equal to the sum of the contributions of the hyperplanes where point particles interact. For the first time, this fact was noted in [Pet5] for the stochastic dynamics corresponding to the Boltzmann–Grad limit of a system of hard spheres (see Chapter 4).

The solutions of the stochastic Boltzmann hierarchy coincide with the solutions of the ordinary Boltzmann hierarchy outside the hyperplanes of lower dimension where point particles interact. As is known, the solutions of the ordinary Boltzmann hierarchy are products of one-particle correlation functions, provided that the initial correlation functions are also products of one-particle correlation functions. A one-particle correlation function is a solution of the nonlinear Boltzmann equation [Pet5]. In other words, the solutions of the ordinary Boltzmann hierarchy satisfy the chaos condition.

It follows from the arguments presented above that the solutions of the stochastic Boltzmann hierarchy also satisfy the chaos condition outside the hyperplanes of lower dimension where point particles interact because they coincide there with the solutions of the ordinary Boltzmann hierarchy.

Note that the stochastic Boltzmann hierarchy is obtained from the stochastic dynamics in the same way as the BBGKY hierarchy is obtained from the Hamilton dynamics. The ordinary Boltzmann hierarchy is obtained directly from the Boltzmann equations [Pet5] or from the BBGKY hierarchy for a system of hard spheres in the Boltzmann–Grad limit, where the boundary conditions are not taken into account [CIP], and it is likely that there is no dynamics corresponding to it.

As is known, Kac [Kac1, Kac2] proposed to use a special Markov process in the momentum space and obtained from the corresponding Kolmogorov–Itô–Liouville equation in the mean-field approximation a hierarchy whose solutions satisfy the chaos condition .

In Chapter 5, it was shown that this Markov process in the momentum space can be obtained by certain averaging with respect to coordinates from stochastic dynamics in the phase space that corresponds to a system of hard spheres. The Kolmogorov–Itô–Liouville equation can also be obtained by averaging with respect to coordinates from the Kolmogorov–Itô–Liouville equation for the distribution function in the phase space. The results of Kac were thus justified. In the present work, we generalize the results of Kac to stochastic dynamics considered.

Note that the Boltzmann equation in the momentum space can be obtained directly from the stochastic Boltzmann hierarchy for initial correlation functions that depend only on momenta and satisfy the chaos condition (see Chapter 5) [LaPe2]. Indeed, outside the hyperplanes of lower dimension where point particles interact, the solutions of the ordinary hierarchy and the stochastic hierarchy coincide, whereas the solutions of the ordinary hierarchy do not depend on coordinates and satisfy the chaos conditions, and one-particle correlation function is a solution of the Boltzmann equation. In this case, the mean-field approximation is not used.

Thus, we have generalized and reproduced the aforementioned results concerning the calculation of averages of observables, a new concept of correlation functions, stochastic hierarchy, chaos, and an analog of the Kac dynamics in the momentum space for our stochastic dynamics with arbitrary differential scattering cross section.

6.2 Stochastic dynamics

6.2.1 Functional average

Consider point particles with unit mass in the three-dimensional space $\mathbb{R}^3$ and denote by $x_1 = (q_1, p_1), \dots, x_N = (q_N, p_N)$ their phase points; $(x)_N = (x_1, \dots, x_N) = x$ at initial time $t = 0$.

Define their stochastic dynamics for negative time $-t$, $t > 0$, as follows: Particles move as free ones until $q_i - p_i\tau = q_j - p_j\tau$, $0 \le \tau \le t$, $i, j \in \{1, \dots, N\}$. Then these two particles collide, their momenta become

$$p_i^* = p_i - \eta_{ij}\eta_{ij}\cdot(p_i - p_j), \quad p_j^* = p_j + \eta_{ij}\eta_{ij}\cdot(p_i - p_j),$$

$$\eta_{ij} \subset S_+^2(\eta_{ij}|\eta_{ij}\cdot(p_i - p_j) \ge 0), \quad |\eta_{ij}| = 1; \tag{6.2.1}$$

if $\eta_{ij} \subset S_-^2(\eta_{ij} \mid \eta_{ij}\cdot(p_i - p_j) \le 0)$, then $p_i^* = p_i$, $p_j^* = p_j$; here, $\eta_{ij}\cdot(p_i - p_j)$ is the scalar product of vectors η_{ij} and $(p_i - p_j)$.

At time $-t$, their phase points are

$$x_i(-t) = (q_i - p_i\tau - p_i^*(t - \tau), p_i^*),$$

$$x_j(-t) = (q_j - p_j\tau - p_j^*(t - \tau), p_j^*), \quad \eta_{ij} \subset S_+^2, \tag{6.2.2}$$

if $\eta_{ij} \subset S_-^2$, $x_i(-t) = (q_i - p_it, p_i)$, $x_j(-t) = (q_j - p_jt, p_j)$.

Particles scatter elastically, but η_{ij} are random vectors with probability density

$$\frac{Q(\eta_{ij}\cdot(p_i - p_j))}{\eta_{ij}\cdot(p_i - p_j)},$$

where $Q(\eta_{ij} \cdot (p_i - p_j))$ is known as a *cross section*.

If $\eta_{ij} \subset S^2_-(\eta_{ij} \mid \eta_{ij} \cdot (p_i - p_j) \leq 0)$, then particles continue to move freely even in the case $q_i - p_i\tau = q_j - p_j\tau$.

For positive time $t > 0$, it is necessary to put in (6.2.2) $(+\tau)$ instead of $(-\tau)$ and S^2_- and S^2_+ instead of S^2_+ and S^2_-, respectively. We neglect the case where three or more particles collide at the same point.

The above-introduced stochastic dynamics defines a trajectory with fixed random vectors in the phase space:

$$\begin{aligned} X(-t) &= (x(-t))_N = (x_1(-t), \dots, x_N(-t)) \\ &= (x_1(-t,(x)_N), \dots, x_N(-t,(x)_N)) = X(-t,(x)_N) = X(-t,x). \end{aligned}$$

The trajectory $X(-t,x)$ obviously possesses the group property $X(-t_1 - t_2, x) = X(-t_1, X(-t_2, x)) = X(-t_2, X(-t_1, x))$.

We define the operator $S_N(-t)$ as the operator of shift along the trajectory:

$$\begin{aligned} S(-t) f_N(x_1, \dots, x_N) &= f_N(x_1(-t), \dots, x_N(-t)) \\ &= f_N(x_1(-t,(x)_N), \dots, x_N(-t,(x)_N)). \end{aligned} \tag{6.2.3}$$

Let $f_N(x_1, \dots, x_N)$ be a real symmetric continuously differentiable normalized function and let $\varphi_N(x_1, \dots, x_N)$ be a real symmetric test function.

We consider an infinitesimal time $-\Delta t$ and introduce the following functional:

$$\begin{aligned} &(S_N(-\Delta t) f_N, \varphi_N) \\ &\quad = \int dx_1 \dots dx_N\, f_N(q_1 - p_1\Delta t, p_1, \dots, q_N \\ &\qquad - p_N\Delta t, p_N)\varphi_N(q_1, p_1, \dots, q_N, p_N) \\ &\qquad + \sum_{i<j=1}^{N} \int dx_1 \dots dx_N \int_0^{\Delta t} d\tau \int_{S^2_+} d\eta_{ij} \frac{Q(\eta_{ij} \cdot (p_i - p_j))}{\eta_{ij} \cdot (p_i - p_j)} \\ &\qquad \times \eta_{ij} \cdot (p_i - p_j)\delta(q_i - p_i\tau - q_j + p_j\tau) \\ &\qquad \times [f_N(q_1 - p_1\Delta t, p_1, \dots, q_i - p_i\tau - p_i^*(\Delta t - \tau), p_i^*, \dots, q_j \\ &\qquad - p_j\tau - p_j^*(\Delta t - \tau), p_j^*, \dots, q_N - p_N\Delta t, p_N) \\ &\qquad - f_N(q_1 - p_1\Delta t, p_1, \dots, q_i - p_i\Delta t, p_i, \dots, q_j \\ &\qquad - p_j\Delta t, p_j, \dots, q_N - p_N\Delta t, p_N)]\varphi_N(q_1, p_1, \dots, q_N, p_N) \end{aligned}$$

$$= \int dx_1 \dots dx_N \Big[f_N(x_1, \dots, x_N) - \Delta t \sum_{i=1}^{N} p_i \cdot \frac{\partial}{\partial q_i} f_N(x_1, \dots, x_N) \Big]$$

$$\times \varphi_N(x_1, \dots, x_N) - \int dx_1 \dots dx_N \Delta t \int_{S_+^2} d\eta_{ij} \delta(q_i - q_j)$$

$$\times Q(\eta_{ij} \cdot (p_i - p_j)) \Big[f_N(q_1, p_1, \dots, q_i, p_i^*, \dots, q_j, p_j^*, \dots, q_N, p_N)$$

$$- f_N(q_1, p_1, \dots, q_i, p_i, \dots, q_j, p_j, \dots, q_N, p_N) \Big]$$

$$\times \varphi_N(q_1, p_1, \dots, q_N, p_N). \tag{6.2.4}$$

Here the operator $S_N(-\Delta t)$ is defined according to the stochastic dynamics as follows: for $q_i - p_i \tau = q_j - p_j \tau$ and $\eta_{ij} \in S_+^2$

$$S_N(-\Delta t) f_N(x_1, \dots, x_i, \dots, x_j, \dots, x_N)|_{q_i - p_i \tau = q_j - p_j \tau}$$
$$= f_N(q_1 - p_1 \Delta t, p_1, \dots, q_i - p_i \tau - p_i^*(\Delta t - \tau), p_i^*, \dots, q_j$$
$$- p_j \tau - p_j^*(\Delta t - \tau), p_j^*, \dots, q_N - p_N \Delta t, p_N),$$

$$S_N(-\Delta t) f_N(x_1, \dots, x_i, \dots, x_j, \dots, x_N)|_{q_i - p_i \tau = q_j - p_j \tau}$$
$$= f_N(q_1 - p_1 \Delta t, p_1, \dots, q_i - p_i \Delta t, p_i, \dots, q_j$$
$$- p_j \Delta t, p_j, \dots, q_N - p_N \Delta t, p_N) \tag{6.2.5}$$

for $\eta_{ij} \in S_-^2$, and

$$S_N(-\Delta t) f_N(x_1, \dots, x_N) = f_N(q_1 - p_1 \Delta t, p_1, \dots, q_N - p_N \Delta t, p_N)$$

for $q_i - \tau p_i \neq q_j - p_j \tau$, for all $i, j \in \{1, \dots, N\}$, $0 \leq \tau \leq \Delta t$.

We now represent the functional average as follows:

$$\big(S_N(-\Delta t) f_N, \varphi_N\big)$$
$$= \int dx_1 \dots dx_N \bar{S}_N(-\Delta t) f_N(x_1, \dots, x_N) \varphi_N(x_1, \dots, x_N). \tag{6.2.6}$$

The introduced operator $\bar{S}_N(-\Delta t)$ is the usual operator of evolution in the theory of Markov processes, and it was obtained as result of a specific averaging procedure with respect to random vectors η_{ij} that takes into account the contribution of the hypersurfaces of lower dimension $q_i - p_i \tau = q_j - p_j \tau$, $0 \leq \tau \leq t$, $i, j \in \{1, \dots, N\}$ where the stochastic particles interact.

The functional average (6.2.6) is defined for an arbitrary test function and determines the result of the action of the operator $\bar{S}_N(-\Delta t)$ on the function $f_N(x_1,\dots,x_N)$, i.e., $\bar{S}_N(-\Delta t)f_N(x_1,\dots,x_N)$, as a generalized function:

$$\begin{aligned}
&\bar{S}_N(-\Delta t)f_N(x_1,\dots,x_N)\\
&\quad= f_N(q_1-p_1\Delta t,p_1,\dots,q_N-p_N\Delta t,p_N)\\
&\qquad+\sum_{i<j=1}^{N}\int_0^{\Delta t}d\tau\int_{S_+^2}d\eta_{ij}\,Q(\eta_{ij}\cdot(p_i-p_j))\delta(q_i-p_i\tau-q_j+p_j\tau)\\
&\qquad\times\big[f_N(q_1-p_1\Delta t,p_1,\dots,q_i-p_i\tau-p_i^*(\Delta t-\tau),p_i^*,\dots,\\
&\qquad\qquad q_j-p_j\tau-p_j^*(\Delta t-\tau),p_j^*,\dots,q_N-\\
&\qquad-p_N\Delta t,p_N)-f_N(q_1-p_1\Delta t,p_1,\dots,q_i-p_i\Delta t,p_i,\dots,\\
&\qquad\qquad q_j-p_j\Delta t,p_j,\dots,q_N-p_N\Delta t,p_N)\big]\\
&\quad= f_N(q_1-p_1\Delta t,p_1,\dots,q_N-p_N\Delta t,p_N)\\
&\qquad+\sum_{i<j=1}^{N}\int_0^{\Delta t}d\tau\int_{S^2}d\eta_{ij}\,Q(\eta_{ij}\cdot(p_i-p_j))\delta(q_i-p_i\tau-q_j+p_j\tau)\\
&\qquad\times S_N(-\Delta t)f_N(x_1,\dots,x_N)\\
&\quad=\bar{f}_N(\Delta t,x_1,\dots,x_N).
\end{aligned}\tag{6.2.7}$$

In (6.2.7) we have used the fact that

$$Q(\eta_{ij}\cdot(p_i-p_j))|_{\eta_{ij}\in S_-^2}=-Q(\eta_{ij}\cdot(p_i-p_j))|_{\eta_{ij}\in S_+^2},$$

and relation (6.2.5).

Note that there is no contradiction between definition (6.2.3), (6.2.5) of

$$S_N(-\Delta t)f_N(x_1,\dots,x_N)=f_N(x_1(-\Delta t),\dots,x_N(-\Delta t))$$

and (6.2.7) corresponding to the usual function $S(-\Delta t)f_N(x_1,\dots,x_N)$. Formula (6.2.7) simply defines the function $\bar{S}_N(-\Delta t)f_N(x_1,\dots,x_N)$ as a generalized function, and the averages of the function $\bar{S}_N(-\Delta t)\times f_N(x_1,\dots,x_N)$ (6.2.7) with respect to the observable $\varphi_N(x_1,\dots,x_N)$ should be calculated as the following functional:

$$\begin{aligned}
(\bar{f}_N(\Delta t),\varphi_N)&=(\bar{S}_N(-\Delta t)f_N,\varphi_N)\\
&=\int dx_1\dots dx_N\,\bar{S}_N(-\Delta t)f_N(x_1,\dots,x_N)\varphi_N(x_1,\dots,x_N)\\
&=(S_N(-\Delta t)f_N,\varphi_N).
\end{aligned}\tag{6.2.8}$$

Thus, numerically, the state $S_N(-\Delta t) f_N(x_1, \dots, x_N)$ is given by

$$S_N(-\Delta t) f_N(x_1, \dots, x_N) = f_N(x_1(-\Delta t), \dots, x_N(-\Delta t))$$
$$= f_N(x_1(-\Delta t, (x)_N), \dots, x_1(-\Delta t, (x)_N)).$$

When we calculate the average of $S_N(-\Delta t) f_N(x_1, \dots, x_N)$ with respect to the observable $\varphi_N(x_1, \dots, x_N)$, we use the generalized function $\bar{S}_N(-\Delta t) f_N(x_1, \dots, x_N) = \bar{f}_N(\Delta t, x_1, \dots, x_N)$ given by formula (6.2.7) and calculate the average $(\bar{S}_N(-\Delta t) f_N, \varphi_N)$ as functional (6.2.8) that coincides with (6.2.4).

Note again that functional (6.2.4) is the average of the observable $\varphi_N(x_1, \dots, x_N)$ over the state

$$S_N(-\Delta t) f_N(x_1, \dots, x_N) = f_N(x_1(-\Delta t), \dots, x_N(-\Delta t)),$$

where φ_N is real symmetric test function and $f_N \geq 0$ is also a real symmetric continuously differentiable function normalized so that

$$\int f_N(x_1, \dots, x_N) dx_1 \dots dx_N = 1. \tag{6.2.9}$$

We stress that, in functionals (6.2.4), (6.2.6), and (6.2.8), the contributions of the hyperplanes of lower dimension $q_i - p_i \tau = q_j - p_j \tau$, $0 \leq \tau \leq t$, $1 \leq i < j \leq N$, where stochastic particles interact are taken into account; they are equal to the second term on the right-hand side of (6.2.4).

6.2.2. Infinitesimal operator

It follows from (6.2.8) that

$$\frac{\partial \bar{S}_N(-t)}{\partial t}\Big|_{t=0} f_N(x_1, \dots, x_N)$$
$$= -\sum_{i=1}^{N} p_i \cdot \frac{\partial}{\partial q_i} f_N(x_1, \dots, x_N) + \sum_{i<j=1}^{N} \int_{S_+^2} d\eta_{ij} Q(\eta_{ij} \cdot (p_i - p_j)) \delta(q_i - q_j)$$
$$\left[f_N(x_1, \dots, x_i^*, \dots, x_j^*, \dots, x_N) - f_N(x_1, \dots, x_i, \dots, x_j, \dots, x_N) \right]$$
$$= \bar{\mathcal{H}}_N f_N(x_1, \dots, x_N), \tag{6.2.10}$$

$$x_i^* = (q_i, p_i^*), x_j^* = (q_j, p_j^*).$$

We define formally the group of operators $\bar{S}_N(-t)$ at arbitrary time $-t$ as follows:

$$\bar{S}_N(-t) = \lim_{n \to \infty} \prod_{i=1}^{n} \bar{S}_N(-\Delta t_i), \quad \sum_{i=1}^{n} \Delta t_i = t, \tag{6.2.11}$$

where the operator $\bar{S}_N(-\Delta t)$ for infinitesimal Δt is defined according to (6.2.7), and the infinitesimal generator of the group $\bar{S}_N(-t)$ is defined by (6.2.10) and is equal to $\bar{\mathcal{H}}_N$.

Now define the state $\bar{f}_N(t, x_1, \ldots, x_N)$ at arbitrary time $t > 0$ as follows:

$$\bar{f}_N(t) = \bar{f}_N(t, x_1, \ldots, x_N) = \bar{S}_N(-t) f_N(x_1, \ldots, x_N)$$
$$= \lim_{n\to\infty} \prod_{i=1}^{n} \bar{S}_N(-\Delta t_i) f_N(x_1, \ldots, x_N), \tag{6.2.12}$$
$$\sum_{i=1}^{n} \Delta t_i = t, \quad \bar{f}_N(t + \Delta t) = \bar{S}_N(-\Delta t)\bar{f}_N(t),$$

i.e., $\bar{S}_N(-\Delta t)\bar{f}_N(t)$ is defined by formula (6.2.7) with $\bar{f}_N(t, x_1, \ldots, x_N)$ instead of $f_N(x_1, \ldots, x_N)$.

We define the functional average (with infinitesimal Δt) of the state $\bar{f}_N(t + \Delta t)$:

$$
\begin{aligned}
&(\bar{f}_N(t + \Delta t), \varphi_N) \\
&\quad = (\bar{S}_N(-\Delta t)\bar{f}_N(t), \varphi_N) \\
&\quad = \int dx_1 \ldots dx_N \, \bar{f}_N(t, q_1 - p_1\Delta t, p_1, \ldots, q_N - p_N\Delta t, p_N) \\
&\qquad \times \varphi_N(q_1, p_1, \ldots, q_N, p_N) \\
&\qquad + \sum_{i<j=1}^{N} \int dx_1 \ldots dx_N \int_0^{\Delta t} d\tau \int_{S_+^2} d\eta_{ij}\, Q(\eta_{ij}\cdot(p_i - p_j)) \\
&\qquad \times \delta(q_i - p_i\tau - q_j + p_j\tau) \\
&\qquad \times [\bar{f}_N(t, q_1 - p_1\Delta t, p_1, \ldots, q_i - p_i\tau - p_i^*(\Delta t - \tau), p_i^*, \ldots, \\
&\qquad\qquad q_j - p_j\tau - p_j^*(\Delta t - \tau), p_j^*, \ldots, q_N - p_N\Delta t, p_N) \\
&\qquad - \bar{f}_N(t, q_1 - p_1\Delta t, p_1, \ldots, q_i - p_i\Delta t, p_i, \ldots, q_j - p_j\Delta t, p_j, \ldots, \\
&\qquad\qquad q_N - p_N\Delta t, p_N)\varphi_N(x_1, \ldots, x_N)] \\
&\quad = \int \bar{S}_N(-\Delta t)\bar{f}_N(t, x_1, \ldots, x_N)\varphi_N(x_1, \ldots, x_N) dx_1 \ldots dx_N \\
&\quad = \int \bar{f}_N(t + \Delta t, x_1, \ldots, x_N)\varphi_N(x_1, \ldots, x_N) dx_1 \ldots dx_N.
\end{aligned} \tag{6.2.13}
$$

If follows from (6.2.13) that

$$
\begin{aligned}
&\bar{S}_N(-\Delta t)\bar{f}_N(t, x_1, \dots, x_N)\\
&\quad = \bar{f}_N(t + \Delta t, x_1, \dots, x_N)\\
&\quad = \bar{f}_N(t, q_1 - p_1 \Delta t, p_1, \dots, q_N - p_N \Delta t, p_N)\\
&\qquad + \sum_{i<j=1}^{N} \int_0^{\Delta t} d\tau \int_{S_+^2} d\eta_{ij}\, Q(\eta_{ij} \cdot (p_i - p_j))\delta(q_i - p_i\tau - q_j + p_j\tau)\\
&\qquad\qquad \times [\bar{f}_N(t, q_1 - p_1\Delta t, p_1, \dots, q_i - p_i\tau - p_i^*(\Delta t - \tau), p_i^*, \dots,\\
&\qquad\qquad\qquad q_j - p_j\tau - p_j^*(\Delta t - \tau), p_j^*, \dots, q_N - p_N\Delta t, p_N)\\
&\qquad\qquad - \bar{f}_N(t, q_1 - p_1\Delta t, p_1, \dots, q_i - p_i\Delta t, p_i, \dots,\\
&\qquad\qquad\qquad q_j - p_j\Delta t, p_j, \dots, q_N - p_N\Delta t, p_N)]. \qquad (6.2.14)
\end{aligned}
$$

From (6.2.14) we obtain the differential equation for the state $\bar{f}_N(t, x_1, \dots, x_N)$:

$$
\begin{aligned}
&\frac{\partial}{\partial t}\bar{f}_N(t, x_1, \dots, x_N)\\
&\quad = -\sum_{i=1}^{N} p_i \cdot \frac{\partial}{\partial q_i}\bar{f}_N(t, x_1, \dots, x_N)\\
&\qquad + \sum_{i<j=1}^{N} \int_{S_+^2} d\eta_{ij}\, Q(\eta_{ij} \cdot (p_i - p_j))\delta(q_i - q_j)\\
&\qquad\qquad \times \Big[\bar{f}_N(t, x_1, \dots, x_i^*, \dots, x_j^*, \dots, x_N)\\
&\qquad\qquad\qquad - \bar{f}_N(t, x_1, \dots, x_i, \dots, x_j, \dots, x_N)\Big]\\
&\quad = \bar{\mathcal{H}}_N \bar{f}_N(t, x_1, \dots, x_N), \qquad (6.2.15)
\end{aligned}
$$

$$
x_i^* = (q_i, p_i^*), \quad x_j^* = (q_j, p_j^*),
$$

with the initial condition

$$
\bar{f}_N(t, x_1, \dots, x_N)_{|t=0} = f_N(x_1, \dots, x_N).
$$

Equation (6.2.15) is the *Kolmogorov–Itô–Liouville equation* for $\bar{f}_N(t, x_1, \dots, x_N)$. It was derived from the functional average (6.2.4), (6.2.13). We stress again that, in the functional average (6.2.4), (6.2.13), the contribution of the hyperplanes $q_i - p_i\tau = q_j - p_j\tau$, $1 \le i < j \le N$, is taken into account. This contribution is expressed by the second terms in (6.2.4) and (6.2.13) . Note that equation (6.2.15) defines the derivative of the function $\bar{f}_N(t, x_1, \dots, x_N)$ in the sense of generalized functions.

We suppose that the function $Q(\eta_{ij} \cdot (p_i - p_j))$ satisfies the condition

$$\int \bar{f}_N(t, x_1, \dots, x_N) dx_1 \dots dx_N = \int f_N(0, x_1, \dots, x_N) dx_1 \dots dx_N.$$

6.2.2 Infinitesimal operator with fixed random vectors

We can also differentiate the function $f_N(x_1(-t), \dots, x_N(-t))$ in the sense of pointwise convergence, i.e., we can differentiate $f_N(x_1(-t), \dots, x_N(-t))$ with respect to time along the trajectory $(x_1(-t), \dots, x_N(-t))$ with fixed parameters η_{ij}. Denote the function $f_N(x_1(-t), \dots, x_N(-t))$ with fixed parameters η_{ij} by $\tilde{f}_N(t, x_1, \dots, x_N)$. Repeating word for word our calculation from Chapters 2–4, we obtain the equation

$$\begin{aligned}
\frac{\partial \tilde{f}_N}{\partial t}&(t, x_1, \dots, x_N) \\
&= -\sum_{i=1}^{N} p_i \cdot \frac{\partial}{\partial q_i} \tilde{f}_N(t, x_1, \dots, x_N) \\
&\quad + \sum_{i<j=1}^{N} \Theta(\eta_{ij} \cdot (p_i - p_j)) \eta_{ij} \cdot (p_i - p_j) \delta(q_i - q_j) \\
&\qquad \times \Big[\tilde{f}_N(t, x_1, \dots, x_i^*, \dots, x_j^*, \dots, x_N) \\
&\qquad\quad - \tilde{f}_N(t, x_1, \dots, x_i, \dots, x_j, \dots, x_N) \Big] \\
&= \tilde{\mathcal{H}}_N \tilde{f}_N(t, x_1, \dots, x_N),
\end{aligned} \tag{6.2.16}$$

$$f_N(t, x_1, \dots, x_N)_{|t=0} = f_N(x_1, \dots, x_N),$$

with the known boundary condition according to which if $q_i = q_j$, then, in the first term of (6.2.16), the momenta (p_i, p_j) should be replaced by (p_i^*, p_j^*) with $\eta_{ij} \cdot (p_i - p_j) \geq 0$, $\Theta(\alpha) = 1$, $\alpha > 0$, $\Theta(\alpha) = 0$, $\alpha < 1$.

We now present a new derivation of equation (6.2.16). Equation (6.2.16) and the infinitesimal operator $\tilde{\mathcal{H}}_N$ can be obtained from the following functional average:

$$\begin{aligned}
&\int dx_1 \dots dx_N f_N(q_1 - p_1 \Delta t, p_1, \dots, q_N - p_N \Delta t, p_N) \varphi_N(q_1, p_1, \dots, q_N, p_N) \\
&\quad + \sum_{i<j=1}^{N} \int dx_1 \dots dx_N \int_0^{\Delta t} d\tau (\eta_{ij} \cdot (p_i - p_j)) \Theta(\eta_{ij} \cdot (p_i - p_j)) \\
&\quad \times \delta(q_i - p_i \tau - q_j + p_j \tau)
\end{aligned}$$

$$\times [f_N(q_1 - p_1\Delta t, p_1, \dots, q_i - p_i\tau - p_i^*(\Delta t - \tau), p_i^*, \dots,$$
$$q_j - p_j\tau - p_j^*(\Delta t - \tau), p_j^*, \dots, q_N - p_N\Delta t, p_N)$$
$$- f_N(q_1 - p_1\Delta t, p_1, \dots, q_i - p_i\Delta t, p_i, \dots,$$
$$q_j - p_j\Delta t, p_j, \dots, q_N - p_N\Delta t, p_N)]\varphi_N(x_1, \dots, x_N)$$
$$= \int dx_1 \dots dx_N \tilde{S}_N(-\Delta t) f_N(x_1, \dots, x_N)\varphi_N(x_1, \dots, x_N)$$
$$= (\tilde{f}_N(\Delta t), \varphi_N). \tag{6.2.17}$$

Equality (6.2.17) holds for an arbitrary test function $\varphi_N(x_1, \dots, x_N)$, and $\tilde{S}_N(-\Delta t) f_N(x_1, \dots, x_N)$ is determined from (6.2.17) as follows:

$$\tilde{S}_N(-\Delta t) f_N(x_1, \dots, x_N) = \tilde{f}_N(\Delta t, x_1, \dots, x_N)$$
$$= f_N(q_1 - p_1\Delta t, p_1, \dots, q_N - p_N\Delta t, p_N)$$
$$+ \sum_{i<j=1}^{N} \int_0^{\Delta t} d\tau \Theta(\eta_{ij} \cdot (p_i - p_j))\eta_{ij} \cdot (p_i - p_j)\delta(q_i - p_i\tau - q_j + p_j\tau)$$
$$\times [f_N(q_1 - p_1\Delta t, p_1, \dots, q_i - p_i\tau - p_i^*(\Delta t - \tau), p_i^*, \dots,$$
$$q_j - p_j\tau - p_j^*(\Delta t - \tau), p_j^*, \dots, q_N - p_N\Delta t, p_N)$$
$$- f_N(q_1 - p_1\Delta t, p_1, \dots, q_i - p_i\Delta t, p_i, \dots, q_j - p_j\Delta t,$$
$$p_j, \dots, q_N - p_N\Delta t, p_N)], \tag{6.2.18}$$

$$\frac{\partial \tilde{S}_N(-t)}{\partial t}\Big|_{t=0} f_N(t, x_1, \dots, x_N) = \tilde{\mathcal{H}}_N f_N(x_1, \dots, x_N).$$

We formally define the group of operators $\tilde{S}_N(-t)$ at arbitrary time $-t$ as follows:

$$\tilde{S}_N(-t) = \lim_{n\to\infty} \prod_{i=1}^{n} \bar{S}_N(-\Delta t_i), \quad \sum_{i=1}^{n} \Delta t_i = t,$$

where the operator $\bar{S}_N(-\Delta t)$ for infinitesimal Δt is defined according to (6.2.18), and the infinitesimal generator of the group $\tilde{S}_N(-t)$ is equal to $\tilde{\mathcal{H}}_N$. Using (6.2.18) and the definition of $\tilde{S}_N(-t)$, one can obtain a distribution function at arbitrary time t, namely, $\tilde{f}_N(t, x_1, \dots, x_N) = \tilde{S}_N(-t) f_N(x_1, \dots, x_N)$. Let us suppose that the distribution function is already obtained at time t; then $\tilde{f}_N(t + \Delta t, x_1, \dots, x_N)$ is defined through $\tilde{f}_N(t, x_1, \dots, x_N)$ as follows:

$$\tilde{f}_N(t + \Delta t, x_1, \dots, x_N)$$
$$= \tilde{S}_N(-\Delta t)\tilde{f}_N(t, x_1, \dots, x_N)$$

$$
\begin{aligned}
&= \tilde{f}_N(t, q_1 - p_1\Delta t, p_1, \dots, q_N - p_N\Delta t, p_N) \\
&\quad + \sum_{i<j=1}^{N} \int_0^{\Delta t} d\tau \Theta(\eta_{ij}\cdot(p_i - p_j))\eta_{ij}\cdot(p_i - p_j)\delta(q_i - p_i\tau - q_j + p_j\tau) \\
&\qquad \times [\tilde{f}_N(t, q_1 - p_1\Delta t, p_1, \dots, q_i - p_i\tau - p_i^*(\Delta t - \tau), p_i^*, \dots, \\
&\qquad\quad q_j - p_j\tau - p_j^*(\Delta t - \tau), p_j^*, \dots, q_N - p_N\Delta t, p_N) \\
&\qquad\quad - \tilde{f}_N(t, q_1 - p_1\Delta t, p_1, \dots, q_i - p_i\Delta t, p_i, \dots, q_j - p_j\Delta t, \\
&\qquad\quad p_j, \dots, q_N - p_N\Delta t, p_N)]. \qquad (6.2.19)
\end{aligned}
$$

Or in terms of averages with test functions $\varphi_N(x_1, \dots, x_N)$

$$
\begin{aligned}
&(\tilde{f}_N(t + \Delta t), \varphi_N) \\
&= \int dx_1 \dots dx_N \tilde{f}_N(t, q_1 - p_1\Delta t, p_1, \dots, q_N - p_N\Delta t, p_N)\varphi_N(x_1, \dots, x_N) \\
&\quad + \sum_{i<j=1}^{N} \int dx_1 \dots dx_N \bigg\{ \int_0^{\Delta t} d\tau \Theta(\eta_{ij}\cdot(p_i - p_j))\eta_{ij}\cdot(p_i - p_j) \\
&\qquad \times \delta(q_i - p_i\tau - q_j + p_j\tau) \\
&\qquad \times \big[\tilde{f}_N(t, q_1 - p_1\Delta t, p_1, \dots, q_i - p_i\tau - p_i^*(\Delta t - \tau), p_i^*, \dots, \\
&\qquad\qquad q_j - p_j\tau - p_j^*(\Delta t - \tau), p_j^*, \dots, q_N - p_N\Delta t, p_N) \\
&\qquad\quad - \tilde{f}_N(t, q_1 - p_1\Delta t, p_1, \dots, q_i - p_i\Delta t, p_i, \dots, q_j - p_j\Delta t, p_j, \dots, \\
&\qquad\qquad q_N - p_N\Delta t, p_N)\big]\bigg\}\varphi_N(x_1, \dots, x_N). \qquad (6.2.20)
\end{aligned}
$$

The differential equation (6.2.16) follows from (6.2.19) and (6.2.20).

6.2.3 Duality principle

Due to the importance of the problem, this subsection repeats in many respects an analogous subsection from Chapter 4. We now explain in what sense the functions $\tilde{S}_N(-\Delta t)f_N(x_1, \dots, x_N)$, $\bar{S}_N(-\Delta t)f_N(x_1, \dots, x_N)$ given by formulas (6.2.18) and (6.2.7) are equivalent to the function $S_N(-\Delta t)f_N(x_1, \dots, x_N)$ given by formulas (6.2.3) and (6.2.5). We have

$$
\begin{aligned}
S_N(-\Delta t)f_N(x_1, \dots, x_N) &= f_N(x_1(-\Delta t), \dots, x_N(-\Delta t) \\
&= f_N(q_1 - p_1\Delta t, p_1, \dots, q_N - p_N\Delta t, p_N)
\end{aligned}
$$

if $q_i - p_i\tau \neq q_j - p_j\tau$ for all $i, j \in \{1, \ldots, N\}$ and $0 \leq \tau \leq \Delta t$,

$$\begin{aligned} S_N(-\Delta t) f_N(x_1, \ldots, x_N) &= f_N(x_1(-\Delta t), \ldots, x_N(-\Delta t)) \\ &= f_N(q_1 - p_1\Delta t, p_1, \ldots, q_i - p_i\tau - p_i^*(\Delta t - \tau), p_i^*, \ldots, \\ &\qquad q_j - p_j\tau - p_j^*(\Delta t - \tau), p_j^*, \ldots, q_N - p_N\Delta t, p_N) \end{aligned} \tag{6.2.5'}$$

if $q_i - p_i\tau = q_j - p_j\tau$, $0 \leq \tau \leq \Delta t$, $\eta_{ij} \subset S_+^2$, and

$$\begin{aligned} S_N(-\Delta t) f_N(x_1, \ldots, x_N) &= f_N(x_1(-\Delta t), \ldots, x_N(-\Delta t)) \\ &= f_N(q_1 - p_1\Delta t, p_1, \ldots, q_i - p_i\Delta t, p_i, \ldots, \\ &\qquad q_j - p_j\Delta t, p_j, \ldots, q_N - p_N\Delta t, p_N) \end{aligned}$$

if $q_i - p_i\tau = q_j - p_j\tau$, $0 \leq \tau \leq \Delta t$, $\eta_{ij} \subset S_-^2$, $i, j \in \{1, \ldots, N\}$.

Thus, the functions $S_N(-\Delta t) f_N(x_1, \ldots, x_N)$ and $S_N(-\Delta t) f_N(t, x_1, \ldots, x_N)$ are numerically given by (6.2.5′) in the entire phase space $(x_1, \ldots, x_N)$. Expressions (6.2.18) and (6.2.19) for $\tilde{S}_N(-\Delta t) f_N(x_1, \ldots, x_N)$ and $\tilde{S}_N(-\Delta t) \tilde{f}_N(t, x_1, \ldots, x_N)$ determine how to integrate the functions $S_N(-\Delta t) f_N(x_1, \ldots, x_N)$, $S_N(-\Delta t) \times f_N(t, x_1, \ldots, x_N)$ with the test function $\varphi_N(x_1, \ldots, x_N)$, but with fixed random vectors η_{ij}, $1 \leq i < j \leq N$, and to take into account the contributions of the hyperplanes $q_i - p_i\tau = q_j - p_j\tau$, $1 \leq i < j \leq N$.

Expressions (6.2.7) and (6.2.14) for the functions $\bar{S}_N(-\Delta t) f_N(x_1, \ldots, x_N)$ and $\bar{S}_N(-\Delta t) \bar{f}_N(t, x_1, \ldots, x_N)$ determine how to integrate these functions with test functions $\varphi_N(x_1, \ldots, x_N)$, to average with respect to the random vectors η_{ij}, and to take into account the contribution of the hyperplanes $q_i - p_i\tau = q_j - p_j\tau$.

We can now formulate the *principle of duality for the distribution function* $f_N(t, x_1, \ldots, x_N)$. Assume that the initial distribution function $f_N(0, x_1, \ldots, x_N) \equiv f_N(x_1, \ldots, x_N)$ is symmetric, continuously differentiable, and normalized to one in the phase space. Then the distribution function (6.2.3), (6.2.5′) $f_N(t, x_1, \ldots, x_N) = S_N(-t) f_N(x_1, \ldots, x_N)$ is a well-defined continuously differentiable function everywhere outside the hyperplanes of lower dimension where particles interact. But in the functional average with some observable $\varphi_N(x_1, \ldots, x_N)$, which is a real symmetric smooth test function, we consider some definite generalized function $\tilde{f}_N(t, x_1, \ldots, x_N)$, $\bar{f}_N(t, x_1, \ldots, x_N)$ corresponding to $S(-t) f_N(x_1, \ldots, x_N) = f_N(t, x_1, \ldots, x_N)$ and calculate the contribution of these hyperplanes of lower dimension where particles interact. Calculating the functional averages $(\tilde{f}(t), \varphi_N)$ or $(\bar{f}_N(t), \varphi_N)$ we use $\tilde{f}_N(t)$ or $\bar{f}_N(t)$ instead of $f_N(t)$. We cannot calculate $(\tilde{f}_N(t), \varphi_N)$ or $(\bar{f}_N(t), \varphi_N)$ directly for arbitrary finite time t; we have the explicit formulas (6.2.4) and (6.2.17) only for infinitesimal Δt. For arbitrary time t we use formulas (6.2.14) and (6.2.19) for the definition of $\bar{f}_N(t + \Delta t)$ and $\tilde{f}_N(t + \Delta t)$ through already defined $\bar{f}_N(t)$, $\tilde{f}_N(t)$ and then calculate $(\bar{f}_N(t + \Delta t), \varphi_N)$ according to (6.2.13), and $(\tilde{f}_N(t + \Delta t), \varphi_N)$ according to (6.2.20).

The duality principle defines the generalized functions $\bar{f}_N(t, x_1, \ldots, x_N) = \bar{S}_N(-t) f_N(x_1, \ldots, x_N)$ or $\tilde{f}_N(t, x_1, \ldots, x_N) = \tilde{S}_N(-t) f_N(x_1, \ldots, x_N)$ through the usual function $f_N(t, x_1, \ldots, x_N) = S_N(-t) f_N(x_1, \ldots, x_N)$ by formulas (6.2.7), (6.2.14) or (6.2.18), (6.2.19). If one has to consider the distribution function as a usual function (numerically), then one should take $f_N(t, x_1, \ldots, x_N) = S_N(-t) f_N(x_1, \ldots, x_N)$; if one has to calculate the functional average (6.2.13) or (6.2.20) with observable $\varphi_N(x_1, \ldots, x_N)$ then one should take the generalized function $\bar{f}_N(t, x_1, \ldots, x_N) = \bar{S}_N(-t) f_N(x_1, \ldots, x_N)$ or the generalized function $\tilde{f}_N(t, x_1, \ldots, x_N) = \tilde{S}_N(-t) f_N(x_1, \ldots, x_N)$.

It has been shown in Chapters 3 and 4 that the differential equation for the function $f_N(t, x_1, \ldots, x_N)$ in the sense of pointwise convergence has the form

$$\begin{aligned} \frac{\partial f_N}{\partial t}(t, x_1, \ldots, x_N) &= -\sum_{i=1}^{N} p_i \cdot \frac{\partial}{\partial q_i} f_N(t, x_1, \ldots, x_N) \\ &\quad + \sum_{i<j=1}^{N} \Theta(\eta_{ij} \cdot (p_i - p_j)) \delta(\Delta t - \tau_{ij})_{|\Delta t=0} \\ &\qquad \times \Big[f_N(t, x_1, \ldots, x_i^*, \ldots, x_j^*, \ldots, x_N) \\ &\qquad\quad - f_N(t, x_1, \ldots, x_i, \ldots, x_j, \ldots, x_N) \Big], \quad q_i = q_j, \end{aligned} \tag{6.2.21}$$

with the above-described boundary condition in the Poisson bracket, according to which (p_i, p_j) should be replaced by (p_i^*, p_j^*) if $q_i - q_j = 0$ and $\big[f_N(t + 0, x_1, \ldots, x_N) = f_N(t, x_1, \ldots, x_i^*, \ldots, x_j^*, \ldots x_N)\big]$, $q_i = q_j$, $\eta_{ij} \cdot (p_i - p_j) > 0$, $\big[f_N(t + 0, x_1, \ldots, x_N) = f_N(t, x_1, \ldots, x_N)\big]$, $q_i = q_j$, $(\eta_{ij} \cdot (p_i - p_j) \leq 0)$.

Here, τ_{ij} is the time of collision. In the coordinate system where the first component of the vectors (q_i, p_i, q_j, p_j) is directed along the vector η_{ij}, the time of collision τ_{ij} is defined as follows:

$$\tau_{ij} = \frac{q_i^1 - q_j^1}{p_i^1 - p_j^1}.$$

Then the (i, j)-th term in (6.2.23) can be expressed as follows:

$$\begin{aligned} &\Theta(p_i^1 - p_j^1) \delta(q_i^1 - q_j^1)(p_i^1 - p_j^1) \\ &\quad \times [f_N(t, x_1, \ldots, x_i^*, \ldots, x_j^*, \ldots, x_N) \\ &\qquad - f_N(t, x_1, \ldots, x_i, \ldots, x_j, \ldots, x_N)]|_{(q_i^2 = q_j^2), (q_i^3 = q_j^3)}. \end{aligned}$$

This term is different from zero on the first axis $q_i^1 - q_j^1$ (with respect to the vector $q_i - q_j$, i.e., for $q_i^2 - q_j^2 = 0$, $q_i^3 - q_j^3 = 0$) and, regarded as a generalized function in the three-dimensional space, is equal to (for analogous calculations, see [GGV] pp. 48–56)

$$\begin{aligned}
&\Theta(p_i^1 - p_j^1)\delta(q_i^1 - q_j^1)\delta(q_i^2 - q_j^2)\delta(q_i^3 - q_j^3)(p_i^1 - p_j^1) \\
&\qquad \times [f_N(t, x_1, \ldots, x_i^*, \ldots, x_j^*, \ldots, x_N) \\
&\qquad\qquad - f_N(t, x_1, \ldots, x_i, \ldots, x_j, \ldots, x_N)] \\
&\quad = \Theta(\eta_{ij} \cdot (p_i - p_j))\delta(q_i - q_j)\eta_{ij} \cdot (p_i - p_j) \\
&\qquad\quad \times [f_N(t, x_1, \ldots, x_i^*, \ldots, x_j^*, \ldots, x_N) \\
&\qquad\qquad - f_N(t, x_1, \ldots, x_i, \ldots, x_j, \ldots, x_N)]. \qquad (6.2.22)
\end{aligned}$$

The expression obtained does not depend on the choice of a coordinate system because $\delta(q_i - q_j)$ and $\eta_{ij} \cdot (p_i - p_j)$ are invariant under rotation. Substituting (6.2.22) in (6.2.21) we obtain (6.2.16).

Recall that the operator

$$\begin{aligned}
&-\sum_{i=1}^{N} p_i \cdot \frac{\partial}{\partial q_i} f_N(t, x_1, \ldots, x_N) + \sum_{i<j=1}^{N} \Theta(\eta_{ij} \cdot (p_i - p_j))\delta(\Delta t - \tau_{ij})|_{\Delta t = 0} \\
&\qquad \times \Big[f_N(t, x_1, \ldots, x_i^*, \ldots, x_j^*, \ldots, x_N) \\
&\qquad\qquad - f_N(t, x_1, \ldots, x_i, \ldots, x_j, \ldots, x_N) \Big]|_{q_i = q_j} \\
&\quad = \mathcal{H}_N f_N(x_1, \ldots, x_N) \qquad (6.2.23)
\end{aligned}$$

with one-dimensional δ-functions is equivalent to the operator

$$-\sum_{i=1}^{N} p_i \cdot \frac{\partial}{\partial q_i} f_N(t, x_1, \ldots, x_N) = \mathcal{H}_N f_N(t, x_1, \ldots, x_N) \qquad (6.2.24)$$

with the boundary condition according to which, at the points $q_i = q_j$, $i, j \in \{1, \ldots, N\}$, the momenta p_i and p_j should be replaced by p_i^* and p_j^* if $\eta_{ij} \cdot (p_i - p_j) \geq 0$, and (p_i, p_j) do not change if $\eta_{ij} \cdot (p_i - p_j) \leq 0$.

Remark 6.1. As shown in the previous chapters, the boundary condition can be omitted in the infinitesimal operator $\mathcal{H}_N$ (6.2.23) with the one-dimensional δ-function because, due to the δ-function, the boundary conditions are automatically fulfilled. The boundary conditions are necessary in the infinitesimal operator $\mathcal{H}_N$ (6.2.24) without δ-functions.

Thus, we have three expressions for the infinitesimal operator $\mathcal{H}_N$ of the group $S_N(-t)$. The first one $\bar{\mathcal{H}}_N$ (6.2.15) was obtained in a weak sense, and it shows how to take into account, in the functional average, the hypersurfaces of lower dimensions $q_i = q_j, i, j \in \{1, \ldots, N\}$ where particles interact. In the second one $\tilde{\mathcal{H}}_N$ (6.2.16) (also calculated in a weak sense), the averaging with respect to the random vectors η_{ij} was not performed, but the hypersurfaces $q_i = q_j, i, j \in \{1, \ldots, N\}$, were taken into account. In the third one $\mathcal{H}_N$ (6.2.23), (6.2.24) calculated point by point, the infinitesimal operator $\mathcal{H}_N$ of the group $S_N(-t)$ is equal to the operator $-\sum_{i=1}^{N} p_i \cdot \frac{\partial}{\partial q_i}$ (6.2.26) with the boundary conditions at $q_i = q_j, i, j \in \{1, \ldots, N\}$ or with the one–dimensional δ-functions (6.2.23).

All these expressions for the infinitesimal operator $\mathcal{H}_N$ are equivalent, but the first one (6.2.15) and the second one (6.2.16) show how to calculate the average of $\frac{\partial \bar{f}}{\partial t}(t, x_1, \ldots, x_N)$ and $\frac{\partial \tilde{f}_N(t,x_1,\ldots,x_N)}{\partial t}$ with observable $\varphi_N(x_1, \ldots, x_N)$, or they define $\frac{\partial \bar{f}_N(t,x_1,\ldots,x_N)}{\partial t}$ and $\frac{\partial \tilde{f}_N(t,x_1,\ldots,x_N)}{\partial t}$ as generalized functions. The third expression for the infinitesimal operator (6.2.23), (6.2.24) defines $\frac{\partial f_N(t,x_1,\ldots,x_N)}{\partial t} = \mathcal{H}_N f_N(t, x_1, \ldots, x_N)$ in the sense of pointwise differentiation and defines it as a usual function with jumps at $q_i = q_j, i, j \in \{1, \ldots, N\}$, which is expressed in the boundary conditions.

Thus, for the derivative $\frac{\partial f_N(t,x_1,\ldots,x_N)}{\partial t}$ the principle of duality is also formulated as for $f_N(t, x_1, \ldots, x_N)$, and according to it the same $\frac{\partial f_N(t,x_1,\ldots,x_N)}{\partial t}$ is considered as a usual function (6.2.24) or as special generalized functions (6.2.15), (6.2.16) in the functional average.

It is also useful to consider the action of the operator of evolution $S_N(-\Delta t) f_N(x_1, \ldots, x_s, x_{s+1}, \ldots, x_N)$ as a usual function with respect to the variables $(x_1, \ldots, x_s)$ and a generalized function with respect to the variables $(x_{s+1}, \ldots, x_N)$. Namely, one associates the following expressions with the function $S_N(-\Delta t) f_N(x_1, \ldots, x_s, x_{s+1}, \ldots, x_N)$:

$$
\begin{aligned}
&\bar{f}_{s,N-s}(\Delta t, x_1, \ldots, x_s, x_{s+1}, \ldots, x_N) \\
&\quad = \bar{S}_{s,N-s}(-\Delta t) f_N(x_1, \ldots, x_s, x_{s+1}, \ldots, x_N) \\
&\quad = S_s(-\Delta t)((x)_s) S^0_{N-s}(-\Delta t, (x)^s_N) f_N(x_1, \ldots, x_s, x_{s+1}, \ldots, x_N) \\
&\qquad + \sum_{i<j=1}^{N}{}' \int_0^{\Delta t} d\tau \int_{S^2_+} d\eta_{ij}\, Q_{ij}(\eta_{ij} \cdot (p_i - p_j)) \delta(q_i - p_i \tau - q_j + p_j \tau) \\
&\qquad\quad \times \big[f_N(q_1 - p_1 \Delta t, p_1, \ldots, q_i - p_i \tau - p_i^*(\Delta t - \tau), p_i^*, \ldots, \\
&\qquad\qquad\qquad q_j - p_j \tau - p_j^*(\Delta t - \tau), p_j^*, \ldots, q_N - p_N \Delta t, p_N) \\
&\qquad\qquad - f_N(q_1 - p_1 \Delta t, p_1, \ldots, q_i - p_i \Delta t, p_i, \ldots, \\
&\qquad\qquad\qquad q_j - p_j \Delta t, p_j, \ldots, q_N - p_N \Delta t, p_N) \big],
\end{aligned} \tag{6.2.25}
$$

$$\bar{f}_{s,N-s}(t, x_1, \dots, x_s, x_{s+1}, \dots, x_N)$$
$$= \lim_{n\to\infty} \prod_{i=1}^{n} \bar{S}_{s,N-s}(\Delta t_i) f_N(x_1, \dots, x_s, x_{s+1}, \dots, x_N), \quad \sum \Delta t_i = t.$$

In (6.2.25), we have used the following notation: $S^0_{N-s}(-\Delta t, (x)^s_N)$ is the operator of free evolution of $N - s$ particles, $\sum'^{N}_{i<j=1}$ means that summation with respect to $1 \le i < j \le s$ is excluded, and $(x)^s_N = (x_{s+1}, \dots, x_N)$.

By analogy with (6.2.15) one obtains

$$\frac{\partial \bar{f}_{s,N-s}(t, x_1, \dots, x_s, x_{s+1}, \dots, x_N)}{\partial t}$$
$$= \Big(\mathcal{H}_s - \sum_{i=s+1}^{N} p_i \cdot \frac{\partial}{\partial q_i}\Big) \bar{f}_{s,N-s}(t, x_1, \dots, x_s, x_{s+1}, \dots, x_N)$$
$$+ \sum_{i<j=1}^{N}{}' \int_{S^2_+} d\eta_{ij}\, Q_{ij}(\eta_{ij} \cdot (p_i - p_j))\delta(q_i - q_j)$$
$$\times \big[\bar{f}_{s,N-s}(t, x_1, \dots, x_i^*, \dots, x_j^*, \dots, x_N)$$
$$- \bar{f}_{s,N-s}(t, x_1, \dots, x_i, \dots, x_j, \dots, x_N)\big], \qquad (6.2.26)$$

where the operator $\mathcal{H}_s$ is given by formulas (6.2.23) or (6.2.24) with boundary conditions and is calculated in the sense of pointwise convergence (see also (3.2.15′) and (3.3.14)).

6.3 Hierarchy for correlation functions

6.3.1 Derivation of hierarchy from equation for distribution function

We define the following sequence of correlation functions:

$$\bar{F}_s^{(N)}(t, x_1, \dots, x_s)$$
$$= N(N-1)\dots(N-s+1) \int dx_{s+1} \dots dx_N$$
$$\times \bar{f}_N(t, x_1, \dots, x_s, \dots, x_{s+1}, \dots, x_N), \quad 1 \le s \le N. \qquad (6.3.1)$$

Using equation (6.2.15), one can derive the following hierarchy for sequence (6.3.1) (see details in Chapter 3):

$$
\begin{aligned}
&\frac{\partial \bar{F}_s^{(N)}(t, x_1, \dots, x_s)}{\partial t} \\
&\quad = -\sum_{i=1}^{s} p_i \cdot \frac{\partial}{\partial q_i} \bar{F}_s^{(N)}(t, x_1, \dots, x_s) \\
&\qquad + \sum_{i<j=1}^{s} \int_{S_+^2} d\eta_{ij}\, Q(\eta_{ij} \cdot (p_i - p_j))\delta(q_i - q_j) \\
&\qquad\quad \times [\bar{F}_s(t, x_1, \dots, x_i^*, \dots, x_j^*, \dots, x_s) - \bar{F}_s(t, x_1, \dots, x_i, \dots, x_j, \dots, x_s)] \\
&\qquad + \sum_{i=1}^{s} \int dx_{s+1} \int_{S_+^2} d\eta_{is+1} Q(\eta_{is+1} \cdot (p_i - p_{s+1}))\delta(q_i - q_{s+1}) \\
&\qquad\quad \times \big[\bar{F}_{s+1}^{(N)}(t, x_1, \dots, x_i^*, \dots, x_{s+1}^*) \\
&\qquad\qquad - \bar{F}_{s+1}^{(N)}(t, x_1, \dots, x_i, \dots, x_{s+1})\big], \quad 1 \le s \le N. \qquad (6.3.2)
\end{aligned}
$$

Performing the formal thermodynamic limit transition as $N \to \infty$ for sequence (6.3.1) and supposing that one can also perform this limit transition in hierarchy (6.3.2), we obtain the limiting hierarchy

$$
\begin{aligned}
&\frac{\partial \bar{F}_s(t, x_1, \dots, x_s)}{\partial t} \\
&\quad = -\sum_{i=1}^{s} p_i \cdot \frac{\partial}{\partial q_i} \bar{F}_s(t, x_1, \dots, x_s) \\
&\qquad + \sum_{i<j=1}^{s} \int_{S_+^2} d\eta_{ij}\, Q(\eta_{ij} \cdot (p_i - p_j))\delta(q_i - q_j) \\
&\qquad\quad \times \big[\bar{F}_s(t, x_1, \dots, x_i^*, \dots, x_j^*, \dots, x_s) - \bar{F}_s(t, x_1, \dots, x_i, \dots, x_j, \dots, x_s)\big] \\
&\qquad + \sum_{i=1}^{s} \int dx_{s+1} \int_{S_+^2} d\eta_{is+1} Q(\eta_{is+1} \cdot (p_i - p_{s+1}))\delta(q_i - q_{s+1}) \\
&\qquad\quad \times \big[\bar{F}_{s+1}(t, x_1, \dots, x_i^*, \dots, x_{s+1}^*) \\
&\qquad\qquad - \bar{F}_{s+1}(t, x_1, \dots, x_i, \dots, x_{s+1})\big], \quad s \ge 1, \qquad (6.3.3)
\end{aligned}
$$

$$
\bar{F}_s(t, x_1, \dots, x_s) = \lim_{N \to \infty} \bar{F}_s^{(N)}(t, x_1, \dots, x_s).
$$

Note that the correlation functions $\bar{F}_s(t, x_1, \dots, x_s)$ do not depend on any random vector η_{ij}.

The sequence of correlation functions $\bar{F}_s(t, x_1, \ldots, x_s)$, $s \geq 1$, and hierarchy (6.3.3) for it were obtained through the distribution function $\bar{f}_N(t, x_1, \ldots, x_N)$ considered as a definite generalized function.

We have used $\bar{f}_N(t, x_1, \ldots, x_N)$ because it was necessary to take into account contribution of the hyperplanes of lower dimension where the stochastic particles interact. The definition of $\bar{F}_s^{(N)}(t, x_1, \ldots, x_s)$ (6.3.1) completely resembles the definition of correlation functions in the framework of standard statistical mechanics.

The first two terms on the right-hand side of hierarchy (6.3.3) are the result of the action of the infinitesimal operator $\bar{\mathcal{H}}_s$ (of the group $\bar{S}_s(-t)$), which is useful for the functional average. We want to derive a hierarchy for the sequence of correlation functions considered as usual (not generalized) functions. To do this we replace operator the $\bar{\mathcal{H}}_s$ in (6.3.2) by the equivalent operator $\mathcal{H}_s = -\sum_{i=1}^s p_i \cdot \frac{\partial}{\partial q_i}$ (6.2.26) with known boundary condition. One obtains the hierarchy

$$\begin{aligned}
&\frac{\partial F_s(t, x_1, \ldots, x_s)}{\partial t}\\
&\quad = -\sum_{i=1}^{s} p_i \cdot \frac{\partial}{\partial q_i} F_s(t, x_1, \ldots, x_s)\\
&\qquad + \sum_{i=1}^{s} \int dx_{s+1} \int_{S_+^2} d\eta_{is+1} Q(\eta_{is+1} \cdot (p_i - p_{s+1}))\delta(q_i - q_{s+1})\\
&\qquad\quad \times \big[F_{s+1}(t, x_1, \ldots, x_i^*, \ldots, x_s, x_{s+1}^*)\\
&\qquad\qquad - F_{s+1}(t, x_1, \ldots, x_i, \ldots, x_s, x_{s+1}^*)\big], \quad s \geq 1, \qquad (6.3.4)
\end{aligned}$$

with boundary conditions according to which, at $q_i = q_j, i, j \in \{1, \ldots, s\}$, the momenta (p_i, p_j) in the first term on the right-hand side of (6.3.4) should be replaced by (p_i^*, p_j^*).

Hierarchy (6.3.4) can be written in an equivalent form with the operator $\mathcal{H}_s$ (6.2.23) instead of $\mathcal{H}_s$ (6.2.26), namely,

$$\begin{aligned}
&\frac{\partial F_s(t, x_1, \ldots, x_s)}{\partial t}\\
&\quad = -\sum_{i=1}^{s} p_i \cdot \frac{\partial}{\partial q_i} F_s(t, x_1, \ldots, x_s)\\
&\qquad + \sum_{i<j=1}^{s} \Theta(\eta_{ij}(p_i - p_j))\delta(\Delta t - \tau_{ij})|_{\Delta t=0}\\
&\qquad\quad \times \big[F_s(t, x_1, \ldots, x_i^*, \ldots, x_j^*, \ldots, x_s)\\
&\qquad\qquad - F_s(t, x_1, \ldots, x_i, \ldots, x_j, \ldots, x_s)\big]|_{q_i=q_j}
\end{aligned}$$

$$+\sum_{i=1}^{s}\int dx_{s+1}\int_{S_+^2} d\eta_{is+1}Q(\eta_{is+1}\cdot(p_i-p_{s+1}))$$
$$\times\big[F_{s+1}(t,x_1,\dots,x_i^*,\dots,x_s,x_{s+1}^*)$$
$$-F_{s+1}(t,x_1,\dots,x_i,\dots,x_s,x_{s+1})\big],\quad s\geq 1, \tag{6.3.5}$$

without boundary condition.

In (6.3.4) and (6.3.5) we have used an equivalent representation of the infinitesimal operator $\mathcal{H}_s$. Note that correlation functions in (6.3.4), (6.3.5) depend on the random vectors $\eta_{ij}, i, j \in \{1,\dots,s\}$.

Remark 6.2. Hierarchy (6.3.4) can be derived directly from the Liouville–Itô equation

$$\frac{\partial}{\partial t}f_N(t,x_1,\dots,x_N) = -\sum_{i=1}^{N} p_i\cdot\frac{\partial}{\partial q_i}f_N(t,x_1,\dots,x_N)$$
$$=\mathcal{H}_N f_N(t,x_1,\dots,x_N) \tag{6.3.6}$$

with boundary condition if one takes into account the boundary condition as in Chapter 3 (Subsection 3.4.2).

It is easy to check that hierarchy (6.3.4) or (6.3.5) with $\mathcal{H}_s$ (6.2.24) or (6.2.23) instead of $\bar{\mathcal{H}}_s$ can also be derived if one uses $\bar{f}_{s,N-s}(t,x_1,\dots,x_s,x_{s+1},\dots,x_N)$ (6.2.25) instead of $\bar{f}_N(t,x_1,\dots, x_s,x_{s+1},\dots,x_N)$ in the definition of correlation function (6.3.1) and equation (6.2.26).

The principle of duality for hierarchies (6.3.3)–(6.3.6) consists of the following: All hierarchies differ only by the infinitesimal operator of the evolution operator $S_s(-t)$ calculated in the sense of pointwise convergence or in the sense of generalized functions. If one has a hierarchy with $\mathcal{H}_s F_s(t,x_1,\dots,x_s)$, i.e., with correlation functions as usual functions, then hierarchy with $\bar{\mathcal{H}}_s\bar{F}_s(t,x_1,\dots,x_s)$, $\tilde{\mathcal{H}}_s\tilde{F}_s(t,x_1,\dots,x_s)$ is associated with the corresponding correlation functions considered as generalized functions and vice versa. One kind of a hierarchy or an other can be obtained by the replacement $\mathcal{H}_s \leftrightarrow \bar{\mathcal{H}}_s$.

In what follows we will use the hierarchy for the sequence of $F_s(t,q_1,\dots,q_s)$, $s\geq 1$, with the infinitesimal operator $\mathcal{H}_s$ (6.2.6), namely

$$\frac{\partial F_s(t,x_1,\dots,x_s)}{\partial t}$$
$$=\mathcal{H}_sF_s(t,x_1,\dots,x_s)+\sum_{i=1}^{s}\int dx_{s+1}\int_{S_+^2}d\eta_{is+1}Q(\eta_{is+1}\cdot(p_i-p_j))\delta(q_i-q_{s+1})$$
$$\times\big[F_{s+1}(t,x_1,\dots,x_i^*,\dots,x_s,x_{s+1}^*)$$

$$- F_{s+1}(t, x_1, \dots, x_i, \dots, x_s, x_{s+1})\Big] \tag{6.3.7}$$

with the initial condition

$$F_s(t, x_1, \dots, x_s)|_{t=0} = F_s(x_1, \dots, x_s)$$

and a known boundary condition. Hierarchies (6.3.4), (6.3.5) can also be obtained from equation (6.2.16) by integrating over $x_{s+1}, \dots, x_N$ and averaging over all η_{ij} excluding those η_{ij} with $i, j \in \{1, \dots, s\}$. One obtains hierarchy (6.3.6) with $\tilde{\mathcal{H}}_s$ instead of $\mathcal{H}_s$ (see details in Chapter 3). Then one uses the duality principle and replaces $\tilde{\mathcal{H}}_s$ by $\mathcal{H}_s$. Note that the correlation functions $F_s(t, x_1, \dots, x_s)$ depend on all random vectors η_{ij} with $i, j \in \{1, \dots, s\}$ that correspond to collisions of particles with numbers $(1, \dots, s)$ and appear on the entire interval $[-t, 0]$.

From the definition of the correlation functions $F_s(t, (x)_s)$, $\bar{F}_s(t, (x)_s)$ it follows that $F_s(t, (x)_s) = \bar{F}_s(t, (x)_s)$ outside the hypersurfaces where stochastic particles interact. This can easily be seen from the definition of $\bar{f}_N(t, (x)_N)$ (6.2.7), (6.2.12) and $\bar{f}_{s,N-s}(t, (x)_s)$ (6.2.25).

6.3.2 Derivation of hierarchy from functional average

We now define the functional average for the following s-particle observable:

$$\varphi_N(x_1, \dots, x_N) = \sum_{i_1 < \dots < i_s} \varphi_s(x_1, \dots, x_s), \tag{6.3.8}$$

where the summation is carried out over all $(i_1 < \dots < i_s) \subset \{1, \dots, N\}$.

It follows from (6.2.13), in view of the symmetry of the functions $\bar{f}_N(t + \Delta t, x_1, \dots, x_N)$ and $\bar{f}_N(t, x_1, \dots, x_N)$ with respect to the variables $(x_1, \dots, x_N)$, that

$$\begin{aligned}
&(\bar{f}_N(t + \Delta t), \varphi_N) \\
&\quad = \int \bar{f}_N(t + \Delta t, x_1, \dots, x_N)\varphi_N(x_1, \dots, x_N) dx_1 \dots x_N \\
&\quad = \Big[\frac{N!}{s!(N-s)!} \int \bar{f}_N(t + \Delta t, x_1, \dots, x_s, x_{s+1}, \dots, x_N) dx_{s+1} \dots dx_N)\Big] \\
&\qquad \times \varphi_s(x_1, \dots, x_s) dx_1 \dots dx_s \\
&\quad = \Big[\frac{N!}{s!(N-s)!} \int \bar{f}_N(t, q_i - p_1 \Delta t, p_1, \dots, q_N - p_N \Delta t, p_N) dx_{s+1} \dots dx_N\Big] \\
&\qquad \times \varphi_s(x_1, \dots, x_s) dx_1 \dots dx_s \\
&\qquad + \frac{N!}{s!(N-s)!} \int \Big\{ \sum_{i<j=1}^{s} \int_0^{\Delta t} d\tau \int_{S_+^2} d\eta_{ij} Q(\eta_{ij} \cdot (p_i - p_j))
\end{aligned}$$

$$\times \delta(q_i - p_i\tau - q_j + p_j\tau)\int \Big[\bar{f}_N(t, q_1 - p_1\Delta t, p_1, \dots,$$

$$q_i - p_i\tau - p_i^*(\Delta t - \tau), p_i^*, \dots, q_j - p_j\tau - p_j^*(\Delta t - \tau), p_j^*, \dots,$$

$$q_N - p_N\Delta t, p_N) - \bar{f}_N(t, q_1 - p_1\Delta t, p_1, \dots, q_i - p_i\Delta t, p_i, \dots,$$

$$q_j - p_j\Delta t, p_j, \dots, q_N - p_N\Delta t, p_N)dx_{s+1}\dots dx_N\Big]\Big\}$$

$$\times \varphi_s(x_1, \dots, x_s)dx_1 \dots dx_s$$

$$+ \frac{N!(N-s)}{s!(N-s)!}\int\Big\{\sum_{i=1}^{s}\int_0^{\Delta t} d\tau \int_{S_+^2} d\eta_{is+1} Q(\eta_{is+1}\cdot(p_i - p_{s+1}))$$

$$\times \delta(q_i - p_i\tau - q_{s+1} + p_{s+1}\tau)\int \Big[\bar{f}_N(t, q_1 - p_1\Delta t, p_1, \dots,$$

$$q_i - p_i\tau - p_i^*(\Delta t - \tau), p_i^*, \dots, q_s - p_s\Delta t, p_s, q_{s+1} - p_{s+1}\tau$$

$$- p_{s+1}^*(\Delta t - \tau), p_{s+1}^*, q_{s+2} - p_{s+2}\Delta t, p_{s+2}, \dots, q_N - p_N\Delta t, p_N)$$

$$- \bar{f}_N(t, q_1 - p_1\Delta t, p_1, \dots, q_s - p_s\Delta t, p_s, q_{s+1} - p_{s+1}\Delta t, p_{s+1},$$

$$q_{s+2} - p_{s+2}\Delta t, p_{s+2}, \dots, q_N - p_N\Delta t, p_N)\Big]dx_{s+1}\dots dx_N\Big\}$$

$$\times \varphi_s(x_1, \dots, x_s)dx_1 \dots dx_s. \tag{6.3.9}$$

By analogy with calculations performed in Chapter 3 one can prove that the contribution of terms with $s + 1 \le i < j \le N$ is equal to zero. We suppose that the functional average on the right-hand side of (5.3.8) is already defined through the correlation functions $\bar{F}_s^{(N)}(t, x_1, \dots, x_s)$, $s \ge 1$. Then the above-obtained formula has the following form:

$$\frac{1}{s!}\int \bar{F}_s^{(N)}(t + \Delta t, x_1, \dots, x_s)\varphi_s(x_1, \dots, x_s)dx_1 \dots dx_s$$

$$= \frac{1}{s!}\int \bar{F}_s^{(N)}(t, q_1 - p_1\Delta t, p_1, \dots, q_s - p_s\Delta t, p_s)\varphi_s(x_1, \dots, x_s)dx_1 \dots dx_s$$

$$+ \frac{1}{s!}\int\Big\{\sum_{i<j=1}^{s}\int_0^{\Delta t} d\tau \int_{S_+^2} d\eta_{ij}\, Q(\eta_{ij}\cdot(p_i - p_j))\delta(q_i - p_i\tau - q_j + p_j\tau)$$

$$\times \Big[\bar{F}_s^{(N)}(t, q_1 - p_1\Delta t, p_1, \dots, q_i - p_i\tau - p_i^*(\Delta t - \tau), p_i^*, \dots,$$

$$q_j - p_j\tau - p_j^*(\Delta t - \tau), p_j^*, \dots, q_s - p_s\Delta t, p_s)$$

$$- \bar{F}_s^{(N)}(t, q_1 - p_1\Delta t, p_1, \dots, q_i - p_i\Delta t, p_i, \dots, q_j - p_j\Delta t, p_j, \dots,$$

$$
q_s - p_s\Delta t, p_s)\Big]\Big\}\varphi_s(x_1,\dots,x_s)dx_1\dots dx_s
$$

$$
+\frac{1}{s!}\int dx_{s+1}\Big\{\sum_{i=1}^{s}\int_0^{\Delta t} d\tau\int_{S_+^2} d\eta_{is+1}Q(\eta_{is+1}\cdot(p_i-p_{s+1}))
$$

$$
\times\,\delta(q_i - p_i\tau - q_{s+1} + p_{s+1}\tau)
$$

$$
\times\int\Big[\bar F_{s+1}^{(N)}(t, q_1 - p_1\Delta t, p_1,\dots, q_i - p_i\tau - p_i^*(\Delta t-\tau), p_i^*,\dots,
$$

$$
q_s - p_s\Delta t, p_s, q_{s+1} - p_{s+1}\tau - p_{s+1}^*(\Delta t - \tau), p_{s+1}^*)
$$

$$
-\bar F_{s+1}^{(N)}(t, q_1 - p_1\Delta t, p_1,\dots, q_i - p_i\Delta t, p_i,\dots, q_s - p_s\Delta t, p_s,
$$

$$
q_{s+1} - p_{s+1}\Delta t, p_{s+1})\Big]\Big\}\varphi_s(x_1,\dots,x_s)dx_1\dots dx_s. \tag{6.3.10}
$$

By differentiating this recurrent formula with respect to Δt one obtains the following equation (in weak sense):

$$
\int\frac{\partial\bar F_s^{(N)}(t,x_1,\dots,x_s)}{\partial t}\varphi_s(x_1,\dots,x_s)dx_1\dots dx_s
$$

$$
=\int\Big\{-\sum_{i=1}^{s} p_i\cdot\frac{\partial}{\partial q_i}\bar F_s^{(N)}(t,x_1,\dots,x_s)
$$

$$
+\sum_{i<j=1}^{s}\int_{S_+^2} d\eta_{ij}\,Q(\eta_{ij}\cdot(p_i-p_j))\delta(q_i-q_j)
$$

$$
\times\Big[\bar F_s^{(N)}(t,x_1,\dots,x_i^*,\dots,x_j^*,\dots,x_s)
$$

$$
-\bar F_s^{(N)}(t,x_1,\dots,x_i,\dots,x_j,\dots,x_s)\Big]\Big\}\varphi_s(x_1,\dots,x_s)dx_1\dots dx_s
$$

$$
+\int\Big\{\sum_{i=1}^{s}\int dx_{s+1}\int_{S_+^2} d\eta_{is+1}Q(\eta_{is+1}(p_i-p_{s+1}))\delta(q_i-q_{s+1})
$$

$$
\times\Big[\bar F_{s+1}^{(N)}(t,x_1,\dots,x_i^*,\dots,x_s,x_{s+1}^*)
$$

$$
-\bar F_{s+1}^{(N)}(t,x_1,\dots,x_i,\dots,x_s,x_{s+1})\Big]\Big\}
$$

$$
\times\,\varphi_s(x_1,\dots,x_s)dx_1\dots dx_s,\quad 1\le s\le N. \tag{6.3.11}
$$

In the obtained hierarchy of equations, the derivative of $\bar F_s^{(N)}(t,x_1,\dots,x_s)$ with respect to time is expressed through $\bar F_s^{(N)}(t,x_1,\dots,x_s)$ and $\bar F_{s+1}^{(N)}(t,x_1,\dots,x_s,x_{s+1})$, and the contributions of the hypersurfaces of lower dimensions where particles interact is taken into account. Obviously, hierarchy (6.3.2) directly follows from (6.3.10).

According to the duality principle, the obtained hierarchy is equivalent to the following hierarchy for correlation functions considered as usual functions at every point of the phase space $(x_1, \dots, x_s)$:

$$\begin{aligned}\frac{\partial F_s^{(N)}(t, x_1, \dots, x_s)}{\partial t} &\\ = \mathcal{H}_s F_s^{(N)}(t, x_1, \dots, x_s) &\\ + \sum_{i=1}^{s} \int dx_{s+1} \int_{S_+^2} d\eta_{is+1} Q(\eta_{is+1} \cdot (p_i - p_{s+1}))\delta(q_i - q_{s+1}) &\\ \times \big[F_{s+1}^{(N)}(t, x_1, \dots, x_i^*, \dots, x_s, x_{s+1}^*) &\\ - F_{s+1}^{(N)}(t, x_1, \dots, x_i, \dots, x_s, x_{s+1})\big], \quad s \geq 1, \quad (6.3.12)\end{aligned}$$

where $\mathcal{H}_s$ is the infinitesimal operator of the group of operators $S_s(-t)$, $-\infty < t < \infty$, calculated in the sense of pointwise convergence. Recall that $\mathcal{H}_s$ can be represented on differentiable functions $f_s(x_1, \dots, x_s)$ as

$$\mathcal{H}_s f_s(x_1, \dots, x_s) = -\sum_{i=1}^{s} p_i \cdot \frac{\partial}{\partial q_i} f_s(x_1, \dots, x_s) \quad (6.3.13)$$

with the boundary condition according to which at points $q_i = q_j$, $i, j \in \{1, \dots, s\}$, the momenta p_i and p_j should be replaced by p_i^* and p_j^* in $\mathcal{H}_s$ if $\eta_{ij} \cdot (p_i - p_j) \geq 0$, and (p_i, p_j) do not change if $\eta_{ij} \cdot (p_i - p_j) \leq 0$, or with the equivalent operator (6.2.23).

One can repeat the above calculation with the functional average for $\tilde{f}_N(t, x_1, \dots, x_N)$ (6.2.20) and obtain hierarchy with the operator $\tilde{\mathcal{H}}_s$ instead of $\mathcal{H}_s$, and then, using the duality principle, replace $\tilde{\mathcal{H}}_s$ by $\mathcal{H}_s$.

It was pointed out in Remark 6.2 that hierarchy (6.3.11) can be obtained if, instead of function (6.2.7), (6.2.14), one uses the function $\bar{f}_{s,N-s}(t, x_1, \dots, x_s, x_{s+1}, \dots, x_N)$ (6.2.25). This is a direct proof of the principle of duality for the stochastic hierarchy, according to which one can replace $\mathcal{H}_s \leftrightarrow \tilde{\mathcal{H}}_s$.

Passing formally to the thermodynamic limit as $N \to \infty$ in hierarchy (6.3.11) and supposing that limit correlation functions exist, one obtains the limit hierarchy

$$\frac{\partial F_s(t, x_1, \dots, x_s)}{\partial t}$$
$$= \mathcal{H}_s F_s(t, x_1, \dots, x_s)$$
$$+ \sum_{i=1}^{s} \int dx_{s+1} \int_{S_+^2} d\eta_{is+1} Q(\eta_{is+1} \cdot (p_i - p_{s+1}))\delta(q_i - q_{s+1})$$
$$\times \big[F_{s+1}(t, x_1, \dots, x_i^*, \dots, x_{s+1}^*) - F_{s+1}(t, x_1, \dots, x_i, \dots, x_{s+1}^*)\big], \quad (6.3.14)$$

$$F_s(t, x_1, \dots, x_s) = \lim_{N \to \infty} F_s^{(N)}(t, x_1, \dots, x_s), \quad s \geq 1,$$

with boundary condition in $\mathcal{H}_s$ and initial data

$$F_s(t, x_1, \dots, x_s)|_{t=0} = F_s(x_1, \dots, x_s). \quad (6.3.15)$$

Note that the correlation functions depend on random vectors η_{ij}, $i, j \in \{1, \dots, s\}$.

Hierarchy (6.3.13) is known as the *stochastic hierarchy with arbitrary scattering cross section.*

Remark 6.3. It is easy to show that hierarchy (6.3.13) can also be deduced in the framework of grand canonical ensemble (see Chapter 3).

6.4 Solutions of the stochastic hierarchy

6.4.1 Abstract form of the stochastic hierarchy

In this section, we investigate hierarchy (6.3.13) with operator $\mathcal{H}_s$ (6.2.24) for correlation functions considered as usual functions.

Denote by $\mathcal{H}$ the direct sum of the infinitesimal operators $\mathcal{H}_s$:

$$\mathcal{H} = \sum_{s=1}^{\infty} \oplus \mathcal{H}_s. \quad (6.4.1)$$

Denote by $\mathcal{A}$ the operator that acts on the sequence of correlation functions

$$F(t) = \big(F_1(t, x_1), \dots, F_s(t, x_1, \dots, x_s), \dots\big) \quad (6.4.2)$$

as follows:

$$(\mathcal{A}F(t))_s(x_1, \dots, x_s)$$
$$= \sum_{i=1}^{s} \int dx_{s+1} \int_{S_+^2} d\eta_{is+1} Q(\eta_{is+1} \cdot (p_i - p_{s+1}))\delta(q_i - q_{s+1})$$
$$\times \big[F_{s+1}(t, x_1, \dots, x_i^*, \dots, x_s, x_{s+1}^*)$$
$$- F_s(t, x_1, \dots, x_i, \dots, x_s, x_{s+1})\big], \quad s \geq 1. \quad (6.4.3)$$

Then hierarchy (6.3.13) can be represented in the abstract form

$$\frac{dF(t)}{dt} = \mathcal{H}F(t) + \mathcal{A}F(t) \tag{6.4.4}$$

with the initial data

$$F(t)|_{t=0} = F(0) = F. \tag{6.4.5}$$

Denote by $S(\pm t)$ the direct sum of the operators $S_s(\pm t)$:

$$S(\pm t) = \sum_{s=1}^{\infty} \oplus S_s(\pm t). \tag{6.4.6}$$

Then solutions of hierarchy (6.4.4) with initial data (6.4.5) can be represented by the series of iterations

$$\begin{aligned} F(t) = \sum_{n=0}^{\infty} \int_0^t dt_1 \ldots \int_0^{t_{n-1}} dt_n S(-t)S(t_1)\mathcal{A}S(-t_1)\ldots \\ \times S(t_n)\mathcal{A}S(-t_n)F(0), \end{aligned} \tag{6.4.7}$$

or componentwise

$$\begin{aligned} F_s(t,(x)_s) = \sum_{n=0}^{\infty} \int_0^t dt_1 \ldots \int_0^{t_{n-1}} dt_n S_s(-t,(x)_s)S_s(t_1,(x)_s) \\ \times \sum_{i=1}^{s} \int dx_{s+1}\delta(q_i - q_{s+1}) \int_{S_+^2} d\eta_{is+1} Q(\eta_{is+1}\cdot(p_i - p_{s+1})) \\ \times \left[S_{s+1}(-t_1,(x)_{s+1}^*) - S_{s+1}(-t_1,(x)_{s+1})\right]\ldots \\ \times S_{s+n-1}(t_{n-1},(x)_{s+n-1}) \sum_{i=1}^{s+n-1} \int dx_{s+n}\delta(q_i - q_{s+n}) \int_{S_+^2} d\eta_{is+n} \\ \times Q(\eta_{is+n}\cdot(p_i - p_{s+n}))\left[S_{s+n}(-t_n,(x)_{s+n}^*) \right. \\ \left. - S_{s+n}(-t_n,(x)_{s+n})\right] F_{s+n}(0,(x)_{s+n}), \end{aligned} \tag{6.4.8}$$

where $(x)_{s+n}^* = (x_1,\ldots,x_i^*,\ldots,x_s,x_{s+n}^*)$ in the i-th term and $S_s(\pm t,(x)_s)$ is the operator of shift along the trajectory $X(\pm t,(x)_s)$ of s particles with initial data $(x)_s$ at $t=0$.

6.4.2 Chaos property

It has been shown in the previous chapters that if phase points $(x)_s$ are outside the hypersurfaces V_{ij} where s particles interact (on the hypersurfaces V_{ij} the vectors

$q_i - q_j$ are parallel to the vectors $p_i - p_j$, $i, j \in \{1, \dots, s\}$), then all operators $S_{s+i}(\pm t, (x)_{s+i})$ in (6.4.8) should be replaced by the operators of free evolution $S^0_{s+i}(\pm t, (x)_{s+i})$, $0 \leq i \leq n$. If the initial correlation functions possess the chaos property

$$F_s(0, x_1, \dots, x_s) = F_1(0, x_1) \dots F_1(0, x_s), \tag{6.4.9}$$

then all correlation functions $F_s(t, x_1, \dots, x_s)$ outside all V_{ij} possess the chaos property

$$F_s(t, x_1, \dots, x_s) = F_1(t, x_1) \dots F_1(t, x_s) \tag{6.4.10}$$

and the one-particle correlation function satisfies the nonlinear Boltzmann equation

$$\begin{aligned}\frac{\partial F_1(t, x_1)}{\partial t} &= -p_1 \frac{\partial}{\partial q_1} F_1(t, x_1) + \int dx_2 \delta(q_1 - q_2) \int_{S^2_+} d\eta_{12} \\ &\times Q(\eta_{12} \cdot (p_1 - p_2)) \big[F_1(t, x_1^*) F_1(t, x_2^*) - F_1(t, x_1) F_1(t, x_2) \big]\end{aligned} \tag{6.4.11}$$

with initial condition $F_1(t, x_1)|_{t=0} = F_1(0, x_1)$.

Note that the corresponding proof of the above-formulated assertion has been performed for the differential scattering cross section $\eta_{ij} \cdot (p_i - p_j)$ of hard spheres in Chapter 4, but it can be repeated word for word for the stochastic dynamics with arbitrary cross section $Q(\eta_{ij} \cdot (p_i - p_j))$ because the hypersurfaces V_{ij} where stochastic particles interact do not depend on the form of differential scattering cross section.

In order to prove the existence of solutions of hierarchy (6.4.4) represented by series (6.4.7), (6.4.8) one needs to impose some restrictions on the differential scattering cross section $Q(\eta_{ij} \cdot (p_i - p_j))$. If

$$Q(\eta_{ij} \cdot (p_i - p_j)) \leq |p_i - p_j|, \quad \eta_{ij} \in S^2_+, \tag{6.4.12}$$

then all results obtained for hard spheres also hold for hierarchies with cross section $Q(\eta_{ij} \cdot (p_i - p_j))$.

Namely, series (6.4.8) is uniformly convergent with respect to $(x)_s$ on compacta and with respect to time on the finite interval $[0, t_0]$ if the sequence of initial functions $F(0)$ belongs to the space $E_{\xi,\beta}$ with the norm

$$\|F(0)\| = \sup_{s \geq 1} \frac{1}{\xi^s} \sup_{(x)_s} e^{\beta \sum_{i=1}^{s} p_i^2} |F_s(0, (x)_s)|, \quad \xi > 0, \quad \beta > 0; \tag{6.4.13}$$

here, the number t_0 depends on Q, ξ, and β, $\xi > 0$, $\beta > 0$.

Series (6.4.8) is uniformly convergent with respect to $(x)_s$ on compacta and with respect to time on an arbitrary interval, i.e., globally in time, if the sequence of initial functions is exponentially decreasing with respect to squared momenta and coordinates i.e., $F(0)$ belongs to $\tilde{E}_{\xi,\beta}$ (for details, see Chapter 4).

The correlation functions $F_s(t, x_1, \dots, x_s)$, $s \geq 1$, are defined in the entire phase space of s particles as usual (not generalized) functions represented by series (6.4.8) and are discontinuous on the hyperplanes V_{ij} where s particles interact. (Outside V_{ij} all operators $S_{s+i}(\pm t, (x)_{s+i}))$ should be replaced by $S^0_{s+i}(\pm t, (x)_{s+i})$, and series (6.4.8) is uniformly convergent for $t \subset [0, t_0]$ and for $(x)_s$ from compacta for $F(0) \in E_{\xi,\beta}$ or globally in time for $F(0) \in \tilde{E}_{\xi,\beta}$).

In series (6.4.8) one can perform integration with respect to $(q_{s+1}, \dots, q_{s+n})$ using δ-functions. In integration with respect to $(p_{s+1}, \dots, p_{s+n})$ one can neglect the hypersurfaces V_{ij} of lower dimension where particles with numbers $(s+1, \dots, s+n)$ interact with each other and with particles with numbers $(1, \dots, s)$ because this is the usual Lebesgue integral. One obtains the following representation for $F_s(t, (x)_s)$:

$$
\begin{aligned}
F_s(t, (x)_s) = {} & \sum_{n=0}^{\infty} \int_0^t dt_1 \dots \int_0^{t_{n-1}} dt_n \, S_s(-t, (x)_s) S_s(t_1, (x)_s) \\
& \times \sum_{i=1}^{s} \int dp_{s+1} \int_{S^2_+} d\eta_{is+1} Q(\eta_{is+1} \cdot (p_i - p_{s+1})) \\
& \times [S_s(-t_1) S^0_1(-t_1)((x)^*_{s+1}) - S_s(-t_1) S^0_1(-t_1)((x)_{s+1})]|_{q_{s+1}=q_i} \cdots \\
& \times S_s(t_{n-1}) S^0_{n-1}(t_{n-1})((x)_{s+n-1}) \\
& \times \sum_{i=1}^{s+n-1} \int dp_{s+n} \int_{S^2_+} d\eta_{is+n} Q(\eta_{is+n} \cdot (p_i - p_{s+n})) \\
& \times [S_s(-t_n) S^0_n(-t_n)((x)^*_{s+n}) - S_s(-t_n) S^0_n(-t_n)((x)_{s+n})] \\
& \times F_{s+n}(0, (x)_{s+n})|_{q_i = q_{s+n}},
\end{aligned}
\tag{6.4.14}
$$

where the operators $S_s(-t_i)$ depend on the phase points with numbers $1, \dots, s$ and the operators $S^0_{s+i}(-t_i)$ depend on the phase points with numbers $s+1, \dots, s+i$.

Remark 6.4. Hierarchy (6.3.3) for the correlation functions $\bar{F}_s(t, x_1, \dots, x_s)$ considered as certain generalized functions (6.3.1) can be solved only formally by using the series of iterations (6.4.8) and replacing the operators $S_{s+i}(-t, x_1, \dots, x_s, x_{s+1}, \dots, x_{s+i})$ by $\bar{S}_{s+i}(-t, x_1, \dots, x_s, x_{s+1}, \dots, x_{s+i})$. However, we are not able yet to rigorously prove the existence of the operators $\bar{S}_{s+i}(-t, x_1, \dots, x_s, x_{s+1}, \dots, x_{s+i})$ and do not know their properties. Therefore, we restrict ourselves to the investigation of hierarchy (6.4.4) for the correlation functions $F_s(t, x_1, \dots, x_s)$ considered as usual functions. We do not know how to rigorously determine $\bar{F}_s(t, x_1, \dots, x_s)$ through $F_s(t, x_1, \dots, x_s)$. It is reasonable to assume that $\bar{F}_s(t, x_1, \dots, x_s)$ can be obtained from $F_s(t, x_1, \dots, x_s)$ represented by series (6.4.14) by replacing the operator $S_s(-t, x_1, \dots, x_s)$ by the operator $\bar{S}_s(-t, x_1, \dots, x_s)$. This also follows, on the formal level, from the definition of $\bar{f}_{s,N-s}(t, x_1, \ldots, x_N)$ and $\bar{f}_N(t, x_1, \ldots, x_N)$.

6.4.3 Spatially homogeneous initial data

Consider initial data $F(0)$ with initial correlation functions independent of positions, i.e.,

$$F_s(0, x_1, \ldots, x_s) = F_s(0, p_1, \ldots, p_s), \quad s \geq 1. \tag{6.4.15}$$

Outside the hypersurfaces V_{ij}, with $S^0_{s+i}(\pm t, (x)_{s+i})$ instead of $S_{s+i}(\pm t, (x)_{s+i})$ (see (6.4.14)), the correlation functions $F_s(t, (x)_s)$ also do not depend on positions, i.e.,

$$F(t, x_1, \ldots, x_s) = F(t, p_1, \ldots, p_s), \tag{6.4.16}$$

because

$$S^0_{s+i}(\pm t, (x)_{s+i}) F_{s+i}(0, p_1, \ldots, p_{s+i}) = F_{s+i}(0, p_1, \ldots, p_{s+i}).$$

Further, if the initial correlation functions have the chaos property

$$F_s(0, p_1, \ldots, p_s) = F_1(0, p_1) \ldots F_1(0, p_s), \quad s \geq 1, \tag{6.4.17}$$

then $F_s(t, x_1, \ldots, x_s)$ have the chaos property (6.4.10) outside the hypersurfaces V_{ij}, and $F_1(t, x_1)$ satisfies the Boltzmann equation (6.4.11) with the initial condition $F_1(t, p_1)|_{t=0} = F_1(0, p_1)$.

Therefore, the solution of the Boltzmann equation (6.4.11) does not depend on positions $F_1(t, x_1) = F_1(t, p_1)$, and satisfies the spatially homogeneous equation

$$\begin{aligned} \frac{\partial F_1(t, p_1)}{\partial t} &= \int dp_2 \int_{S^2_+} d\eta_{12} Q(\eta_{12} \cdot (p_1 - p_2)) \\ &\times [F_1(t, p_1^*) F_1(t, p_2^*) - F_1(t, p_1) F_1(t, p_2)]. \end{aligned} \tag{6.4.18}$$

Thus, using the stochastic dynamics and the stochastic Boltzmann hierarchy, we have deduced the spatially inhomogeneous (6.4.11) and spatially homogeneous (6.4.18) Boltzmann equations without mean-field approximation.

Recall that Kac [Kac1, Kac2] proposed a certain Markov process in the momentum space of N particles defined through the corresponding Kolmogorov–Itô–Liouville equation in mean-field approximation. He derived a hierarchy for the sequence of correlation functions and proved that, in the thermodynamic limit as $N \to \infty$, solutions of the hierarchy have the chaos property if initial data have this property. This means that the correlation functions at arbitrary time, from the interval where solutions of hierarchy exist, is the product of one-particle correlation functions, which satisfies the spatially homogeneous Boltzmann equation [Grn, Szn1, Szn2, Wag].

In the next section we show that the Kac's Markov process in the momentum space and the corresponding Kolmogorov–Itô–Liouville equation can be obtained from our stochastic process in the entire phase space by means of a specific averaging procedure in the space of positions.

6.5 Stochastic process in momentum space

6.5.1 Averaging procedure in spatially homogeneous case

Consider the functional average (6.2.4) in the spatially homogeneous case where the functions f_N and φ_N do not depend on positions:

$$\begin{aligned}
&f_N(q_1, p_1, \ldots, q_N, p_N) = f_N(p_1, \ldots, p_N),\\
&\varphi_N(q_1, p_1, \ldots, q_N, p_N) = \varphi_N(p_1, \ldots, p_N),\\
&f_N(q_1 - p_1\Delta t, p_1, \ldots, q_N - p_N t, p_N) = f_N(p_1, \ldots, p_N),\\
&f_N(q_1 - p_1\Delta t, p_1, \ldots, q_i - p_i\tau - p_i^*(\Delta t - \tau), p_i^*, \ldots,\\
&\qquad q_j - p_j\tau - p_j^*(\Delta t - \tau), p_j^*, \ldots, q_N - p_N\Delta t, p_N)\\
&\quad = f_N(p_1, \ldots, p_i^*, \ldots, p_j^*, \ldots, p_N).
\end{aligned} \tag{6.5.1}$$

For spatially homogeneous f_N and φ_N the functional average $(S_N(-\Delta t)f_N, \varphi_N)$ (6.2.4) is divergent: the first term and the second term are proportional to V^N and V^{N-1}, respectively. Instead of functional (6.2.4), we introduce the following functional:

$$\begin{aligned}
&\lim_{V\to\infty} \frac{1}{V^N} \int_\Lambda dq_1 \int dp_1 \ldots \int_\Lambda dq_N \int dp_N\, f_N(p_1, \ldots, p_N)\varphi_N(p_1, \ldots, p_N)\\
&\quad + \lim_{V\to\infty} \frac{1}{V^{N-1}} \sum_{i<j=1}^{N} \int_\Lambda dq_1 \int dp_1 \ldots \int_{\Lambda_i} dq_i \int dp_i \ldots \int_{\Lambda_j} dq_j \int dp_j \ldots\\
&\quad \times \int_\Lambda dq_N \int dp_N \int_0^{\Delta t} d\tau \int_{S_+^2} d\eta_{ij}\, Q(\eta_{ij}\cdot(p_i - p_j))\delta(q_i - p_i\tau - q_j + p_j\tau)\\
&\quad \times \left[f_N(p_1, \ldots, p_i^*, \ldots, p_j^*, \ldots, p_N) - f_N(p_1, \ldots, p_i, \ldots, p_j, \ldots, p_N)\right]\\
&\quad \times \varphi_N(p_1, \ldots, p_N)\\
&= \int dp_1 \ldots dp_N\, f_N(p_1, \ldots, p_N)\varphi_N(p_1, \ldots, p_N)\\
&\qquad + \Delta t \sum_{i<j=1}^{N} \int dp_1 \ldots dp_N \int_{S_+^2} d\eta_{ij}\, Q(\eta_{ij}\cdot(p_i - p_j))\\
&\qquad \times \left[f_N(p_1, \ldots, p_i^*, \ldots, p_j^*, \ldots, p_N) - f_N(p_1, \ldots, p_i, \ldots,\right.\\
&\qquad\qquad \left. p_j, \ldots, p_N)\right]\varphi_N(p_1, \ldots, p_N)
\end{aligned}$$

$$= \int dp_1 \dots dp_N [\bar{S}_N(-\Delta t) f_N(p_1, \dots, p_N)] \varphi_N(p_1, \dots, p_N)$$

$$= (\bar{S}_N(-\Delta t) f_N, \varphi_N). \tag{6.5.2}$$

By Λ, Λ_i, and Λ_j we denote the spheres centered at the origin and at the points $-p_i\tau$ and $-p_j\tau$, respectively, and having the volume $V = V(\Lambda) = V(\Lambda_i) = V(\Lambda_j)$.

The functional $(S_N(-\Delta t) f_N, \varphi_N)$ was obtained from the functional (6.2.4) by averaging over the configurational space (space of positions) and is the average of the state $S_N(-\Delta t) f_N(p_1, \dots, p_N)$ with respect to the observable $\varphi_N(p_1, \dots, p_N)$.

From (6.5.2) one obtains

$$\begin{aligned} S_N(-\Delta t) & f_N(p_1, \dots, p_N) \\ &= \bar{f}_N(\Delta t, p_1, \dots, p_N) \\ &= f_N(p_1, \dots, p_N) + \Delta t \sum_{i<j=1}^{N} \int_{S_+^2} d\eta_{ij}\, Q(\eta_{ij} \cdot (p_i - p_j)) \\ &\qquad \times \big[f_N(p_1, \dots, p_i^*, \dots, p_j^*, \dots, p_N) \\ &\qquad\qquad - f_N(p_1, \dots, p_i, \dots, p_j, \dots, p_N) \big]. \end{aligned} \tag{6.5.3}$$

Note that the operator $S_N(-\Delta t)$ (6.5.3) is defined on functions that depend only on momenta.

Formula (6.5.3) holds for infinitesimal Δt and defines the operator of evolution $S_N(-\Delta t)$. For $N = 2$, it is true for arbitrary $\Delta t > 0$. For arbitrary $t > 0$, we formally define the group of operators of evolution $\bar{S}_N(-t)$ as follows:

$$S_N(-t) = \lim_{n\to\infty} \prod_{i=1}^{n} S_N(-\Delta t_i), \qquad \sum_{i=1}^{n} \Delta t_i = t, \tag{6.5.4}$$

where $S_N(-\Delta t_i)$ are already defined according to (6.5.3).

The state $f_N(t + \Delta t, p_1, \dots, p_N)$ is defined through $f_N(t, p_1, \dots, p_N)$ as follows:

$$\begin{aligned} f_N(t + \Delta t, p_1, \dots, p_N) &= S_N(-\Delta t) f_N(t, p_1, \dots, p_N) \\ &= f_N(t, p_1, \dots, p_N) + \Delta t \sum_{i<j=1}^{N} \int_{S_+^2} d\eta_{ij}\, Q(\eta_{ij} \cdot (p_i - p_j)) \\ &\qquad \times \big[f_N(t, p_1, \dots, p_i^*, \dots, p_j^*, \dots, p_N) \\ &\qquad\qquad - f_N(t, p_1, \dots, p_i, \dots, p_j, \dots, p_N) \big]. \end{aligned} \tag{6.5.5}$$

The functional average $(f_N(t + \Delta t), \varphi_N)$ is defined by (6.5.2) if, instead of $f_N(p_1, \dots, p_N)$, one takes $f_N(t, p_1, \dots, p_N)$.

6.5.2 Differential equation for spatially homogeneous distribution functions

Using (6.5.5), we obtain the differential equation

$$\begin{aligned}\frac{\partial f_N(t,p_1,\dots,p_N)}{\partial t}\\ &= \sum_{i<j=1}^{N}\int_{S_+^2} d\eta_{ij}\, Q(\eta_{ij}\cdot(p_i-p_j))\big[f_N(t,p_1,\dots,p_i^*,\dots,p_j^*,\dots,p_N)\\ &\qquad - f_N(t,p_1,\dots,p_i,\dots,p_j,\dots,p_N)\big] \qquad (6.5.6)\end{aligned}$$

with initial data

$$f_N(t,p_1,\dots,p_N)|_{t=0} = f_N(0,p_1,\dots,p_N) = f_N(p_1,\dots,p_N).$$

This is the Kolmogorov–Itô–Liouville equation for a certain Markov process in the momentum space.

We will also consider the modification of equation (6.5.6) in the mean-field approximation, namely

$$\begin{aligned}\frac{\partial f_N(t,p_1,\dots,p_N)}{\partial t}\\ &= \frac{1}{N}\sum_{i<j=1}^{N}\int_{S_+^2} d\eta_{ij}\, Q(\eta_{ij}\cdot(p_i-p_j))\big[f_N(t,p_1,\dots,p_i^*,\dots,p_j^*,\dots,p_N)\\ &\qquad - f_N(t,p_1,\dots,p_i,\dots,p_j,\dots,p_N)\big], \qquad (6.5.7)\end{aligned}$$

$$f_N(t,p_1,\dots,p_N)|_{t=0} = f_N(p_1,\dots,p_N).$$

Equation (6.5.7) can be obtained if one considers functional (6.5.2) in the mean-field approximation with additional factor $1/N$.

6.5.3 Hierarchy for correlation functions in mean-field approximation

We define the sequence of correlation functions

$$F_s^{(N)}(t,p_1,\dots,p_s) = \int dp_{s+1}\dots dp_N\, f_N(t,p_1,\dots,p_N), \quad s\ge 1, \qquad (6.5.8)$$

where $f_N(t,p_1,\dots,p_N)$ is a solution of (6.5.7). It is easy to obtain the following hierarchy of integro-differential equations:

$$
\begin{aligned}
&\frac{\partial F_s^{(N)}(t, p_1, \dots, p_N)}{\partial t} \\
&\quad = \frac{1}{N} \sum_{i<j=1}^{s} \int_{S_+^2} d\eta_{ij}\, Q_i(\eta_{ij} \cdot (p_i - p_j))\big[F_s^{(N)}(t, p_1, \dots, p_i^*, \dots, p_j^*, \dots, p_s) \\
&\qquad\quad - F_s^{(N)}(t, p_1, \dots, p_i, \dots, p_j, \dots, p_s)\big] \\
&\qquad + \frac{N-s}{N} \sum_{i=1}^{s} \int dp_{s+1} \int d\eta_{is+1} Q_i(\eta_{is+1} \cdot (p_i - p_{s+1})) \\
&\qquad\quad \times \big[F_{s+1}^{(N)}(t, p_1, \dots, p_i^*, \dots, p_s, p_{s+1}^*) \\
&\qquad\qquad - F_{s+1}^{(N)}(t, p_1, \dots, p_i, \dots, p_s, p_{s+1})\big], \quad 1 \le s \le N. \qquad (6.5.9)
\end{aligned}
$$

Passing formally to the thermodynamic limit as $N \to \infty$ and supposing that the limit correlation functions exist in a certain sense, i.e.,

$$
\lim_{N\to\infty} F_s^{(N)}(t, p_1, \dots, p_N) = F_s(t, p_1, \dots, p_s), \quad s \ge 1,
$$

one obtains the limit hierarchy

$$
\begin{aligned}
&\frac{\partial F_s(t, p_1, \dots, p_s)}{\partial t} \\
&\quad = \sum_{i=1}^{s} \int dp_{s+1} \int d\eta_{is+1} Q_i(\eta_{is+1} \cdot (p_i - p_{s+1})) \\
&\qquad \times \big[F_{s+1}(t, p_1, \dots, p_i^*, \dots, p_s, p_{s+1}^*) \\
&\qquad\quad - F_{s+1}(t, p_1, \dots, p_i, \dots, p_s, p_{s+1})\big], \quad s \ge 1, \qquad (6.5.10)
\end{aligned}
$$

with initial data

$$
F_s(t, p_1, \dots, p_s)|_{t=0} = F_s(p_1, \dots, p_s).
$$

For the justification of the existence of the thermodynamic limit, see [Grn, Szn1, Szn2, Wag]. Consider the initial correlation functions that have the chaos property

$$
F_s(p_1, \dots, p_s) = F_1(p_1) \dots F_1(p_s).
$$

Hierarchy (6.5.10) admits the separation of variables because, on the right-hand side of (6.5.10), we have the sum of s operators acting on each of s variables. As a result, we establish that $F_s(t, p_1, \dots, p_s)$ also have the chaos property, i.e.,

$$
F_s(t, p_1, \dots, p_s) = F_1(t, p_1) \dots F_1(t, p_s),
$$

and the one-particle correlation function $F_1(t, p_s)$ satisfies the spatially homogeneous Boltzmann equation

$$\frac{\partial F_s(t, p_1)}{\partial t} = \int dp_2 \int_{S_+^2} d\eta_{12} Q(\eta_{12} \cdot (p_1 - p_2)) \\ \times [F_1(t, p_1^*) F_1(t, p_2^*) - F_1(t, p_1) F_1(t, p_2)]. \quad (6.5.11)$$

Thus, we have obtained the spatially homogeneous Boltzmann equation by using the stochastic dynamics in the momentum space by averaging the stochastic dynamics in the phase space and in the mean-field approximation. In Section 6.4, we have obtained the spatially inhomogeneous Boltzmann equation and the homogeneous Boltzmann equation by using solutions of the stochastic Boltzmann hierarchy outside the hypersurfaces V_{ij} of lower dimension where point particles interact and without the mean-field approximation.

Chapter 7

Analog of Liouville equation and BBGKY hierarchy for a system of hard spheres with inelastic collisions

7.1 Introduction

It is commonly accepted that systems of hard spheres with inelastic collision are a proper model of granular flow. Statistical mechanics of systems of hard spheres should be a theoretical basis of the theory of granular flow. In attempts to adapt classical statistical mechanics to systems of hard spheres with inelastic collisions, one faces new very difficult problems connected with inelasticity.

First of all, it is necessary to define the density of probability (distribution function) on the phase space because the Jacobian of the transformation induced by a shift along trajectories of hard spheres with inelastic collisions is different from one and is singular (its derivative with respect to time contains δ-functions). It is necessary to deduce the Liouville equation for the defined distribution function and correctly formulate boundary conditions associated with inelasticity. Finally, one should deduce an analog of the BBGKY hierarchy for the corresponding correlation functions.

The above-mentioned problems are solved in the present section. The distribution function is defined as follows:

$$D_N(t,(x)_N) = D_N(0, X(-t,(x)_N))\left[\frac{\partial X(-t,(x)_N)}{\partial (x)_N}\right]^2, \qquad (7.1.1)$$

where $D_N(0,(x)_N)$ is the initial distribution function, $X(-t,(x)_N)$ is the trajectory of N hard spheres at time $-t$ with initial data $(x)_N$ at the initial time $t = 0$, and $\frac{\partial X(-t,(x)_N)}{\partial (x)_N}$ is the corresponding singular Jacobian different from one. The distribution function satisfies the law of conservation of full probability

$$\int D_N(t,(x)_N)d(x)_N = \int D_N(0,(x)_N)d(x)_N \qquad (7.1.2)$$

and the Liouville equation

$$\frac{\partial}{\partial t} D_N(t,(x)_N) = -\sum_{i=1}^{N} p_i \cdot \frac{\partial}{\partial q_i} D_N(t,(x)_N), \tag{7.1.3}$$

$$D_N(t,(x)_N)\,|_{t=0} = D_N(0,(x)_N)$$

with boundary conditions according to which, for $q_i - q_j - a\eta = 0$ (where $|\eta| = 1$ and a is the diameter of the sphere) and $\eta \cdot (p_i - p_j) > 0$, the momenta p_i, p_j should be replaced by

$$p_i^* = p_i + \frac{\varepsilon}{1-2\varepsilon}\eta\,\eta \cdot (p_i - p_j),$$

$$p_j^* = p_j - \frac{\varepsilon}{1-2\varepsilon}\eta\,\eta \cdot (p_i - p_j), \quad \frac{1}{2} < \varepsilon \le 1,$$

in the operator $-\sum_{i=1}^{N} p_i \cdot \frac{\partial}{\partial q_i}$ and in $D_N(t,(x)_N)$. Moreover, for the above-mentioned phase points, the following identity is true:

$$D_N(t,q_1,p_1,\ldots,q_i,p_i,\ldots,q_j,p_j,\ldots,q_N,p_N)\frac{1}{(1-2\varepsilon)^2}$$
$$= \times D_N(t,q_1,p_1,\ldots,q_i,p_i^*,\ldots,q_j,p_j^*,\ldots,q_N,p_N), \tag{7.1.4}$$

where ε is a parameter associated with inelastic collisions $\left(\frac{1}{2} < \varepsilon \le 1\right)$. The momenta do not change if $\eta \cdot (p_i - p_j) < 0$, i.e., $p_i^* = p_i$, $p_j^* = p_j$.

For the sequence of correlation functions:

$$F_s^{(N)}(t,(x)_s)$$
$$= N(N-1)\ldots(N-s+1)\int D_N(t,x_1,\ldots x_s,x_{s+1},\ldots,x_N)dx_{s+1}\ldots dx_N,$$

the following BBGKY hierarchy is deduced:

$$\frac{\partial F_s^{(N)}(t,(x)_s)}{\partial t}$$
$$= -\sum_{i=1}^{s} p_i \cdot \frac{\partial}{\partial q_i} F_s^{(N)}(t,(x)_s)$$
$$+ a^2 \sum_{i=1}^{s} \int dp_{s+1} \int_{S_+^2} d\eta\,\eta \cdot (p_i - p_{s+1})$$
$$\times \left[\frac{1}{(1-2\varepsilon)^2} F_{s+1}^{(N)}(t,q_1,p_1,\ldots,q_i,p_i^*,\ldots,q_s,p_s,q_i - a\eta, p_{s+1}^*)\right.$$
$$\left. - F_{s+1}^{(N)}(t,q_1,p_1,\ldots,q_i,p_i,\ldots,q_s,p_s,q_i + a\eta, p_{s+1})\right], \tag{7.1.5}$$

$$S_2^+(\eta|\eta\cdot(p_i-p_{s+1})\geq 0),\quad F_s^{(N)}(t,(x)_s)_{|_{t=0}}=F_s^{(N)}((x)_s),\quad 1\leq s\leq N.$$

One should add the same boundary condition as for $D_N(t,(x)_N)$ in the Liouville equation.

In the present section, we do not touch the problem of the existence of a solution of hierarchy (7.1.4), and the existence of the thermodynamic limit will be investigated in what follows.

7.2 Trajectories of a system of hard spheres with inelastic collisions

7.2.1 Dynamics

In the three-dimensional space $\mathbb{R}^3$, we consider particles with mass m that are hard spheres with diameter a. Particles move freely until they touch each other and the distance between their centers is equal to a. Then they collide inelastically.

Denote, as in the previous chapters, the position of the center of sphere by $q\in\mathbb{R}^3$ and its momentum by $p\in\mathbb{R}^3$. Let N be the number of particles of the considered system. Particles with numbers i and j collide if $q_i-q_j=a\eta$. If their momenta before collisions are p_i and p_j and $\eta\cdot(p_i-p_j)<0$, then, after an inelastic collision, they become

$$\begin{aligned} p_i^* &= p_i-\varepsilon\eta\,\eta\cdot(p_i-p_j),\\ p_j^* &= p_j+\varepsilon\eta\,\eta\cdot(p_i-p_j), \end{aligned}\tag{7.2.1}$$

where the parameter $\frac{1}{2}<\varepsilon\leq 1$ characterizes an inelastic collision, the unit vector η is directed from the center of the sphere with number i to the center of the sphere with number j, $\eta\cdot(p_i-p_j)$ is the scalar product of vectors η and p_i-p_j. Formulas (7.2.1) define a linear transformation of momenta p_i, p_j. If $\eta\cdot(p_i-p_j)>0$, then, after a collision, we have $p_i^*=p_i$ and $p_j^*=p_j$.

Note that this law of inelastic collision (7.2.1) is true for a real evolution of a system with increasing time t (for the dynamics forward in time). In statistical mechanics, we also need an imaginary evolution of a system with decreasing time (dynamics backward in time). We define the law of inelastic collision with decreasing time as the transformation inverse to (7.2.1). To obtain this desired transformation, we consider (7.2.1) as an equation with respect to p_i and p_j for given p_i^* and p_j^*.

Calculating the scalar product

$$\eta\cdot(p_i^*-p_j^*)=(1-2\varepsilon)\eta\cdot(p_i-p_j),$$

one obtains from (7.2.1) the desired linear transformation inverse to (7.2.1):

$$\begin{aligned} p_i &= p_i^* + \frac{\varepsilon}{1-2\varepsilon}\eta\,\eta\cdot(p_i^* - p_j^*), \\ p_j &= p_j^* - \frac{\varepsilon}{1-2\varepsilon}\eta\,\eta\cdot(p_i^* - p_j^*), \qquad \eta\cdot(p_i^* - p_j^*) > 0. \end{aligned} \tag{7.2.2}$$

In what follows, we need only the dynamics backward in time, and it will be useful to change the notation in (7.2.2) and write the momenta (p_i^*, p_j^*) instead of the momenta (p_i, p_j) and vice versa. Then transformation (7.2.2) takes the following form:

$$\begin{aligned} p_i^* &= p_i + \frac{\varepsilon}{1-2\varepsilon}\eta\,\eta\cdot(p_i - p_j), \\ p_j^* &= p_j - \frac{\varepsilon}{1-2\varepsilon}\eta\,\eta\cdot(p_i - p_j), \qquad \eta\cdot(p_i - p_j) > 0. \end{aligned} \tag{7.2.3}$$

It follows from (7.2.3) that the components of the vectors p_i^* and p_j^* perpendicular to the vector η do not change, and the components parallel to the vector η change according to (7.2.3). It is obvious that the Jacobian of transformation (7.2.3) J can easily be calculated:

$$J = \frac{1}{1-2\varepsilon}. \tag{7.2.4}$$

If $\eta\cdot(p_i - p_j) < 0$, then momenta do not change, i.e., $p_i^* = p_i$ and $p_j^* = p_j$. Let us calculate the kinetic energy after a collision in the backward motion of particles with numbers i and j. According to (7.2.3), we have

$$p_i^{*2} + p_j^{*2} = p_i^2 + p_j^2 + 2\frac{\varepsilon - \varepsilon^2}{(1-2\varepsilon)^2}\left(\eta\cdot(p_i - p_j)\right)^2 \ge p_i^2 + p_j^2 \tag{7.2.5}$$

because $\varepsilon - \varepsilon^2 > 0$ for $\frac{1}{2} < \varepsilon < 1$.

We now calculate the kinetic energy after a collision in the forward motion. According to (7.2.1), one obtains

$$p_i^{*2} + p_j^{*2} = p_i^2 + p_j^2 + 2(-\varepsilon + \varepsilon^2)\left(\eta\cdot(p_i - p_j)\right)^2 \le p_i^2 + p_j^2 \tag{7.2.6}$$

because $-\varepsilon + \varepsilon^2 < 0$ for $\frac{1}{2} < \varepsilon < 1$. From (7.2.5) and (7.2.6), one can see that $p_i^{*2} + p_j^{*2}$ is greater than $p_i^2 + p_j^2$ for the backward motion ($t < 0$), and $p_i^{*2} + p_j^{*2}$ is less than $p_i^2 + p_j^2$ for the forward motion:

$$\begin{aligned} p_i^{*2} + p_j^{*2} &\ge p_i^2 + p_j^2, \quad t < 0, \\ p_i^{*2} + p_j^{*2} &\le p_i^2 + p_j^2, \quad t > 0. \end{aligned} \tag{7.2.7}$$

Thus, in the dynamics with inelastic collisions defined above, the kinetic energy increases for $t < 0$ and decreases for $t > 0$. Only in the case $\eta\cdot(p_i - p_j) = 0$, one has $p_i^{*2} + p_j^{*2} = p_i^2 + p_j^2$, and the kinetic energy is preserved even for $\frac{1}{2} < \varepsilon < 1$.

7.2.2 Trajectory

Denote by $Q_1(-t), \dots, Q_N(-t)$ the positions of hard spheres at time $-t$, $t > 0$, by $P_1(-t), \dots, P_N(-t)$ their momenta, by $q_1, \dots, q_N$ their initial positions, by $p_1, \dots, p_N$ their initial momenta at time $t = 0$, and by $(x)_N = (q_1, p_1, \dots, q_N, p_N)$ the initial phase point. Obviously, we consider only admissible configurations, i.e., $|q_i - q_j| \geq a$ for all $i, j \in \{1, \dots, N\}$. As mentioned above, particles move freely until they touch each other and then collide and their momenta change according to (7.2.3).

We will neglect instantaneous collisions of three or more particles because the set of such initial positions and momenta is of Lebesgue measure zero. Denote by $t_{ij}((x)_N)$ the time of the collision of particles with numbers i and j. Considered as a function of $(x)_N$, $t_{ij}((x)_N)$ is continuously differentiable outside a certain set of Lebesgue measure zero.

The trajectory

$$X(-t,(x)_N) = (Q_1(-t,(x)_N), P_1(-t,(x)_N), \dots, Q_N(-t,(x)_N), P_N(-t,(x)_N)),$$

$$Q_i(-t) \equiv Q_i(-t,(x)_N), \quad P_i(-t) = P_i(-t,(x)_N), \quad i = 1, \dots, N,$$

is constructed as follows: Until the first collision, we have

$$X(-t,(x)_N) = (q_1 - p_1 t, p_1, \dots, q_N - p_N t, p_N). \tag{7.2.8}$$

If, at time $t_{ij}((x)_N)$, the particles with numbers i and j collide, then, for $t > t_{ij}(x)_N$, the trajectory $X(-t,(x)_N)$ is again given by formula (7.2.8), but the positions and momenta of the i-th and j-th particles are given by

$$\begin{aligned} &q_i - p_i t_{ij}(x) - p_i^*(t - t_{ij}(x)), \quad p_i^*, \\ &q_j - p_j t_{ij}(x) - p_j^*(t - t_{ij}(x)), \quad p_j^*, \end{aligned} \tag{7.2.9}$$

where p_i^* and p_j^* are expressed in terms of p_i and p_j according to (7.2.3).

One can continue the trajectory according to (7.2.9) after all collisions if infinitely many collisions on a finite time interval are absent. Then the momenta of all particles involved in these infinite number of collisions coincide and their spheres touch each other. The corresponding set of initial phase points lies on the hyperplanes of lower dimension and has the Lebesgue measure zero.

It is obvious that the trajectory has the group property

$$X(-t_1 - t_2,(x)_N) = X(-t_1, X(-t_2,(x)_N)) = X(-t_2, X(-t_1,(x)_N))$$

and satisfies the following boundary condition: for $q_i - q_j = a\eta$, $\eta \cdot (p_i - p_j) > 0$, $i, j \in \{1, \dots, N\}$,

$$X(-t, q_1, p_1, \dots, q_i, p_i, \dots, q_j, p_j, \dots, q_N, p_N)$$

$$= X(-t, q_1, p_1, \ldots, q_i, p_i^*, \ldots, q_j, p_j^*, \ldots, q_N, p_N); \tag{7.2.10}$$

if $q_i - q_j = a\eta$ but $\eta \cdot (p_i - p_j) < 0$, then the momenta p_i and p_j do not change. This boundary condition means that, after a collision, particles depart and the distance between them increases.

The trajectory $X(-t, (x)_N)$ is a continuously differentiable function almost everywhere with respect to its initial data $(x)_N$ and time on every time interval between collisions. The detailed proof of the above-mentioned properties of the trajectory $X(-t, (x)_N)$ can be found in [PGM3] and [CGP]; one should only make some modification connected with the inelastic character of collisions.

We summarize the above-formulated results in the following theorem:

Theorem 7.1. *The trajectory $X(-t, (x)_N)$ of N hard spheres that inelastically collide exists for arbitrary time $t > 0$, is continuously differentiable with respect to the initial phase points $(x)_N$ and time t on the intervals between collisions, and has the group property for almost all initial $(x)_N$ that belong to a certain domain outside a hypersurface of Lebesgue measure zero.*

Theorem 7.1 asserts that the trajectories $X(-t, (x)_N)$ are well defined between the times of collisions almost everywhere (a.e.) with respect to $(x)_N$. In many respects, the trajectories of our system of hard spheres with inelastic collisions have the same properties as a system of hard spheres with elastic collisions. These properties were formulated in Theorem 7.1. However, the trajectories of hard spheres with inelastic collisions also have certain specific properties different from those in the case of elastic collisions. One of these specific properties is that the map of the phase space induced by the shift along trajectories does not preserve the volume.

According to the definition of trajectories (7.2.8), (7.2.9), the Jacobian

$$\frac{\partial(X_1(-t, (x)_N), \ldots, X_N(-t, (x)_N))}{(\partial x_1, \ldots, \partial x_N)} = \frac{\partial(X(-t, (x)_N))}{\partial(x)_N} \tag{7.2.11}$$

is equal to one if, for the initial point $(x)_N$, there are no collisions until time $-t$, and is equal to

$$\frac{\partial(P_1(-t, (x)_N), \ldots, P_N(-t, (x)_N))}{(\partial p_1, \ldots, \partial p_N)} = \left(\frac{1}{1 - 2\varepsilon}\right)^n \tag{7.2.12}$$

if there are n pair collisions for the initial point $(x)_N$. The Jacobian of transformation (7.2.3) is equal to

$$\frac{\partial(p_i^*, p_j^*)}{\partial(p_i, p_j)} = \frac{1}{1 - 2\varepsilon}.$$

7.3 Evolution operator

7.3.1 Definition of evolution operator

Let $f_N(x_1, \dots, x_N) = f_N((x)_N)$ be a continuous symmetric (permutation invariant) function defined on the phase space $\mathbb{R}^{6N}$ of N particles and equal to zero on the set of forbidden configurations. We define, first formally, an operator $S_N(-t)$ as an operator of shift along the trajectory $X(-t, (x)_N)$ as follows:

$$\begin{aligned}(S_N(-t) f_N)(x_1, \dots, x_N) &= f_N(X_1(-t, (x)_N), \dots, X_N(-t, (x)_N)) \\ &= f_N(X(-t, (x)_N)) \end{aligned} \tag{7.3.1}$$

on admissible configurations, and

$$(S_N(-t) f_N)(x_1, \dots, x_N) = 0$$

on the set of forbidden configurations.

According to the definition of the trajectory $X(-t, (x)_N)$, the function $f_N(X(-t, (x)_N))$ has jumps of momenta at the time of collision $t_{ij}((x)_N)$ because momenta after collisions are different from momenta before collisions and is again a symmetric function.

In classical statistical mechanics of systems of particles with elastic collisions, the function $f_N(X(-t, (x)_N))$ is proportional to the probability density of the considered system at time t in the phase space. It must satisfy the law of conservation of full probability, i.e., full probability must be independent of time. We also need to deduce an equation for $f_N(X(-t, (x)_N))$ and an equation for the sequence of correlation functions. Therefore, we impose some condition on the function $f_N((x)_N)$.

Assume that $f_N((x)_N)$ belongs to the Banach space L_N of functions equal to zero on the set of forbidden configurations, such that $|q_i - q_j| < a$ for at least one pair $i, j \in \{1, \dots, N\}$, and Lebesgue integrable with the norm

$$\|f_N\| = \int |f_N(x_1, \dots, x_N)| dx_1 \dots dx_N = \int |f_N((x)_N)| d(x)_N. \tag{7.3.2}$$

Denote by L_N^0 the subspace of L_N consisting of continuously differentiable functions with compact support that are equal to zero in some neighborhood of the forbidden configuration. The subspace L_N^0 is everywhere dense in L_N.

If $f_N \in L_N^0$, then $f_N(X(-t, (x)_N))$ is a continuously differentiable function with respect to t and $(x)_N$ almost everywhere. Indeed, the trajectory $X(-t, (x)_N)$ is a continuously differentiable function with respect to time t and initial points $(x)_N$ a.e. on time intervals between collisions. Collisions happen if $|Q_i(t, (x)_N) - Q_j(-t, (x)_N)| = a$ for some $i, j \in \{1, \dots, N\}$, but the function $f_N(X(-t, (x)_N))$ is equal to zero in some neighborhood of these hypersurfaces. Outside these hypersurfaces, the trajectories are continuously differentiable with respect to time t and initial points $(x)_N$ a.e., and, therefore, the functions $f_N(X(-t,(x)_N))$ have the same property because $f_N((x)_N) \in L_N^0$. According to definition (7.2.1), $f_N(X(-t,(x)_N))$ is equal to zero on the forbidden configurations together with $f_N((x)_N) \in L_N^0$.

It is obvious that the operator $S_N(-t)$ has the group property

$$S_N(-t_1-t_2) = S_N(-t_1)S_N(-t_2) = S_N(-t_2)S_N(-t_1). \tag{7.3.3}$$

7.3.2 Properties of evolution operator

We consider again $f_N(X(-t,(x)_N))$ with $f_N((x)_N) \in L_N^0$ and show that it is Lebesgue integrable. Indeed, it is continuous with respect to $(x)_N$ a.e. and has compact support, whence

$$\int |f_N(X(-t,(x)_N)|d(x)_N < \infty.$$

We need only to prove that $f_N(X(-t,(x)_N))$ has compact support with respect to $(x)_N$ if $f_N((x)_N)$ has compact support. If $f_N((x)_N)$ has compact support, say $\sum_{i=1}^N(q_i^2+p_i^2) \le R$, $R>0$, then $f_N(X(-t,(x)_N))$ has compact support

$$\sum_{i=1}^N \left[Q_i^2(-t,(x)_N) + P_i^2(-t,(x)_N)\right] \le R$$

with respect to $Q_i, P_i, i = 1,\dots,N$.

If

$$\sum_{i=1}^N P_i^2(-t,(x)_N) \le R,$$

then

$$\sum_{i=1}^N p_i^2 \le R$$

because, at each collision of the i-th and j-th particles at time $0 \le \tau \le t$, one has

$$P_i^{*2}(-\tau,(x)_N) + P_j^{*2}(-\tau,(x)_N) \ge P_i^2(-\tau,(x)_N) + P_j^2(-\tau,(x)_N),$$

and, therefore,

$$R \ge \sum_{i=1}^N P_i^2(-t,(x)_N) \ge \sum_{i=1}^N p_i^2.$$

One has

$$\sum_{i=1}^N Q_i^2(-t,(x)_N) \le R$$

and, therefore,

$$\sum_{i=1}^{N} q_i^2 < r,$$

where $r > 0$ is finite because $Q_i(-t, (x)_N)$ is shifted from q_i at a finite distance by finite $P_i(-\tau, (x)_N), 0 \le \tau \le t$.

Thus, $f_N(X(-t, (x)_N))$ has compact support with respect to $(x)_N$ together with $f_N((x)_N)$.

It the case of elastic collision, the operator $S_N(-t)$ is isometric because Jacobian (7.2.11) is equal to one. In our case of inelastic collision, Jacobian (7.2.11) is different from one for initial $(x)_N$ such that collisions occur.

If $\mathcal{D}$ is some domain in $\mathcal{R}^{6N}$ and $\mathcal{D}_{-t}$ is the image of $\mathcal{D}$ induced by a shift along the trajectories $X(-t, (x)_N)$, then

$$\int_{\mathcal{D}} d(x)_N \neq \int_{\mathcal{D}_{-t}} d(X(-t, (x)_N)) = \int_{\mathcal{D}} \frac{\partial X(-t, (x)_N)}{\partial (x)_N} d(x)_N,$$

because $\frac{\partial(X(-\tau,(x)_N)}{\partial(x)_N} \neq 1$ and $\frac{\partial}{\partial\tau}\frac{\partial(X(-\tau,(x)_N))}{\partial(x)_N}$ is proportional to $\delta(\tau - \tau_l)$ for $(x)_N$ for which collisions occur at time $\tau = \tau_l, 0 \le \tau_l \le t$, and the Jacobian has a jump at time τ_l.

Denote by $V(\mathcal{D})$ and $V(\mathcal{D}_{-t})$ the volumes of the domains $\mathcal{D}$ and $\mathcal{D}_{-t}$, respectively. Then

$$\begin{aligned} V(\mathcal{D}_{-t}) &= \int_{\mathcal{D}} \frac{\partial X(-t, (x)_N)}{\partial (x)_N} d(x)_N \\ &= \int_{\mathcal{D}} \frac{\partial X(0, (x)_N)}{\partial (x)_N} d(x)_N + \int_{\mathcal{D}} \left[\int_0^t \frac{\partial}{\partial \tau} \frac{\partial X(-\tau, (x)_N)}{\partial (x)_N} d\tau \right] d(x)_N \\ &= V(\mathcal{D}) + \int_{\mathcal{D}} \left[\int_0^t \frac{\partial}{\partial \tau} \frac{\partial X(-\tau, (x)_N)}{\partial (x)_N} d\tau \right] d(x)_N. \end{aligned}$$

It follows from these formulas that the contributions of the hypersurfaces

$$|Q_i(-\tau_l, (x)_N) - Q_j(-\tau_l, (x)_N)| = a, \quad i, j \in \{1, \dots, N\},$$

in $V(\mathcal{D}_{-t})$ are finite (for more details, see Appendices A and B).

Nevertheless, the operator $S(-t)$ is "isometric" on L_N^0 in the following sense: Consider the function

$$f_N(X(-t, (x)_N)) \left(\frac{\partial(X(-t, (x)_N))}{\partial (x)_N} \right)^2.$$

Later, we will show that

$$\int f_N(X(-t,(x)_N))\Big(\frac{\partial(X(-t,(x)_N))}{\partial(x)_N}\Big)^2 d(x)_N = \int f_N((x)_N)d(x)_N,$$
$$\int |f_N(X(-t,(x)_N))|\Big(\frac{\partial(X(-t,(x)_N))}{\partial(x)_N}\Big)^2 d(x)_N = \int |f_N((x)_N)|d(x)_N. \tag{7.3.4}$$

It is obvious that

$$\frac{\partial(X(-t,(x)_N))}{\partial(x)_N}\bigg|_{t=0} = 1,$$

and it follows from (7.3.4) that the function

$$D_N(t,(x)_N) = f_N(X(-t,(x)_N))\left(\frac{\partial(X(-t,(x)_N))}{\partial(x)_N}\right)^2, \tag{7.3.5}$$

which is equal to $f_N((x)_N)$ at $t = 0$, may be considered as a probability density in the phase space of systems of hard spheres with inelastic collisions.

Function (7.3.5) has the following "group" property:

$$f_N(X(-t_1-t_2,(x)_N))\left(\frac{\partial(X(-t_1-t_2,(x)_N))}{\partial(x)_N}\right)^2$$
$$= f_N(X(-t_1,X(-t_2,(x)_N))\left(\frac{\partial(X(-t_1,X(-t_2,(x)_N)))}{\partial(x)_N}\right)^2$$
$$= f_N(X(-t_2,X(-t_1,(x)_N))\left(\frac{\partial(X(-t_2,X(-t_1,(x)_N)))}{\partial(x)_N}\right)^2, \tag{7.3.6}$$

$$\frac{\partial(X(-t_1,X(-t_2,(x)_N)))}{\partial(x)_N} = \frac{\partial(X(-t_1,X(-t_2,(x)_N)))}{\partial X(-t_2,(x)_N)}\frac{\partial(X(-t_2,(x)_N))}{\partial(x)_N},$$
$$\frac{\partial(X(-t_2,X(-t_1,(x)_N)))}{\partial(x)_N} = \frac{\partial(X(-t_2,X(-t_1,(x)_N)))}{\partial X(-t_1,(x)_N)}\frac{\partial(X(-t_1,(x)_N))}{\partial(x)_N}.$$

Note that function (7.3.5) is continuously differentiable together with functions $f_N(X(-t,(x)_N))$, $f_N((x)_N) \in L^0_N$. Indeed, the function $f_N(X(-t,(x)_N))$ is continuously differentiable with respect to time t and initial data $(x)_N$ a.e. The Jacobian $\frac{\partial(X(-t,(x)_N))}{\partial(x)_N}$ is a constant function of time on time intervals between collisions and has jumps at times of collisions, but the function $f_N(X(-t,(x)_N))$ is equal to zero in the neighborhood of the times of collisions, and, therefore, the function $D_N(t,(x)_N)$ is continuously differentiable as well as $f_N(X(-t,(x)_N))$. The proof is presented below.

7.3.3 Differential equation for distribution function

Let us show that the function $D_N(t,(x)_N)$ is differentiable with respect to time. It is the product of the two functions $f_N(X(-t,(x)_N))$ and $\frac{\partial X(-t,(x)_N)}{\partial (x)_N}$. We get

$$\frac{\partial}{\partial t} D_N(t,(x)_N) = \Big[\frac{\partial}{\partial t} f_N(X(-t,(x)_N))\Big]\Big(\frac{\partial X(-t,(x)_N)}{\partial (x)_N}\Big)^2 + f_N(X(-t,(x)_N))\Big[\frac{\partial}{\partial t}\Big(\frac{\partial X(-t,(x)_N)}{\partial (x)_N}\Big)^2\Big]. \tag{7.3.7}$$

We now calculate the derivative of $f_N(X(-t,(x)_N))$ for $f_N((x)_N) \in L^0_N$. Using the group property of $S_N(-t)$ (7.3.3), one obtains (for details, see [PGM3] and [CGP])

$$\begin{aligned}
&\frac{\partial}{\partial t} f_N(X(-t,(x)_N)) \\
&= \lim_{\Delta t \to 0}\Big[S_N(-t)\frac{(S_N(-\Delta t) - I) f_N((x)_N)}{\Delta t}\Big] \\
&= S_N(-t)\Big[-\sum_{i=1}^{N} p_i \cdot \frac{\partial}{\partial q_i} f_N((x)_N)\Big] \\
&= -\sum_{i=1}^{N} P_i(-t,(x)_N)\frac{\partial}{\partial Q_i(-t,(x)_N)} f_N(X(-t,(x)_N)),
\end{aligned} \tag{7.3.8}$$

$$\begin{aligned}
\frac{\partial}{\partial t} f_N(X(-t,(x)_N)) &= \lim_{\Delta t\to 0}\frac{S_N(-\Delta t) - I}{\Delta t} S_N(-t) f_N((x)_N) \\
&= -\sum_{i=1}^{N} p_i \cdot \frac{\partial}{\partial q_i} f_N(X(-t,(x)_N)).
\end{aligned}$$

Let us explain the derivation of formulas (7.3.8).

One has

$$\lim_{\Delta t \to 0} \frac{1}{\Delta t}(S_N(-\Delta t) - I) f_N((x)_N) = -\sum_{i=1}^{N} p_i \cdot \frac{\partial}{\partial q_i} f_N((x)_N)$$

on the set $|q_i - q_j| > a$, $i,j \in \{1,\dots,N\}$, because $f_N \in L^0_N$, and it is equal to zero on some neighborhood of forbidden configuration. Note that terms with $\frac{\partial}{\partial t} P_i(-t,(x)_N)$ is absent in the first formula (7.3.8) because $P_i(-t,(x)_N)$ has jumps only at $|Q_i(-t,(x)_N) - Q_j(-t,(x)_N)| = 0$ where $f_N(X(-t,(x)_N))$ is equal to zero.

The trajectory $X(-t,(x)_N)$ at $q_i - q_j = a\eta$, $\eta\cdot(p_i - p_j) > 0$, has a jump at $t = +0$, $X(-0,(x)_N) - X(0,(x)_N) = (x)^*_N - (x)_N$, $(x)^*_N = (q_1, p_1, \dots, q_i, p^*_i, \dots, q_j,$

$p_j^*, \dots, q_N, p_N)$, but the function $f_N \in L_N^0$ is equal to zero in some neighborhood of such points; therefore, $f_N((x)_N^*) = f_N((x)_N) = 0$, i.e., the function $f_N(X(-t, x))$ does not have a jump at $t = +0$. At time $t > 0$, we have $X(-t, (x)_N) = X(-t, (x)_N^*)$ and $f_N(X(-t, (x)_N)) = f_N(X(-t, (x)_N^*))$ for $q_i - q_j = a\eta$, $\eta \cdot (p_i - p_j) > 0$.

Note that $f_N(X(-t, (x)_N))$ may be different from zero with respect to $(x)_N$ on this neighborhood of forbidden configuration where $f_N((x)_N)$ is equal to zero. Therefore,

$$\lim_{\Delta t \to 0} \frac{1}{\Delta t}(S_N(-\Delta t) - I) f_N(X(-t, (x)_N))$$

$$= -\sum_{i=1}^{N} p_i \cdot \frac{\partial}{\partial q_i} f_N(X(-t, (x)_N))$$

with a boundary condition according to which, at $q_i - q_j = a\eta$, $\eta \cdot (p_i - p_j) > 0$, $i, j \in \{1, \dots, N\}$, the momenta p_i and p_j in the expression $-\sum_{i=1}^{N} p_i \cdot \frac{\partial}{\partial q_i} f_N(X(-t, (x)_N))$ should be replaced by p_i^* and p_j^*.

Thus, we obtain two expressions for $\frac{\partial}{\partial t} f_N(X(-t, (x)_N))$, namely

$$\frac{\partial}{\partial t} f_N(X(-t, (x)_N))$$

$$= -\sum_{i=1}^{N} P_i(-t, (x)_N) \frac{\partial}{\partial Q_i(-t, (x)_N)} f_N(X(-t, (x)_N)) \tag{7.3.8a}$$

and

$$\frac{\partial}{\partial t} f_N(X(-t, (x)_N)) = -\sum_{i=1}^{N} p_i \cdot \frac{\partial}{\partial q_i} f_N(X(-t, (x)_N))$$

$$= -\sum_{j=1}^{N} \left[\frac{\partial f_N(X(-t, (x)_N))}{\partial Q_j(-t, (x)_N)} \sum_{i=1}^{N} p_i \cdot \frac{\partial}{\partial q_i} Q_j(-t, (x)_N) \right.$$

$$\left. + \frac{\partial f_N(X(-t, (x)_N))}{\partial P_j(-t, (x)_N)} \sum_{i=1}^{N} p_i \cdot \frac{\partial}{\partial q_i} P_j(-t, (x)_N) \right]. \tag{7.3.8b}$$

The right-hand side of (7.3.8a) is a continuous function with respect to $(x)_N$ a.e. because $f_N((x)_N) \in L_N^0$ and $P_i(-t, (x)_N)$, $i = 1, \dots, N$, are continuous functions of time and $(x)_N$ a.e. on time intervals between collisions and have jumps only at times of collisions, but the function $f_N(X(-t, (x)_N))$ is equal to zero in some neighborhood of times of collisions. $Q_i(-t, (x)_N)$, $i = 1, \dots, N$, are continuous functions of time and $(x)_N$ a.e. on time intervals between collisions.

The right-hand side of (7.3.8b) is also a continuous function with respect to $(x)_N$ a.e. because $(f_N(x)_N) \in L_N^0$, and $Q_j(-t, (x)_N)$, $P_j(-t, (x)_N)$, $1 \le j \le N$, are continuously differentiable functions with respect to $(x)_N$ a.e. on time intervals between collisions, but the functions $\frac{\partial f_N(X(-t,(x)_N))}{\partial Q_j(-t,(x)_N)}$ and $\frac{\partial f_N(X(-t,(x)_N))}{\partial P_i(-t,(x)_N)}$ are equal to zero in some neighborhood of times of collisions.

Note that, on the right-hand side of (7.3.8b) we have the *following boundary conditions*:

$$\text{for} \quad q_i - q_j = a\eta, \quad |\eta| = 1, \quad \eta \cdot (p_i - p_j) > 0, \quad i, j \in \{1, \ldots, N\},$$

the expressions $-\sum_{i=1}^{N} p_i \cdot \frac{\partial}{\partial q_i} f_N(X(-t,(x)_N))$, $f_N(X(-t,(x)_N))$ should be replaced by:

$$-\sum_{i=1}^{N} p_i \cdot \frac{\partial}{\partial q_i} f_N(X(-t,(x)_N))|_{p_i=p_i^*, p_j=p_j^*},$$
$$f_N(X(-t,(x)_N))|_{p_i=p_i^*, p_j=p_j^*}. \tag{7.3.9}$$

For $\eta \cdot (p_i - p_j) < 0$, on the contrary, momenta do not change. These boundary conditions follow from the definition of the trajectory at $q_i - q_j = a\eta$, $\eta \cdot (p_i - p_j) > 0$ (7.3.10), namely $X(-t,(x)_N) = X(-t,(x)_N) \;|_{p_i=p_i^*,\; p_j=p_j^*}$, and the fact that $f_N((x)_N) \in L_N^0$.

On the right-hand side of (7.3.8a), analogous boundary conditions are absent because the function $f_N(X(-t,(x)_N))$ is equal to zero for $|Q_i(-t,(x)_N) - Q_j(-t,(x)_N)| = a$, $i, j \in \{1, \ldots, N\}$, and the term $\frac{\partial}{\partial t} P_i(-t,(x)_N) \frac{\partial f_N(X(-t,(x)_N))}{\partial P_i(-t,(x)_N)}$ is equal to zero because $f_N(X(-t,(x)_N))$ is equal to zero where $P_i(-t,(x)_N)$ have jumps, $i = 1, \ldots, N$.

At first sight, according to the boundary condition (7.3.9) for $q_i - q_j = a\eta$, $\eta \cdot (p_i - p_j) > 0$, there are jumps on the right-hand side of (7.3.8b) because the momenta (p_i, p_j) are replaced by (p_i^*, p_j^*). Let us show that this is not true.

As mentioned above, the right-hand side of (7.3.8a) is a continuous function of $(x)_N$ a.e. on the entire phase space of admissible configurations, i.e., $|q_i - q_j| \geq a$ for all pairs $i, j \in \{1, \ldots, N\}$. The right-hand side of (7.3.8b), together with the boundary condition (7.3.9), identically coincides with the right-hand side of (7.3.8a) and, therefore, it is also a continuous function of $(x)_N$ a.e. on the entire phase space of admissible configurations.

We now consider the second term on the right-hand side of (7.3.7) and show that it is equal to zero. Indeed, the Jacobian $\frac{\partial X(-t,(x)_N)}{\partial (x)_N}$ is a constant function of t for a given fixed $(x)_N$ and has jumps at times of collisions. Therefore, $\frac{\partial}{\partial t}\frac{\partial X(-t,(x)_N)}{\partial (x)_N}$ is equal to zero on time intervals between collisions; however, the function $f(X(-t,(x)_N))$ is equal to zero in the neighborhood of times of collisions, and, as a result,

$$f_N(X(-t,(x)_N))\left[\frac{\partial}{\partial t}\left(\frac{\partial X(-t,(x)_N)}{\partial (x)_N}\right)^2\right] \equiv 0,$$

i.e., the second term on the right-hand side of (7.3.7) is equal to zero.

Taking into account the results obtained above, we get

$$\begin{aligned}
&\frac{\partial}{\partial t} D_N(t,(x)_N) \\
&\quad = \left[\frac{\partial}{\partial t} f_N(X(-t,(x)_N))\right]\left(\frac{\partial X(-t,(x)_N)}{\partial (x)_N}\right)^2 \\
&\quad = \left[-\sum_{i=1}^{N} P_i(-t,(x)_N)\frac{\partial}{\partial Q_i(-t,(x)_N)} f_N(X(-t,(x)_N))\right] \\
&\qquad\quad \times \left(\frac{\partial X(-t,(x)_N)}{\partial (x)_N}\right)^2 \\
&\quad = \left[-\sum_{i=1}^{N} p_i\cdot\frac{\partial}{\partial q_i} f_N(X(-t,(x)_N))\right]\left(\frac{\partial X(-t,(x)_N)}{\partial (x)_N}\right)^2 \\
&\quad = -\sum_{i=1}^{N} p_i\cdot\frac{\partial}{\partial q_i}\left[f_N(X(-t,(x)_N))\left(\frac{\partial X(-t,(x)_N)}{\partial (x)_N}\right)^2\right] \\
&\quad = -\sum_{i=1}^{N} p_i\cdot\frac{\partial}{\partial q_i} D_N(t,(x)_N).
\end{aligned} \tag{7.3.10}$$

The last equality in (7.3.10) follows from the fact that the Jacobian $\frac{\partial X(-t,(x)_N)}{\partial (x)_N}$ is constant (piecewise constant) everywhere with respect to $(x)_N$, excluding points $(x)_N$ at which there are collisions at time t, but the function $f_N(X(-t,(x)_N))$ is equal to zero in the neighborhood of these points, and, therefore,

$$f_N(X(-t,(x)_N))\left(-\sum_{i=1}^{N} p_i\cdot\frac{\partial}{\partial q_i}\frac{\partial X(-t,(x)_N)}{\partial (x)_N}\right)$$

is equal to zero at such points. Taking into account that

$$D_N(t,(x)_N) = f_N(X(-t,(x)_N))\left(\frac{\partial X(-t,(x)_N)}{\partial (x)_N}\right)^2,$$

for $\eta \in S_+^2$ $(\eta\cdot(p_i - p_j) > 0)$ one obtains

$$\begin{aligned}
&D_N(-t, x_1,\dots,q_i,p_i,\dots,q_i - a\eta, p_j,\dots,x_N) \\
&\quad = f_N(X(-t,x_1,\dots,q_i,p_i,\dots,q_i - a\eta,p_j,\dots,x_N)) \\
&\qquad \times \left(\frac{\partial X(-t,x_1,\dots,q_i,p_i,\dots,q_i - a\eta,p_j,\dots,x_N)}{\partial (x)_N}\right)^2
\end{aligned}$$

$$
\begin{aligned}
&= f_N(X(-t, x_1, \ldots, q_i, p_i^*, \ldots, q_i - a\eta, p_j^*, \ldots, x_N)) \\
&\quad \times \left(\frac{\partial X(-t-0, x_1, \ldots, q_i, p_i^*, \ldots, q_i - a\eta, p_j^*, \ldots, x_N)}{\partial (x)_N^*} \right. \\
&\quad \times \left. \frac{\partial X(-0, x_1, \ldots, q_i, p_i^*, \ldots, q_i - a\eta, p_j^*, \ldots, x_N)}{\partial (x)_N} \right)^2 \\
&= f_N(X(-t-0, x_1, \ldots, q_i, p_i^*, \ldots, q_i - a\eta, p_j^*, \ldots, x_N)) \\
&\quad \times \left(\frac{\partial X(-t-0, x_1, \ldots, q_i, p_i^*, \ldots, q_i - a\eta, p_j^*, \ldots, x_N)}{\partial (x)_N^*} \right)^2 \frac{1}{(1-2\varepsilon)^2} \\
&= D_N(-t, x_1, \ldots, q_i, p_i^*, \ldots, q_i - a\eta, p_j^*, \ldots, x_N) \frac{1}{(1-2\varepsilon)^2}. \qquad (7.3.11)
\end{aligned}
$$

In deducing (7.3.11), we have taken into account that $f_N(X(-t,(x)_N)) = f_N(X(-t,(x)_N^*))$ for $q_i - q_j = a\eta$, $\eta \cdot (p_i - p_j) > 0$, used the fact that the Jacobian $\frac{\partial X(-t,(x)_N)}{\partial (x)_N}$ can be calculated as the product of Jacobians on consecutive time intervals between collisions, and separated the Jacobian that corresponds to collisions of the i-th and j-th particles at time $t = +0$. The last Jacobian is equal to $\frac{1}{1-2\varepsilon}$.

We must add the following boundary condition to (7.3.10): for

$$
q_i - q_j = a\eta, \quad |\eta| = 1, \quad \eta \cdot (p_i - p_j) > 0, \quad i, j \in \{1, \ldots, N\},
$$

the expressions

$$
-\sum_{i=1}^{N} p_i \cdot \frac{\partial}{\partial q_i} D_N(t,(x)_N), \quad D_N(t,(x)_N)
$$

should be replaced by

$$
\begin{aligned}
&-\frac{1}{(1-2\varepsilon)^2} \sum_{i=1}^{N} p_i \cdot \frac{\partial}{\partial q_i} D_N(t,(x)_N) \mid_{p_i = p_i^*, p_j = p_j^*}, \\
&\frac{1}{(1-2\varepsilon)^2} D_N(t,(x)_N) \mid_{p_i = p_i^*, p_j = p_j^*},
\end{aligned} \qquad (7.3.12)
$$

and for $\eta \cdot (p_i - p_j) < 0$ the momenta (p_i, p_j) do not change.

The boundary condition (7.3.12) for $D_N(t,(x)_N)$ follows directly from the boundary condition (7.3.9) for $f_N(X(-t,(x)_N))$, $f_N((x)_N) \in L_N^0$, and from equality (7.3.11).

The results obtained can be summarized in the following fundamental theorem:

Theorem 7.2. *The probability density on the phase space of a system of hard spheres with inelastic collisions $D_N(t,(x)_N)$ is a differentiable function with respect to time t*

and $(x)_N$ *a.e. and satisfies the Liouville equation*

$$\frac{\partial}{\partial t} D_N(t,(x)_N) = -\sum_{i=1}^{N} p_i \cdot \frac{\partial}{\partial q_i} D_N(t,(x)_N) \tag{7.3.13}$$

with the boundary condition (7.3.12) and the initial condition

$$D_N(t,(x)_N)|_{t=0} = D_N(0,(x)_N) = f_N((x)_N)$$

$$\left(\left.\frac{\partial X(-t,(x)_N)}{\partial (x)_N}\right|_{t=0} = 1 \text{ because } X(-t,(x)_N)|_{t=0} = (x)_N \right).$$

Note that the right-hand side of (7.3.13), together with the boundary conditions (7.3.12), is a continuous function on the phase space of admissible configurations a.e., as it follows from the second expression on the right-hand side of (7.3.10). Indeed, we have already shown that

$$-\sum_{i=1}^{N} P_i(-t,(x)_N) \frac{\partial}{\partial Q_i(-t,(x)_N)} f_N(X(-t,(x)_N))$$

is a continuous function of $(x)_N$ on admissible configurations of the phase space a.e. The Jacobian $\frac{\partial X(-t,(x)_N)}{\partial (x)_N}$ has jumps only at the points $(x)_N$ for which there are collisions at time t, but the multiplier written above is equal to zero in the neighborhood of these points, and, therefore, the expression

$$-\sum_{i=1}^{N} P_i(-t,(x)_N) \frac{\partial}{\partial Q_i(-t,(x)_N)} f_N(X(-t,(x)_N)) \left(\frac{\partial X(-t,(x)_N)}{\partial (x)_N} \right)^2 \tag{7.3.14}$$

has the desired property of continuity. The expression

$$-\sum_{i=1}^{N} p_i \cdot \frac{\partial}{\partial q_i} D_N(t,(x)_N),$$

together with the boundary condition (7.3.12), coincides with this continuous function and has the same of continuity a.e. on the entire phase space of admissible configurations.

Remark 7.1. One can impose some additional conditions on functions $f_N((x)_N) \in L_N^0$ in order to make the function $D_N(t,(x)_N)$ continuous everywhere on admissible configurations in the phase space.

Namely, we restrict ourselves to functions $f_N((x)_N) \in L_N^0$ that are also equal to zero in the neighborhoods of the hyperplanes where three or more particles collide instantaneously, times of collisions become infinite, and the number of collisions on a finite time interval is infinite. Obviously, this set of functions is again everywhere dense in L_N. We continue denoting it by L_N^0. The functions $D_N(t,(x)_N)$ that correspond to such $f_N((x)_N)$ are continuous with respect to $(x)_N$ everywhere on the phase space and at every time t.

7.4 Equation for a sequence of correlation functions

7.4.1 Definition of correlation functions

We use the commonly accepted definition of correlation function. Namely, correlation functions $F_s^{(N)}(t, x_1, \dots, x_s)$ within the framework of the canonical ensemble are defined through the probability density $D_N(t, x_1, \dots, x_N)$ as follows:

$$F_s^{(N)}(t, x_1, \dots, x_s)$$
$$= N(N-1)\dots(N-s+1)$$
$$\times \int D_N(t, x_1, \dots, x_s, x_{s+1}, \dots, x_N) dx_{s+1} \dots dx_N, \quad 1 \leq s \leq N. \quad (7.4.1)$$

In (7.4.1), we integrate over the entire phase space of particles with numbers $s+1, \dots, N$, but the function $D_N(t, (x)_N)$ is equal to zero on forbidden configurations, and the integration in (7.4.1) is actually carried out over the admissible configuration $|q_i - q_j| \geq a$, $i, j \in \{1, \dots, N\}$. It is assumed that the initial probability density $D_N(0, (x)_N) = f_N((x)_N)$ is normalized to unity:

$$\int D_N(0, (x)_N) d(x)_N = \int f_N((x)_N) d(x)_N = 1.$$

7.4.2 Equation for correlation functions

In order to deduce equations for correlation functions, we differentiate both sides of (7.4.1) with respect to time and use in the right-hand side of (7.4.1) the Liouville equation (7.3.12). One obtains

$$\frac{\partial}{\partial t} F_s^{(N)}(t, x_1, \dots, x_s) = N(N-1)\dots(N-s+1)$$
$$\times \int \left(-\sum_{i=1}^{N} p_i \cdot \frac{\partial}{\partial q_i} D_N(t, x_1, \dots, x_s, x_{s+1}, \dots, x_N) \right) dx_{s+1} \dots dx_N. \quad (7.4.2)$$

We are now in the same situation as for a system of N hard spheres with elastic collisions, and we obtain the following hierarchy of equations:

$$\frac{\partial}{\partial t} F_s^{(N)}(t, x_1, \dots, x_s)$$
$$= -\sum_{i=1}^{s} p_i \cdot \frac{\partial}{\partial q_i} F_s^{(N)}(t, x_1, \dots, x_s)$$
$$+ a^2 \sum_{i=1}^{s} \int dp_{s+1} \int_{S^2} d\eta\, \eta \cdot (p_i - p_{s+1})$$

$$\times F_{s+1}^{(N)}(t, x_1, \dots, x_s, q_i - a\eta, p_{s+1})$$
$$+ \frac{1}{2}a^2 \int dx_{s+1} \int dp_{s+2} \int_{S^2} d\eta \, \eta \cdot (p_{s+1} - p_{s+2})$$
$$\times F_{s+2}^{(N)}(t, x_1, \dots, x_{s+1}, q_{s+1} - a\eta, p_{s+2}), \quad 1 \leq s \leq N, \tag{7.4.3}$$

where η is the unit vector and S^2 is the unit sphere.

We now split the spheres S^2 in the second and the third term on the right-hand side of (7.4.3) into two parts, namely, $S^2 = S_+^2 \cup S_-^2$, where

$$S_+^2 = (\eta \in \mathbb{R}^3 | \, \|\eta\| = 1, \eta \cdot (p_i - p_{s+1}) > 0),$$
$$S_-^2 = (\eta \in \mathbb{R}^3 | \, \|\eta\| = 1, \eta \cdot (p_i - p_{s+1}) < 0), \quad i = 1, \dots, s$$

and

$$S_+^2 = (\eta \in \mathbb{R}^3 | \, \|\eta\| = 1, \eta \cdot (p_{s+1} - p_{s+2}) > 0),$$
$$S_-^2 = (\eta \in \mathbb{R}^3 | \, \|\eta\| = 1, \eta \cdot (p_{s+1} - p_{s+2}) < 0).$$

It follows from (7.3.11) that the correlation functions satisfy the following boundary conditions:

$$F_{s+1}^{(N)}(t, x_1, \dots, q_i, p_i, \dots, x_s, q_i - a\eta, p_{s+1})$$
$$= F_{s+1}^{(N)}(t, x_1, \dots, q_i, p_i^*, \dots, x_s, q_i - a\eta, p_{s+1}^*) \frac{1}{(1-2\varepsilon)^2},$$

$$F_{s+2}^{(N)}(t, x_1, \dots, x_s, q_{s+1}, p_{s+1}, q_{s+1} - a\eta, p_{s+2})$$
$$= F_{s+2}^{(N)}(t, x_1, \dots, x_s, q_{s+1}, p_{s+1}^*, q_{s+1} - a\eta, p_{s+2}^*) \frac{1}{(1-2\varepsilon)^2} \tag{7.4.4}$$

for $\eta \cdot (p_i - p_{s+1}) > 0$ and $\eta \cdot (p_{s+1} - p_{s+2}) > 0$, respectively.

Let us show that the third term on the right-hand side of (7.4.3) is equal to zero. To this end, we represent it as follows, using (7.4.4):

$$\frac{a^2}{2} \int dq_{s+1} \int dp_{s+1} \int dp_{s+2} \bigg[\int_{S_+^2} d\eta \, |\eta \cdot (p_{s+1} - p_{s+2})|$$
$$\times F_{s+2}^{(N)}(t, x_1, \dots, q_{s+1}, p_{s+1}^*, q_{s+1} - a\eta, p_{s+2}^*) \frac{1}{(1-2\varepsilon)^2}$$
$$- \int_{S_-^2} d\eta \, |\eta \cdot (p_{s+1} - p_{s+2})|$$

$$\times F_{s+2}^{(N)}(t, x_1, \dots, q_{+1}, p_{s+1}, q_{s+1} - a\eta, p_{s+2})\Bigg]. \tag{7.4.5}$$

In the first term of (7.4.5), we use the momenta p_{s+1}^* and p_{s+2}^* as new variables of integration. Taking into account that

$$\eta \cdot (p_{s+1}^* - p_{s+2}^*) = \frac{1}{1-2\varepsilon}\eta \cdot (p_{s+1} - p_{s+2}) < 0$$

for $\eta \cdot (p_{s+1} - p_{s+2}) > 0$, $\frac{1}{2} < \varepsilon < 1$, we get $\eta \in S_-^2$ with respect to the variables p_{s+1}^*, p_{s+2}^*. We also have

$$dp_{s+1}dp_{s+2}\left|\frac{1}{1-2\varepsilon}\right| = dp_{s+1}^*dp_{s+2}^*.$$

Note that we use the constant Jacobian in the momentum space equal to $\left|\frac{1}{1-2\varepsilon}\right|$ and take into account that the linear transformation (7.2.3) maps the domain $\eta\cdot(p_{s+1}-p_{s+2}) > 0$ into the domain $\eta \cdot (p_{s+1}^* - p_{s+2}^*) < 0$.

Therefore, the first term is equal to

$$\frac{a^2}{2}\int dq_{s+1}\int dp_{s+1}^*\int dp_{s+2}^*\int_{S_-^2} d\eta|\eta \cdot (p_{s+1}^* - p_{s+2}^*)|$$
$$\times F_{s+2}^{(N)}(t, x_1, \dots, q_{s+1}, p_{s+1}^*, q_{s+1} - a\eta, p_{s+2}^*),$$

and it cancels the second term.

We now split the spheres S_2 into two parts, namely $S_+^2 = (\eta \mid \eta \cdot (p_i - p_{s+1}) > 0)$ and $S_-^2 = (\eta \mid \eta\cdot(p_i - p_{s+1}) < 0)$, replace the vector $\eta \in S_-^2$ by the vector $-\eta \in S_+^2$, and use for $\eta \in S_+^2$ the boundary conditions (7.4.4)

$$F_{s+1}^{(N)}(t, x_1, \dots, q_i, p_i, \dots, q_i - a\eta, p_{s+1})$$
$$= F_{s+1}^{(N)}(t, x_1, \dots, q_i, p_i^*, \dots, q_i - a\eta, p_{s+1})\frac{1}{(1-2\varepsilon)^2}$$

in the second term on the right-hand side of (7.4.3).

Finally, in view of this, hierarchy (7.4.3) takes the form

$$\frac{\partial}{\partial t}F_s^{(N)}(t, x_1, \dots, x_s)$$
$$= -\sum_{i=1}^{s} p_i \cdot \frac{\partial}{\partial q_i}F_s^{(N)}(t, x_1, \dots, x_s)$$
$$+ a^2\sum_{i=1}^{s}\int dp_{s+1}\int_{S_+^2} d\eta\,\eta \cdot (p_i - p_{s+1})$$

$$\times \Bigg[\frac{1}{(1-2\varepsilon)^2} F_{s+1}^{(N)}(t, x_1, \dots, q_i, p_i^*, \dots, q_i - a\eta, p_{s+1}^*)$$

$$- F_{s+1}^{(N)}(t, x_1, \dots, q_i, p_i, \dots, q_i + a\eta, p_{s+1}) \Bigg], \quad 1 \le s \le N \tag{7.4.6}$$

with the same boundary condition at $q_i - q_j = a\eta$ for

$$-\sum_{i=1}^{s} p_i \cdot \frac{\partial}{\partial q_i} F_s^{(N)}(t, x_1, \dots, x_s), \quad F_s^{(N)}(t, x_1, \dots, x_s)$$

as for the Liouville equation (7.3.12) for $D_s(t, x_1, \dots, x_s)$. (In the term $F_{s+1}(t, x_1, \dots, q_i, p_i, \dots, q_i - a\eta, p_{s+1})$ with $\eta \in S_-^2$, one uses the new $\eta' = -\eta$, $\eta' \in S_+^2$.) We also have the initial condition

$$F_s^{(N)}(t, x_1, \dots, x_s)|_{t=0} = F_s^{(N)}(0, x_1, \dots, x_s), \quad s \ge 1.$$

We consider the equation for $F_1^{(N)}(t, x_1)$

$$\frac{\partial}{\partial t} F_1^{(N)}(t, x_1)$$

$$= -p_1 \frac{\partial}{\partial q_1} F_1^{(N)}(t, x_1) + a^2 \int dp_2 \int_{S_+^2} d\eta\, \eta \cdot (p_1 - p_2)$$

$$\times \left[\frac{1}{(1-2\varepsilon)^2} F_2^{(N)}(t, q_1, p_1^*, q_1 - a\eta, p_2^*) - F_2^{(N)}(t, q_1, p_1, q_1 + a\eta, p_2) \right]$$

and integrate it with respect to x_1 over the entire phase space. Using the same tricks as in the proof of the fact that the term with F_{s+2} is zero and assuming that

$$\lim_{|q_1| \to \infty} F_1^{(N)}(t, q_1, p_1) = 0,$$

one obtains

$$\frac{\partial}{\partial t} \int F_1^{(N)}(t, x_1) dx_1 = 0. \tag{7.4.7}$$

This means that $\int F_1^{(N)}(t, x_1) dx_1$ does not depend on t, i.e.,

$$\int F_1^{(N)}(t, x_1) dx_1 = \int F_1^{(N)}(0, x_1) dx_1.$$

Taking into account that, according to definition (7.4.1),

$$F_1^{(N)}(t, (x)_1) = N \int D_N(t, x_1, x_2, \dots, x_N) dx_2 \dots dx_N$$

one obtains the law of conservation of full probability

$$\int D_N(t, x_1, \dots, x_N) dx_1 \dots dx_N = \int D_N(0, x_1, \dots, x_N) dx_1 \dots dx_N. \quad (7.4.8)$$

We summarize the results obtained above in the following theorem:

Theorem 7.3. *The sequence of correlation functions $F_s^{(N)}(t, x_1, \dots, x_s)$ (7.4.1), $1 \le s \le N$, satisfies the hierarchy of equations (7.4.6) with boundary and initial conditions, and the probability density $D_N(t, x_1, \dots, x_N)$ (7.3.5) satisfies the law of conservation of full probability (7.4.8).*

Remark 7.2. If one introduces the probability density by the formula

$$D_N(t, (x)_N) = f_N(X(-t, (x)_N) \left| \left[\frac{\partial X(-t, (x)_N)}{\partial (x)_N} \right] \right|^n \quad (7.4.9)$$

for $n \ge 1$, $n \ne 2$, then the sequence of correlation functions (7.4.1) satisfies the following hierarchy:

$$\begin{aligned}
&\frac{\partial}{\partial t} F_s^{(N)}(t, x_1, \dots, x_s) \\
&\quad = -\sum_{i=1}^{s} p_i \cdot \frac{\partial}{\partial q_i} F_s^{(N)}(t, x_1, \dots, x_s) \\
&\quad + a^2 \sum_{i=1}^{s} \int dp_{s+1} \int_{S_+^2} d\eta\, \eta \cdot (p_i - p_{s+1}) \\
&\quad\quad \times \Bigg[\left| \frac{1}{(1-2\varepsilon)^n} \right| F_{s+1}^{(N)}(t, x_1, \dots, q_i, p_i^*, q_i - a\eta, p_{s+1}^*) \\
&\quad\quad\quad - F_{s+1}^{(N)}(t, x_1, \dots, q_i, p_i, \dots, q_i + a\eta, p_{s+1}) \Bigg] \\
&\quad + \frac{1}{2} a^2 \int dx_{s+1} \int dp_{s+1} \int_{S_+^2} d\eta\, \eta \cdot (p_{s+1} - p_{s+2}) \\
&\quad\quad \times \Bigg[\left| \frac{1}{(1-2\varepsilon)^n} \right| F_{s+2}^{(N)}(t, x_1, \dots, q_{s+1}, p_{s+1}^*, q_{s+1} - a\eta, p_{s+2}^*) \\
&\quad\quad\quad - F_{s+2}^{(N)}(t, x_1, \dots, q_{s+1}, p_{s+1}, q_{s+1} + a\eta, p_{s+2}) \Bigg]. \quad (7.4.10)
\end{aligned}$$

The third term on the right-hand side of (7.4.10) is different from zero because the calculation used in the case $n = 2$ is not true for $n \ne 2$. After the change of the integration variables from (p_{s+1}, p_{s+2}) to (p^*_{s+1}, p^*_{s+2}), the multiplier $\left|\frac{1}{(1-2\varepsilon)^{n-2}}\right|$ remains in the first term with F_{s+2}, and this term does not cancel the second term.

For the probability density $D_N(t,(x)_N)$ (7.4.9) with $n \neq 2$, the law of conservation of full probability (7.4.8) is not true. This means that, for the system of hard spheres with inelastic collisions, the unique "candidate" for the probability density is the function $D_N(t,(x)_N)$ (7.4.9) with $n = 2$, i.e., $D_N(t,(x)_N)$ defined according to (7.3.5).

7.4.3 Boundary conditions for correlation functions

According to the boundary conditions for the function $D_N(t,(x)_N)$ and for the Liouville equation (7.3.11), we have the boundary conditions (7.4.4) for the correlation functions and the following boundary conditions for the BBGKY hierarchy (7.4.6): in the expression

$$-\sum_{i=1}^{s} p_i \cdot \frac{\partial}{\partial q_i} F_s^{(N)}(t, x_1, \dots, x_s)$$

for $q_i - q_j - a\eta = 0$, $\eta \in S^2_+$, $i, j \in \{1, \dots, s\}$, the momenta p_i and p_j should be replaced by p_i^* and p_j^* (1.3), and $F_s^{(N)}$ by $\frac{1}{(1-2\varepsilon)^2} F_s^{(N)}$.

At first sight, the momenta p_i, p_{s+1}, $i = 1, \dots, s$, in η, $(p_i - p_{s+1})$ in the first term of the integral on the right-hand side of equation (7.4.6) should also be replaced by p_i^* and p^*_{s+1}. This is not true. The reason is that, under the integral sign in (7.4.2), the behavior of the integrand on hypersurfaces of lower dimension can be neglected. We prefer to explain this assertion by a very simple example of a system of two spheres (rods) in the one-dimensional case.

We have the Liouville equation

$$\frac{\partial D_2(t, x_1, x_2)}{\partial t} = -\left(p_1 \frac{\partial}{\partial q_1} + p_2 \frac{\partial}{\partial q_2}\right) D_2(t, x_1, x_2)$$

with the following boundary condition: for $q_1 - q_2 - a\eta = 0$, $\eta \cdot (p_1 - p_2) > 0$,

$$D_2(t, q_1, p_1, q_2, p_2) = \frac{1}{(1-2\varepsilon)^2} D_2(t, q_1, p_1^*, q_2, p_2^*),$$

$$\left(-p_1 \frac{\partial}{\partial q_1} - p_2 \frac{\partial}{\partial q_2}\right) D_2(t, q_1, p_1, q_2, p_2)$$

$$= \frac{1}{(1-2\varepsilon)^2} \left(-p_1{}^* \frac{\partial}{\partial q_1} - p_2^* \frac{\partial}{\partial q_2}\right) D_2(t, q_1, p_1^*, q_2, p_2^*).$$

Following [LaPe1] and taking into account that $\frac{\partial}{\partial t} D_2(t, x_1, x_2)$ is different from zero on the admissible configurations $|q_1 - q_2| \geq a$ and continuous with respect to

(x_1, x_2), we get

$$\begin{aligned}
&\int \frac{\partial}{\partial t} D_2(t, x_1, x_2) dq_2 dp_2 \\
&\quad = \int \left(-p_1 \frac{\partial}{\partial q_1} - p_2 \frac{\partial}{\partial q_2}\right) D_2(t, q_1, p_1, q_2, p_2) dq_2 dp_2 \\
&\quad = \lim_{\varepsilon \to 0} \int \left\{\left(\int_{-\infty}^{q_1 - a - \varepsilon} dq_2 + \int_{q_1 + a + \varepsilon}^{\infty} dq_2\right)\right. \\
&\qquad\qquad \left. \times \left(-p_1 \frac{\partial}{\partial q_1} - p_2 \frac{\partial}{\partial q_2}\right) D_2(t, q_1, p_1, q_2, p_2)\right\} dp_2.
\end{aligned}$$

We now calculate the following integrals:

$$\begin{aligned}
&\left(\int_{-\infty}^{q_1 - a - \varepsilon} dq_2 + \int_{q_1 + a + \varepsilon}^{\infty} dq_2\right)\left(-p_2 \frac{\partial}{\partial q_2} D_2(t, q_1, p_1, q_2, p_2)\right) \\
&\qquad = -p_2 D_2(t, q_1, p_1, q_1 - a - \varepsilon, p_2) + p_2 D_2(t, q_1, p_1, q_1 + a + \varepsilon, p_2),
\end{aligned}$$

$$\begin{aligned}
&\left(\int_{-\infty}^{q_1 - a - \varepsilon} dq_2 + \int_{q_1 + a + \varepsilon}^{\infty} dq_2\right)\left(-p_1 \frac{\partial}{\partial q_1} D_2(t, q_1, p_1, q_2, p_2)\right) \qquad (7.4.11) \\
&\qquad = -p_1 \frac{\partial}{\partial q_1}\left(\int_{-\infty}^{q_1 - a - \varepsilon} dq_2 + \int_{q_1 + a + \varepsilon}^{\infty} dq_2\right) D_2(t, q_1, p_1, q_2, p_2) \\
&\qquad\quad + p_1 D_2(t, q_1, p_1, q_1 - a - \varepsilon, p_2) - p_1 D_2(t, q_1, p_1, q_1 + a + \varepsilon, p_2).
\end{aligned}$$

Passing to the limit as $\varepsilon \to 0$ in (7.4.11) and using the formulas obtained above, the continuity of the function $D_2(t, x_1, x_2)$ on admissible configurations, and the fact that p_1 and p_2 are fixed and independent of ε, we obtain

$$\begin{aligned}
\frac{\partial F_1(t, q_1, p_1)}{\partial t} &= \lim_{\varepsilon \to 0} \int \left\{\left(\int_{-\infty}^{q_1 - a - \varepsilon} dq_2 + \int_{q_1 + a + \varepsilon}^{\infty} dq_2\right)\right. \\
&\qquad \left. \times \left(-p_1 \frac{\partial}{\partial q_1} - p_2 \frac{\partial}{\partial q_2}\right) D_2(t, q_1, p_1, q_2, p_2)\right\} dp_2 \\
&= -p_1 \frac{\partial}{\partial q_1} F_1(t, q_1, p_1) + \int dp_2 \Big\{(p_1 - p_2)\Big[D_2(t, q_1, p_1, q_1 - a, p_2) \\
&\qquad - D_2(t, q_1, p_1, q_1 + a, p_2)\Big]\Big\}.
\end{aligned}$$

Consider the following two cases: $p_1 - p_2 > 0$ and $p_1 - p_2 < 0$. In the first case, according to the boundary condition, we have

$$D_2(t, q_1, p_1, q_1 - a, p_2) = \frac{1}{(1-2\varepsilon)^2} D_2(t, q_1, p_1^*, q_1 - a, p_2^*),$$

$$D_2(t, q_1, p_1, q_1 + a, p_2) = D_2(t, q_1, p_1, q_1 + a, p_2).$$

In the second case, one has

$$D_2(t, q_1, p_1, q_1 - a, p_2) = D_2(t, q_1, p_1, q_1 - a, p_2),$$

$$D_2(t, q_1, p_1, q_1 + a, p_2) = \frac{1}{(1-2\varepsilon)^2} D_2(t, q_1, p_1^*, q_1 + a, p_2^*).$$

Denote by η the unit inner vector of the sphere (rod) $|q_2 - q_1| = a$ centered at q_1 so that $\eta = +1$ at the point $q_2 = q_1 - a$ and $\eta = -1$ at the point $q_2 = q_1 + a$.

In the first case, we have $(p_1 - p_2) = \eta \cdot (p_1 - p_2) > 0$, $\eta = +1$, and

$$\begin{aligned}(p_1 - p_2)\Big[D_2(t, q_1, p_1, q_1 - a, p_2) - D_2(t, q_1, p_1, q_1 + a, p_2)\Big] \\ = \eta(p_1 - p_2)\Big[\frac{1}{(1-2\varepsilon)^2} D_2(t, q_1, p_1^*, q_1 - a\eta, p_2^*) \\ - D_2(t, q_1, p_1, q_1 + a\eta, p_2)\Big]\Big|_{\eta=1}.\end{aligned}$$

In the second case, we have $-(p_1 - p_2) = (p_1 - p_2)\eta$, $\eta \cdot (p_1 - p_2) > 0$, $\eta = -1$, and

$$\begin{aligned}&(p_1 - p_2)\Big[D_2(t, q_1, p_1, q_1 - a, p_2) - D_2(t, q_1, p_1, q_1 + a, p_2)\Big] \\ &= (p_1 - p_2)\Big[D_2(t, q_1, p_1, q_1 + a\eta, p_2) - \frac{1}{(1-2\varepsilon)^2} D_2(t, q_1, p_1^*, q_1 - a\eta, p_2^*)\Big] \\ &= \eta(p_1 - p_2)\Big[\frac{1}{(1-2\varepsilon)^2} D_2(t, q_1, p_1^*, q_1 - a\eta, p_2^*) \\ &\qquad - D_2(t, q_1, p_1, q_1 + a\eta, p_2)\Big]\Big|_{\eta=-1}.\end{aligned}$$

Denote by S_+^1 the vector η for which $\eta \cdot (p_1 - p_2) > 0$. For $(p_1 - p_2) > 0$, S_+^1 consists of the vector $\eta = +1$; for $(p_1 - p_2) < 0$, S_+^1 consists of the vector $\eta = -1$. Denote $D_2(t, q_1, p_1, q_2, p_2) = F_2(t, q_1, p_1, q_2, p_2)$. Finally, we obtain the equations

$$\frac{\partial F_1(t,q_1,p_1)}{\partial t}$$
$$= -p_1 \frac{\partial}{\partial q_1} F_1(t,q_1,p_1) + \int dp_2 \sum_{\eta \subset S_+^1} \eta \cdot (p_1 - p_2)$$
$$\times \left[\frac{1}{(1-2\varepsilon)^2} F_2(t,q_1,p_1^*,q_1 - a\eta, p_2^*) - F_2(t,q_1,p_1,q_1 + a\eta, p_2) \right], \tag{7.4.12}$$

$$\frac{\partial F_2(t,q_1,p_1,q_2,p_2)}{\partial t} = \left(-p_1 \frac{\partial}{\partial q_1} - p_2 \frac{\partial}{\partial q_2} \right) F_2(t,q_1,p_1,q_2,p_2),$$

and the boundary condition for the second equation is the same as for $D_2(t,q_1,p_1,q_2,p_2)$.

Equations (7.4.12) are, in fact, hierarchy (7.4.6) for $N = 2$.

Analogous calculations were performed for one-dimensional point particles in [PeC1] on a formal level.

7.4.4 Grand canonical ensemble

As is known (see [PGM3] and [CGP]), in grand canonical ensemble one has a sequence of nonnormalized distribution functions $D_N(t,(x)_N)$, $N \geq 0$, $D_0 = 1$, that satisfy the Liouville equation (7.3.13) with the boundary condition (7.3.12). The sequence of correlation functions is defined as follows:

$$F_s(t,(x)_s)$$
$$= \frac{1}{\Xi} \sum_{n=0}^{\infty} \int \frac{1}{n!} D_{s+n}(t,x_1,\ldots,x_s,x_{s+1},\ldots,x_{s+n}) dx_{s+1} \ldots dx_{s+n}, \tag{7.4.13}$$
$$s \geq 1,$$

where Ξ is the grand partition function:

$$\Xi = 1 + \sum_{n=1}^{\infty} \int \frac{1}{n!} D_n(t,x_1,\ldots,x_n) dx_1 \ldots dx_n$$
$$= 1 + \sum_{n=1}^{\infty} \int \frac{1}{n!} D_n(0,x_1,\ldots,x_n) dx_1 \ldots dx_n. \tag{7.4.14}$$

In (7.4.14), we have used the law of conservation of full probability (7.4.8).

By repeating the procedure of deducing hierarchy (7.4.6) for canonical ensemble [Subsection 7.4.1], for grand canonical ensemble one obtains the hierarchy

$$\begin{aligned}
&\frac{\partial F_s(t, x_1, \dots, x_s)}{\partial t} \\
&\quad = -\sum_{i=1}^{s} p_i \cdot \frac{\partial}{\partial q_i} F_s(t, x_1, \dots, x_s) \\
&\quad\quad + a^2 \sum_{i=1}^{s} \int dp_{s+1} \int_{S_+^2} d\eta \, \eta \cdot (p_i - p_{s+1}) \\
&\quad\quad \times \bigg[\frac{1}{(1-2\varepsilon)^2} F_{s+1}(t, x_1, \dots, q_i, p_i^*, \dots, q_i - a\eta, p_{s+1}^*) \\
&\quad\quad - F_{s+1}(t, x_1, \dots, q_i, p_i, \dots, q_i - a\eta, p_{s+1}) \bigg], \quad s \geq 1, \qquad (7.4.15)
\end{aligned}$$

with the same boundary conditions as for canonical ensemble.

Appendix A

In this appendix, we present two very simple examples that explain why

$$\int f(X(-t,x)) \frac{\partial (X(-t,x))}{\partial x} dx \neq \int f(x) dx.$$

Consider the interval $[0, \infty)$. On this interval, we define the following map T:

$$\begin{aligned}
T(x) &= x \quad \text{if} \quad 0 \leq x \leq 1, \\
T(x) &= 2x \quad \text{if} \quad x > 1.
\end{aligned} \qquad (A.1)$$

Let us show that the Jacobian $\frac{dT(x)}{dx}$ is defined as follows:

$$\begin{aligned}
\frac{dT(x)}{dx} &= 1 \quad \text{if} \quad 0 \leq x < 1, \\
\frac{dT(x)}{dx} &= \delta(x-1) \quad \text{if} \quad x = 1, \\
\frac{dT(x)}{dx} &= 2 \quad \text{if} \quad 1 < x < \infty.
\end{aligned} \qquad (A.2)$$

We calculate $\frac{dT(x)}{dx}$ as a distribution (generalized function). Let $\varphi(x)$ be a test function. Then

$$\begin{aligned}
\int_0^\infty \frac{dT(x)}{dx} \varphi(x) dx &= -\int_0^\infty T(x) \varphi'(x) dx \\
&= -\int_0^1 x \varphi'(x) dx - \int_1^\infty 2x \varphi'(x) dx
\end{aligned}$$

$$= -\varphi(1) + \int_0^1 \varphi(x)dx + 2\varphi(1) + 2\int_1^\infty \varphi(x)dx$$
$$= \varphi(1) + \int_0^1 \varphi(x)dx + 2\int_1^\infty \varphi(x)dx$$
$$= \int_0^\infty \delta(x-1)\varphi(x)dx + \int_0^1 1\cdot\varphi(x)dx + \int_1^\infty 2\cdot\varphi(x)dx. \tag{A.3}$$

This formula gives $\frac{dT(x)}{dx}$, as stated before in (A.2).

Now consider the following integral with an arbitrary smooth function $f(x)$defined on $[0,\infty)$:

$$\int_0^\infty f(T(x))\frac{dT(x)}{dx}dx = \int_0^1 1\cdot f(x)dx + f(1) + \int_1^\infty 2\cdot f(2x)dx$$
$$= f(1) + \int_0^1 f(x)dx + \int_2^\infty f(x)dx. \tag{A.4}$$

It follows from (A.4) that there is a finite contribution of the "hypersurface" $x = 1$ where the map $T(x)$ is discontinuous and the interval $1 \le x \le 2$ is absent, i.e., is lost in the map $T(x)$. Assume that $f(1) = 0$. Then (A.4) is reduced to the following final formula:

$$\int_0^\infty f(T(x))\frac{dT(x)}{dx}dx = \int_0^1 f(x)dx + \int_2^\infty f(x)dx$$
$$= \int_0^\infty f(x)dx - \int_1^2 f(x)dx. \tag{A.5}$$

It follows from (A.5) that

$$\int_0^\infty f(T(x))\frac{dT(x)}{dx}dx < \int_0^\infty f(x)dx \tag{A.6}$$

for a positive "distribution" $f(x) \ge 0$ different from zero on the interval $(1, 2]$.

Consider the second example with the map

$$T(x) = x,\ \ 0 \le x \le 1, \quad T(x) = \frac{1}{2}x, \quad x > 1.$$

For $T(x)$, one obtains

$$\frac{dT(x)}{dx} = 1, \quad 0 \le x \le 1,$$
$$\frac{dT(x)}{dx} = -\frac{1}{2}\delta(x-1), \quad x = 1,$$

and

$$\frac{dT(x)}{dx} = \frac{1}{2}, \quad x > 1.$$

It is easy to check that

$$\int_0^\infty f(T(x))\frac{dT(x)}{dx}dx = -\frac{1}{2}f(1) + \int_0^1 f(x)dx + \int_{\frac{1}{2}}^\infty f(x)dx.$$

If $f(x) \geq 0$ and $f(1) = 0$, then

$$\int_1^\infty f(T(x))\frac{dT(x)}{dx}dx = \int_0^\infty f(x)dx + \int_{\frac{1}{2}}^1 f(x)dx > \int_0^\infty f(x)dx. \tag{A.7}$$

These two examples show that, for a discontinuous map, there may be "loss" or "gain" of domains.

These simple examples can help us to understand why

$$\int f_N(X(-t,(x)_N))\frac{\partial(X(-t,(x)_N))}{\partial(x)_N}d(x)_N \neq \int f((x)_N)d(x)_N. \tag{A.8}$$

The reason is that the map $X(-t,(x)_N)$ is discontinuous and, after collisions $\frac{\partial(X(-t,(x)_N))}{\partial(x)_N} \neq 1$, on the left-hand side the contributions of the hypersurfaces where collisions occur may be finite; some domains in the phase space may be "lost" in the map induced by a shift along the trajectories $X(-t,(x)_N)$ or may be "gained" in the map induced by a shift along the trajectories $X(t,(x)_N)$.

We have shown in Sec. 3 that

$$\int f_N(X(-t,(x)_N))\left(\frac{\partial(X(-t,(x)_N))}{\partial(x)_N}\right)^2 = \int f_N((x)_N)d(x)_N$$

for $f_N((x)_N) \subset L_1^0$, which means that the additional multiplier $\frac{\partial(X(-t,(x)_N))}{\partial(x)_N}$, different from 1 after collisions, compensates for the "loss" of domains in the phase space. Note that $f_N((x)_N) \subset L_1^0$ is equal to zero on hypersurfaces where collisions occur, and, therefore, the contributions of these hypersurfaces are equal to zero.

Appendix B

In deriving formulas (7.3.11), we did not take into account that, for some p_i and p_j, the momenta after collisions p_i^* and p_j^* are equal to p_i and p_j, and

$$\frac{\partial X(+0,(x)_N)}{\partial(x)_N} = 1.$$

For example, this is the case if $\eta \cdot (p_i - p_j) = 0$. These momenta belong to hypersurfaces of lower dimension, and one can neglect them because $D_N(t,(x)_N) \subset L_N$ for $f_N((x)_N) \subset L_N^0$.

If one considers the generalized functions $f_N((x)_N)$ concentrated, e.g., on the hypersurfaces $p_1 = \ldots = p_N = p$ and having compact support with respect to $(q)_N$, then

$$\frac{\partial X(-t,(x)_N)}{\partial (x)_N} = 1$$

and, in $D_N(t, x_1, \ldots, q_i, p_i^*, \ldots, q_i - a\eta, p_j^*, \ldots, x_N)$, $\eta \cdot (p_i - p_j) > 0$, we have the momenta $p_i^* = p_i$ and $p_j^* = p_j$.

For such initial distribution functions $f_N((x)_N)$, the hierarchy for the correlation functions (7.4.15) reduces to the following one:

$$\frac{\partial}{\partial t} F_s(t, x_1, \ldots, x_s) = -\sum_{i=1}^{s} p_i \cdot \frac{\partial}{\partial q_i} F_s(t, x_1, \ldots, x_s), \quad s \geq 1. \tag{B.1}$$

The second and the third term on the right-hand side of (7.4.3) are equal to zero because $p_i - p_{s+1} = 0$ and $p_{s+1} - p_{s+2} = 0$.

Hierarchy (B.1) has the stationary solution

$$F_s(t, x_1, \ldots, x_s) = \prod_{i=1}^{s} (p_i - p) \prod_{i<j=1}^{s} \Theta(|q_i - q_j| = a).$$

Chapter 8

Solution of BBGKY hierarchy for a system of hard spheres with inelastic collisions

8.1 Introduction

We consider an analog of the BBGKY hierarchy for systems of hard spheres with inelastic collisions. It is commonly accepted that these systems are proper models of granular flow. We continue the investigation of these systems begun in Chapter 7. We use the same notation.

In Chapter 7, the Liouville equation for distribution functions of systems of finite numbers N of hard spheres with inelastic collisions was investigated. It was proved that the distribution function is defined as follows:

$$D_N(t,(x)_N) = \left[\frac{\partial X(-t,(x)_N)}{\partial (x)_N}\right]^2 S_N(-t,(x)_N) D_N(0,(x)_N), \quad N \geq 1, \quad (8.1.1)$$

where $D_N(0,(x)_N)$ is the initial distribution function, $S_N(-t,(x)_N)$ is the operator of shift along the trajectory $X(-t,(x)_N)$ of N spheres with initial data $(x)_N$, and $\frac{\partial X(-t,(x)_N)}{\partial (x)_N}$ is the Jacobian. It was shown that only the distribution functions (8.1.1), with squared Jacobian, satisfy the law of conservation of full probability

$$\int D_N(t,(x)_N) d(x)_N = \int D_N(t,(x)_N) dx_N, \quad (8.1.2)$$

the Liouville equation

$$\frac{\partial}{\partial t} D_N(t,(x)_N) = -\sum_{i=1}^{N} p_i \cdot \frac{\partial}{\partial q_i} D_N(t,(x)_N) = \mathcal{H}_N D_N(t,(x)_N) \quad (8.1.3)$$

and specific boundary conditions according to which, at $q_i - q_j = a\eta$, $\eta \cdot (p_i - p_j) > 0$ $\big(|\eta| = 1$ and a is the diameter of hard spheres$\big)$, the momenta p_i, p_j should be replaced by

$$p_i^* = p_i + \frac{\varepsilon}{1-2\varepsilon} \eta\, \eta \cdot (p_i - p_j),$$

$$p_j^* = p_j - \frac{\varepsilon}{1-2\varepsilon} \eta\, \eta \cdot (p_i - p_j),$$

and

$$D_N(t, x_1, \ldots, q_i, p_i, \ldots, q_i - a\eta, p_j, \ldots, x_N)$$
$$= \frac{1}{(1-2\varepsilon)^2} D_N(t, x_1, \ldots, q_i, p_i^*, \ldots, q_i - a\eta, p_j^*, \ldots, x_N). \qquad (8.1.4)$$

The parameter ε in (8.1.4), $\frac{1}{2} < \varepsilon < 1$, characterizes inelasticity. The corresponding sequence of correlation functions satisfies an analog of the BBGKY hierarchy, namely,

$$\frac{\partial}{\partial t} F_s(t, (x)_s)$$
$$= -\sum_{i=1}^{s} p_i \cdot \frac{\partial}{\partial q_i} F_s(t, (x)_s)$$
$$+ a^2 \sum_{i=1}^{s} \int dp_{s+1} \int_{S_2^+} d\eta\, \eta \cdot (p_i - p_{s+1})$$
$$\times \left[\frac{1}{(1-2\varepsilon)^2} F_{s+1}(t, x_1, \ldots, q_i, p_i^*, \ldots, x_s, q_i - a\eta, p_{s+1}^*) \right.$$
$$\left. - F_{s+1}(t, x_1, \ldots, q_i, p_i, \ldots, x_s, q_i + a\eta, p_{s+1}) \right], \quad s \geq 1, \qquad (8.1.5)$$

with the same boundary condition in the first term on the right-hand side of (8.1.6) as conditions (8.1.4) and (8.1.5) for $D_N(t, (x)_N)$ and with the initial conditions

$$F_s(t, (x)_s)|_{t=0} = F_s(0, (x)_s), \quad s \geq 1. \qquad (8.1.6)$$

In the present paper, we consider hierarchy (8.1.6) with initial data (8.1.7), i.e., the Cauchy problem for hierarchy (8.1.6) in the Banach space L_1 of sequences of integrable symmetric functions

$$f = (f_1(x_1), \ldots, f_s((x)_s), \ldots) \qquad (8.1.7)$$

equal to zero on forbidden configurations where $|q_i - q_j| < a$ for at least one pair $i, j \in \{1, \ldots, s\}$ with the norm

$$\|f\| = \sum_{s=0}^{\infty} \|f_s\|, \qquad \|f_s\| = \int |f_s((x)_s)|\, d(x)_s. \qquad (8.1.8)$$

The corresponding group of evolution operators $U(t)$, $t \geq 0$, bounded and strongly continuous in L_1 has been constructed:

$$U(t) = e^{\int dx} J(-t) S(-t) e^{-\int dx}; \qquad (8.1.9)$$

here,

$$\Big(\int dxf\Big)_s((x)_s) = \int dx_{s+1} f_{s+1}(x_1, \dots, x_s, x_{s+1}) \tag{8.1.10}$$

is a bounded operator in L_1, $\|\int dx\| \leq 1$, $J(-t)$ is the direct sum of the operators of multiplication by the squared Jacobian, namely,

$$(J(-t)f)_s((x)_s) = \left[\frac{\partial X(-t,(x)_s)}{\partial (x)_s}\right]^2 f_s((x)_s),$$

and $S(-t)$ is the direct sum of the operators $S_s(-t,(x)_s)$ of shift along the trajectory $X(-t,(x)_s)$:

$$\big(S(-t)f\big)_s((x)_s) = \Big(S_s(-t)f_s\Big)((x)_s) = f_s\Big(X(-t,(x)_s)\Big).$$

It is proved that the group $U(t)$ is strongly differentiable on the everywhere dense set $L_1^0 \subset L_1$ that consists of finite sequences $f \in L_1$ of differentiable functions equal to zero in a certain neighborhood of forbidden configurations and having compact support. The infinitesimal generator B of the group $U(t)$ coincides on L_1^0 with the operator on the right-hand side of hierarchy (8.1.6).

Denoting the sequence of correlation functions $\big(F_1(t,x_1),\dots,F_s(t,(x)_s),\dots)\big)$ by $F(t)$, we consider hierarchy (8.1.6) as an abstract evolution equation in L_1 for the sequence $F(t)$, namely

$$\begin{aligned} \frac{dF(t)}{dt} &= BF(t), \\ F(t)|_{t=0} &= F(0), \end{aligned} \tag{8.1.11}$$

and show that the Cauchy problem (8.1.12) for hierarchy (8.1.6) has the unique solution

$$F(t) = U(t)F(0), \qquad F(0) \in L_1, \tag{8.1.12}$$

which is strong for $F(0) \in L_1^0$ and generalized for arbitrary $F(0) \in L_1$.

8.2 Solution of hierarchy for correlation functions

8.2.1 Solution formula

As is known (see Chapter 7), the correlation functions defined by the formulas

$$\begin{aligned} &F_s(t,x_1,\dots,x_s) \\ &\quad = \frac{1}{\Xi}\sum_{n=0}^{\infty}\frac{1}{n!}\int dx_{s+1}\dots dx_{s+n}\left[\frac{\partial X(-t,x_1,\dots,x_s,x_{s+1},\dots,x_{s+n})}{\partial(x_1,\dots,x_s,x_{s+1},\dots,x_{s+n})}\right]^2 \\ &\qquad\quad \times S_{s+n}(-t,x_1,\dots,x_s,x_{s+1},\dots,x_{s+n}) \end{aligned}$$

$$\times D_{s+n}(0, x_1, \ldots, x_s, x_{s+1}, \ldots, x_{s+n})$$

$$= \frac{1}{\Xi} \sum_{n=0}^{\infty} \frac{1}{n!} \int d(x)^s_{s+n} \left[\frac{\partial X(-t, (x)_{s+n})}{\partial (x)_{s+n}} \right]^2$$

$$\times S_{s+n}(-t, (x)_{s+n}) D_{s+n}(0, (x)_{s+n}), \quad s \geq 1, \tag{8.2.1}$$

where

$$\Xi = \sum_{n=0}^{\infty} \frac{1}{n!} \int d(x)_n \left[\frac{\partial X(-t, (x)_n)}{\partial (x)_n} \right]^2 S_n(-t, (x)_n) D_n(0, (x)_n)$$

$$= \sum_{n=0}^{\infty} \frac{1}{n!} \int d(x)_n D_n(0, (x)_n).$$

Recall that we use the same notation as in Chapter 7, namely,

$$(x)_{s+n} = (x_1, \ldots, x_s, x_{s+1}, \ldots, x_{s+n}),$$

$$d(x)^s_{s+n} = dx_{s+1} \ldots dx_{s+n},$$

$$S^2_+ = (\eta \in \mathbb{R}^3 \mid \|\eta\| = 1, \eta \cdot (p_i - p_j) > 0),$$

$S_{s+n}(-t, (x)_{s+n})$ is the operator of shift along the trajectory $X_{s+n}(-t, (x)_{s+n})$, $D_{s+n}(0, (x)_{s+n})$ is the initial distribution function, $\frac{\partial X(-t,(x)_{s+n})}{\partial (x)_{s+n}}$ is the Jacobian. Expression (8.2.1) are formal solutions of the hierarchy is the grand partition function

$$\frac{\partial F_s(t, (x)_s)}{\partial t}$$

$$= -\sum_{i=1}^{s} p_i \cdot \frac{\partial}{\partial q_i} F_s(t, (x)_s) + a^2 \sum_{i=1}^{s} \int dp_{s+1} \int_{S^2_+} d\eta \eta \cdot (p_i - p_{s+1})$$

$$\times \left[\frac{1}{(1-2\varepsilon)^2} F_{s+1}(t, q_1, p_1, \ldots, q_i, p^*_i, \ldots, q_s, p_s, q_i - a\eta, p^*_{s+1}) \right.$$

$$\left. - F_{s+1}(t, q_1, p_1, \ldots, q_i, p_i, \ldots, q_s, p_s, q_i + a\eta, p_{s+1}) \right], \quad s \geq 1, \tag{8.2.2}$$

with corresponding boundary and initial conditions.

A correlation function satisfies the following boundary condition: at $q_i - q_j = a\eta$, $\eta \cdot (p_i - p_j) > 0$, the momenta p_i and p_j in the first term on the right-hand side of (8.2.2) should be replaced by

$$p^*_i = p_i + \frac{\varepsilon}{1-2\varepsilon} \eta \, \eta \cdot (p_i - p_j), \qquad p^*_j = p_j - \frac{\varepsilon}{1-2\varepsilon} \eta \, \eta \cdot (p_i - p_j)$$

in $-\sum_{i=1}^{s} p_i \cdot \frac{\partial}{\partial q_i}$ and $F_s(t,(x)_s)$ should be replaced by

$$\frac{1}{(1-2\varepsilon)^2} F_s(t, x_1, \dots, q_i, p_i^*, \dots, q_i - a\eta, p_j^*, \dots, x_s).$$

In (8.2.1), the correlation functions $F_s(t,(x)_s)$ are expressed via the initial distribution functions $D_{s+n}(0,(x)_{s+n})$, $n \geq 0$. We transform (8.2.1) so that $F_s(t,(x)_s)$ are expressed via the initial correlation functions $F_{s+n}(0,(x)_{s+n})$, $n \geq 0$. For this purpose, we use (8.2.1) in the case $t = 0$, where $S(0,(x)_{s+n}) = I$ (I is the identity operator) and

$$\frac{\partial X(0,(x)_{s+n})}{\partial (x)_{s+n}} = 1.$$

We obtain

$$F_s(0,(x)_s) = \frac{1}{\Xi} \sum_{n=0}^{\infty} \frac{1}{n!} \int D_{s+n}(0,(x)_{s+n}) d(x)_{s+n}^{s}. \tag{8.2.3}$$

Denote the following sequences by $F(0)$, $D(0)$, and f:

$$\begin{aligned} F(0) &= (F_1(0,x_1), \dots, F_s(0,(x)_s), \dots), \\ D(0) &= (D_1(0,x_1), \dots, D_s(0,(x)_s), \dots), \\ f &= (f_1(x_1), \dots, f_s((x)_s), \dots). \end{aligned} \tag{8.2.4}$$

Let $\int dx$ denote the following operator:

$$\left(\int dx f\right)_s ((x)_s) = \int f_{s+1}((x)_s, x_{s+1}) dx_{s+1}. \tag{8.2.5}$$

Formulas (8.2.3) can be represented as follows:

$$F(0) = \frac{1}{\Xi} e^{\int dx} D(0), \tag{8.2.6}$$

and we have $D(0) = \Xi e^{-\int dx} F(0)$.

Denote by $J(t)$, as in Introduction, the direct sum of operators of multiplication of sequences (8.2.4) by $\left[\frac{\partial X(-t,(x)_s)}{\partial (x)_s}\right]^2$, namely,

$$(J(-t)f)_s((x)_s) = \left[\frac{\partial X(-t,(x)_s)}{\partial (x)_s}\right]^2 f_s((x)_s)$$

and by $S(-t)$ the direct sum of the operators $S_s(-t,(x)_s)$, i.e.,

$$(S(-t)f)_s((x)_s) = (S_s(-t,(x)_s)f_s)(x_s) = f_s(X(-t,(x)_s)).$$

In terms of these operators, formulas (8.2.1) can be represented as follows:

$$F(t) = \frac{1}{\Xi} e^{\int dx} J(-t) S(-t) D(0),$$
$$F_s(t,(x)_s) = \frac{1}{\Xi} (e^{\int dx} J(-t) S(-t) D(0))_s (x)_s. \tag{8.2.7}$$

Finally, expressing $D(0)$ in (8.2.7) through $F(0)$, according to (8.2.6), we get

$$F(t) = e^{\int dx} J(-t) S(-t) e^{-\int dx} F(0) = U(t) F(0),$$
$$U(t) = e^{\int dx} J(-t) S(-t) e^{-\int dx}, \tag{8.2.8}$$

or componentwise

$$F_s(t,(x)_s) = \sum_{n=0}^{\infty} \sum_{k=0}^{n} \frac{(-1)^k}{(n-k)!k!} \int \left[\frac{\partial X(-t,(x)_{s+n-k})}{\partial (x)_{s+n-k}} \right]^2$$
$$\times S_{s+n-k}(-t,(x)_{s+n-k}) F_{s+n}(0,(x)_{s+n}) d(x)_{s+n}^{s}. \tag{8.2.9}$$

$U(t)$ is the evolution operator of hierarchy (8.2.2). These formulas have been obtained on formal level. In the next subsection, we present the justification of these formulas.

8.2.2 Convergence of series

Suppose that sequence f (8.2.4) consists of integrable symmetric functions $\int |f_s((x)_s)| = \|f_s\| < \infty$ equal to zero on forbidden configurations and having the norm

$$\|f\| = \sum_{s=0}^{\infty} \|f_s\| < \infty. \tag{8.2.10}$$

This means that f belongs to the Banach space L_1 consisting of sequences of integrable symmetric functions equal to zero on forbidden configurations with norm $\|f\|$ and componentwise linear operations.

In Chapter 7, we have proved that

$$\int \left[\frac{\partial X(-t,(x)_s)}{\partial (x)_s} \right]^2 \left| S(-t,(x)_s) f_s((x)_s) \right| d(x)_s = \int \left| f_s((x)_s) \right| d(x)_s.$$

This means that

$$\| J(-t) S(-t) f \| = \|f\|. \tag{8.2.11}$$

If is obvious that

$$\left\| \int dx f \right\| \le \|f\|,$$

so that

$$\left\| \int dx \right\| \leq 1, \qquad \|e^{\pm \int dx}\| \leq e, \qquad e^{\pm \int dx} e^{\mp \int dx} = 1. \tag{8.2.12}$$

In view of (8.2.9)–(8.2.11), it follows from (8.2.8) that

$$\|U(t) f\| = \|e^{\int dx} J(-t) S(-t) e^{-\int dx} f\| \leq e^2 \|f\|$$

for arbitrary $f \subset L_1$, which means that the operator of evolution $U(t)$ is bounded in the space L_1, i.e.,

$$\|U(t)\| \leq e^2. \tag{8.2.13}$$

8.2.3 Group property

We have proved in Chapter 7 that the operator $S(-t)$ has the group property

$$S(-t_1 - t_2) = S(-t_1) S(-t_2) = S(-t_2) S(-t_1) \tag{8.2.14}$$

for arbitrary $t_1 > 0$ and $t_2 > 0$.

It was also proved in [PeC1] that the operator $J(-t)$ has the following property:

$$\begin{aligned} \left[\frac{\partial X(-t_1 - t_2, (x)_s)}{\partial (x)_s} \right]^2 &= \left[\frac{\partial X(-t_1, X(-t_2, (x)_s))}{\partial X(-t_2, (x)_s)} \frac{\partial X(-t_2, (x)_s)}{\partial (x)_s} \right]^2 \\ &= \left[\frac{\partial X(-t_2, X(-t_1, (x)_s))}{\partial X(-t_1, (x)_s)} \frac{\partial X(-t_1, (x)_s)}{\partial (x)_s} \right]^2. \end{aligned} \tag{8.2.15}$$

This equality follows from the fact that the Jacobian

$$\frac{\partial X(-t_1 - t_2, (x)_s)}{\partial (x)_s}$$

is equal to the product of the Jacobians that correspond to the consecutive time intervals $[0, t_2]$, $[t_2, t_2 + t_1]$ or $[0, t_1]$, $[t_1, t_1 + t_2]$.

We now show that the product of the operators $J(-t)$ and $S(-t)$ possesses the group property

$$\begin{aligned} J(-t_1 - t_2) S(-t_1 - t_2) &= J(-t_1) S(-t_1) J(-t_2) S(-t_2) \\ &= J(-t_2) S(-t_2) J(-t_1) S(-t_1). \end{aligned} \tag{8.2.16}$$

Let us prove (8.2.15) in the s-particle subspace. We have

$$\left[\frac{\partial X(-t_1, (x)_s)}{\partial (x)_s} \right]^2 S_s(-t_1, (x)_s) \left\{ \left[\frac{\partial X(-t_2, (x)_s)}{\partial (x)_s} \right]^2 S_s(-t_2, (x)_s) \right\}$$

$$= \left[\frac{\partial X(-t_1,(x)_s)}{\partial (x)_s}\right]^2 \left[\frac{\partial X(-t_2, X(-t_1,(x)_s))}{\partial X(-t_1,(x)_s)}\right]^2 S_s(-t_1-t_2,(x)_s)$$

$$= \left[\frac{\partial X(-t_1-t_2,(x)_s)}{\partial (x)_s}\right]^2 S_s(-t_1-t_2,(x)_s).$$

We have used the fact that

$$S_s(-t,(x)_s)\Big[f_s((x)_s)g_s((x)_s)\Big] = \Big[S_s(-t,(x)_s)f_s((x)_s)\Big]\Big[S_s(-t,(x)_s)g_s((x)_s)\Big].$$

These equalities are equivalent to the following ones:

$$J(-t_1)S(-t_1)J(-t_2)S(-t_2) = J(-t_1-t_2)S(-t_1-t_2).$$

By analogy, one can prove that

$$J(-t_2)S(-t_2)J(-t_1)S(-t_1) = J(-t_1-t_2)S(-t_1-t_2).$$

Thus, equality (8.2.15) is proved.

Using (8.2.13)–(8.2.15), we can prove that the operator $U(t)$ possesses the group property

$$U(t_1+t_2) = U(t_1)U(t_2) = U(t_2)U(t_1). \tag{8.2.17}$$

Indeed,

$$\begin{aligned} U(t_1+t_2) &= e^{\int dx} J(-t_1-t_2)S(-t_1-t_2)e^{-\int dx} \\ &= e^{\int dx} J(-t_1)S(-t_1)J(-t_2)S(-t_2)e^{-\int dx} \\ &= e^{\int dx} J(-t_1)S(-t_1)e^{-\int dx}e^{\int dx} J(-t_2)S(-t_2)e^{-\int dx} \\ &= U(t_1)U(t_2) \end{aligned}$$

because

$$e^{-\int dx}e^{\int dx} = I.$$

By analogy, we get

$$U(t_1+t_2) = U(t_2)U(t_1),$$

and (8.2.16) is proved.

8.2.4 Strong continuity of the group

We have proved that the evolution operators $U(t)$, $t \geq 0$, are bounded in L_1 according to (8.2.12) and possess the group property according to (8.2.16).

We now prove that the evolution operator $U(t)$ is strongly continuous, i.e.,

$$\lim_{\Delta t \to 0} \|U(t+\Delta t)f - U(t)f\| = 0, \quad f \in L_1. \tag{8.2.18}$$

We follow [PGM3] and [CGP] with some modifications.

In view of the boundedness of $U(t)$ and its group property, it is sufficient to prove that

$$\lim_{\Delta t \to 0} \|U(\Delta t) f - f\| = 0. \tag{8.2.19}$$

The operator $U(\Delta t)$ is the product of the operators $e^{\int dx}$, $J(-\Delta t)$, $S(-\Delta t)$, $e^{-\int dx}$, where the operators $e^{\pm \int dx}$ are bounded in L_1. This means that, in order to prove (8.2.18), it is sufficient to prove the strong continuity of the group $J(-t)S(-t)$ for $f \in L_1^0$. This property follows from the fact that

$$\begin{aligned} &e^{-\int dx} J(-\Delta t) S(-\Delta t) e^{-\int dx} f - f \\ &\quad = e^{\int dx} \Big[J(-\Delta t) S(-\Delta t) e^{-\int dx} f - e^{-\int dx} f \Big] \end{aligned}$$

and that $e^{-\int dx} f \in L_1^0$ for $f \in L_1^0$. Recall that the subspace L_1^0 consists of finite sequences of functions $f_s((x)_s) \in L_s^0$. The functions $f_s((x)_s) \in L_s^0$ are continuously differentiable, have compact support, and are equal to zero in a certain neighborhood of forbidden configurations. The subspace L_s^0 is everywhere dense in L_s as well as L_1^0 in L_1.

Note that the strong continuity of $J(-t)S(-t)$ was proved in [see (7.3.3)]. Indeed, it was proved that the functions

$$[J(-t)S(-t)f]_s((x)_s), \; f_s((x)_s) \in L_s^0,$$

are continuous in t, $t \geq 0$, uniformly in $(x)_s$ on compacta. Therefore,

$$\lim_{\Delta t \to 0} \int \Big| J(-\Delta t) S(-\Delta t) f_s((x)_s) - f_s((x)_s) \Big| d(x)_s = 0$$

because the integrand has compact support and tends to zero as $\Delta t \to 0$ uniformly in $(x)_s$ on compacta.

Taking into account that $f \in L_1^0$ is a finite sequence, we prove that

$$\lim_{\Delta t \to 0} \Big\| U(\Delta t) f - f \Big\| = 0, \quad f \in L_1^0.$$

Taking into account the boundedness of $U(t)$ and the fact that L_1^0 is everywhere dense in L_1, we get

$$\lim_{\Delta t \to 0} \Big\| U(\Delta t) f - f \Big\| = 0$$

for arbitrary $f \in L_1$. Thus, the strong continuity of the evolution operator $U(t)$ (8.2.17) is proved.

We summarize the results obtained above in the following theorem:

Theorem 8.1. *The evolution operators $U(t)$, $t \geq 0$, are a group of bounded strongly continuous operators in L_1.*

8.3 Infinitesimal generator of the group and a solution of the BBGKY hierarchy

8.3.1 Infinitesimal generator

As is known, the group of bounded strongly continuous operators $U(t)$ in L_1 is strongly differentiable, and its infinitesimal generator is defined on an everywhere dense set in L_1. We now proceed to the determination of this infinitesimal generator. We follow [PGM3] and [CGP] with some modifications.

Theorem 8.2. *The infinitesimal generator B of the group $U(t)$ is closed, and its spectrum is concentrated on the imaginary axis. On the set L_1^0 everywhere dense in L_1, B coincides with the operator $B = -\mathcal{H} + \left[\int dx, \mathcal{H}\right]$, or componentwise*

$$\begin{aligned}
&(Bf)_s(x_1,\dots,x_s)\\
&= -\mathcal{H} f_s(x_1,\dots,x_s) + a^2 \sum_{i=1}^{s} \int dp_{s+1} \int_{S_+^2} d\eta\, \eta\cdot(p_i - p_{s+1})\\
&\quad\times\Bigg[\frac{1}{(1-2\varepsilon)^2} f_{s+1}(x_1,\dots,q_i,p_i^*,\dots,x_s,q_i - a\eta, p_{s+1}^*)\\
&\quad - f_{s+1}(x_1,\dots,q_i,p_i,\dots,x_s,q_i + a\eta, p_{s+1})\Bigg],
\end{aligned}\tag{8.3.1}$$

$$\mathcal{H} f_s(x_1,\dots,x_s) = -\sum_{i=1}^{s} p_i \cdot \frac{\partial}{\partial q_i} f_s(x_1,\dots,x_s),$$

with the following boundary condition:

If $q_i - q_j = a\eta$, $\eta\cdot(p_i - p_{s+1}) > 0$, then the momenta p_i and p_j should be replaced by p_i^* and p_j^*, and $f_{s+2}(x_1,\dots,x_s,q_{s+1},p_{s+1},q_{s+1} - a\eta, p_{s+2})$ should be replaced by $\frac{1}{(1-2\varepsilon)^2} f_{s+2}(x_1,\dots,x_s,q_{s+1},p_{s+1}^*,q_{s+1} - a\eta, p_{s+2}^*)$. The operator $\mathcal{H}$ with this boundary condition is the infinitesimal generator of the group $J(-t)S(-t)$.

On the set L_1^0, the operators B and $U(t)$ commute, i.e.,

$$BU(t) = U(t)B$$

and

$$\frac{d}{dt}U(t) = BU(t) = U(t)B.$$

Proof. The proof of Theorem 8.2 coincides completely with the corresponding proof for a system of hard spheres with elastic collisions (see [PGM3] and [CGP]). As for systems of hard spheres with elastic collisions, the crucial point is the identity

$$\left[\int dx, \left[\int dx, \mathcal{H}\right]\right] f = 0, \quad f \in L_1^0.$$

In our case of systems of hard spheres with inelastic collisions, the projection of this identity onto the s-particle subspace has the form

$$\int dx_{s+2} a^2 \int dp_{s+1} \int_{S_+^2} d\eta\, \eta \cdot (p_{s+1} - p_{s+2})$$

$$\times \left[\frac{1}{(1-2\varepsilon)^2} f_{s+2}(x_1, \dots, x_s, q_{s+1}, p^*_{s+1}, q_{s+1} - a\eta, p^*_{s+2}) \right.$$

$$\left. - f_{s+2}(x_1, \dots, x_s, q_{s+1}, p_{s+1}, q_{s+1} + a\eta, p_{s+2} \right]$$

and is equal to zero, as follows from Chapter 7 (relation (7.4.5)).

Note that the functions $(J(-t)S(-t)f_{s+1})(x_1, \dots, x_{s+1})$ are (possibly) different from zero in a neighborhood of forbidden configurations, where the functions $f_{s+1}(x_1, \dots, x_{s+1}) \in L^0_{s+1}$ vanish. This implies that the functions $(U(t)f)_{s+1}(x_1, \dots, x_{s+1})$, $f \in L_1^0$, are (possibly) different from zero in a neighborhood of forbidden configurations and on the hypersurfaces $|q_i - q_{s+1}| = a$, $i = 1, \dots, s$. This means that the second term in the relation [see (8.3.1)]

$$\left(\frac{d}{dt} U(t) f\right)_s (x_1, \dots, x_s) = (BU(t)f)_s(x_1, \dots, x_s)$$

is different from zero.

8.3.2 Existence of solutions of the BBGKY hierarchy

The BBGKY hierarchy (8.2.2) is the evolution equation for the infinite sequence $F(t)$ of correlation functions $F(t) = (F_1(t, x_1), \dots, F_s(t, x_1, \dots, x_s), \dots)$. This equation has the form

$$\frac{dF(t)}{dt} = BF(t) = -\mathcal{H}F(t) + \left[\mathcal{H}, \int dx\right] F(t) \tag{8.3.2}$$

with the initial condition

$$F(t)|_{t=0} = F(0). \tag{8.3.3}$$

The operator B is the infinitesimal generator of the group $U(t)$ (8.2.8).

One can consider the BBGKY hierarchy as the abstract evolution equation (8.3.2) in the Banach space L_1 with initial data (8.3.3), where $F(0) \in L_1$. Then it follows from Theorem 8.2 that

$$F(t) = U(t)F(0) \tag{8.3.4}$$

is the strong solution of the Cauchy problem for the BBGKY hierarchy (8.3.2) with initial data $F(0) \in L_1^0$. For general initial data $F(0) \in L_1^0$, $F(t) = U(t)F(0)$ is a generalized solution in the following sense: Strong solutions exist for $F(0) \in L_1^0$ and are represented by (8.3.4). The set L_1^0 is everywhere dense in L_1, and, for arbitrary $F(0) \in L_1^0$, there exists a sequence $F(0)^i \in L_1^0$ that converges strongly to $F(0)$. It follows from the boundedness of $U(t)$ that the sequence $U(t)F(0)^i$ also converges to $U(t)F(0)$, and, in this sense, $F(t) = U(t)F(0)$ is a generalized solution of the BBGKY hierarchy.

The above results can be summarized in the following theorem:

Theorem 8.3. *The Cauchy problem for the BBGKY hierarchy (8.3.2) has a solution in L_1 given by (8.3.4). For initial data $F(0) \in L_1^0 \subset L_1$, this solution is strong. For arbitrary initial data $F(0) \in L_1$, it is a generalized solution.*

8.3.3 States of infinite systems

Solutions of hierarchy (8.2.2) or (8.3.2) constructed above describe states of finite systems because the average number $\bar{N}$ of particles corresponding to the state $F(t) \in L_1$, $F(0) \in L_1$, is finite, i.e.,

$$\bar{N} = \int F_1(t, x_1) dx_1 < \infty. \tag{8.3.5}$$

This means that solutions of hierarchy (8.2.2) or (8.3.2) for initial data $F(0) \in L_1$ cannot describe states of an infinite system, i.e., a system consisting of infinite average number of particles located in the entire phase space on admissible configurations.

Usually, in the case of elastic collisions, perturbations of equilibrium states of infinite systems, i.e., Gibbs states for given temperature and density, are considered as initial data for an infinite system. We hope to realize this approach in the case of an infinite system of hard spheres with inelastic collisions.

As pointed out in Chapter 7, hierarchy (8.2.2) or (8.3.2) has the stationary solution

$$F_s((x)_s) = \prod_{i=1}^{s} \delta(p_i - p) \prod_{i<j=1}^{s} \Theta(|q_i - q_j| - a), \quad s \geq 1, \tag{8.3.6}$$

where p is an arbitrary momentum. It would be natural to consider some perturbation of sequence (8.3.6) as initial data for an infinite system; for example,

$$F_s((x)_s) = \prod_{i=1}^{s} \frac{1}{(2\pi\beta)^{\frac{3}{2}}} e^{-\beta(p_i - p)^2} \prod_{i<j=1}^{s} \Theta(|q_i - q_j| - a), \quad s \geq 1, \tag{8.3.7}$$

where β, $\beta > 0$, is some positive parameter. Sequence (8.3.7) can be considered as a perturbation of sequence (8.3.6) because it converges to the latter as $\beta \to 0$ in the sense of generalized functions.

8.4 Stochastic Boltzmann hierarchy for granular flow

8.4.1 Stochastic dynamics for hard spheres with inelastic collisions

The stochastic dynamics for N hard spheres with inelastic collisions can be obtained in full analogy with the stochastic dynamics for hard spheres with elastic collisions in the Boltzmann–Grad limit when the diameter a tends to zero. This dynamics for negative time is defined as follows: point particles move as free ones until their positions coincide. If the positions of i-th and j-th particles coincide at time $-\tau$, $\tau > 0$, then their momenta change and after collision for negative time they become

$$\begin{aligned} p_i^* &= p_i + \frac{\varepsilon}{1-2\varepsilon}\eta_{ij}\, \eta_{ij}\cdot(p_i - p_j), \\ p_j &= p_j - \frac{\varepsilon}{1-2\varepsilon}\eta_{ij}\, \eta_{ij}\cdot(p_i - p_j), \quad i, j \in \{1, \dots, N\}. \end{aligned} \tag{8.4.1}$$

If $\eta \in S^2_-$, the momenta do not change. The unit vector for $\eta \in S^2_+$ is a random one and is uniformly distributed on the sphere S^2,

$$Q_i(-t) = q_i - p_i\tau - p_i^*(t-\tau),$$

$$Q_j(-t) = q_j - p_j\tau - p_j^*(t-\tau),$$

$$q_i - p_i\tau = q_j - p_j\tau, \quad \tau > 0, \quad t > 0.$$

This means that, after collision, particles move as free ones with momenta p_i^*, p_j^* until the next collision.

For positive time $t > 0$, one should put ε instead $\frac{\varepsilon}{1-2\varepsilon}$ in (8.4.1) and the sign "+" instead of "−" before momenta in (8.4.2).

8.4.2 Stochastic trajectories and operator of evolution

Stochastic trajectories with inelastic collisions are defined for fixed random vectors η_{ij} in full analogy with stochastic trajectories for elastic collisions. Namely, $X_i(-t,(x)_N) = X_i(-t, x_1, \dots, x_N)$, $X_i(-t,(x)_N)|_{t=0} = (x)_i$ have the semigroup property and differ from the trajectories of free particles on hypersurfaces of lower dimension where the vectors of difference of initial positions are parallel to the vectors of difference of initial momenta

$$q_i - q_j = \tau(p_i - p_j), \quad i, j \in \{1, \dots, N\} \tag{8.4.2}$$

The operator $S_N(-t)$ is defined as the operator of shift along trajectories. If $D_N(x_1, \dots, x_N) = D_N((x)_N)$ is a continuous symmetric function defined on the phase space of N particles, then

$$S_N(-t)D_N((x)_N) = D_N(X(-t,(x)_N)). \tag{8.4.3}$$

For fixed random vectors, the operators $S_N(-t)$ possess the group of property. The operator of evolution is defined as follows:

$$J_N(-t)S_N(-t)D_N((x)_N) = D_N(t,(x)_N). \tag{8.4.4}$$

where $J_N(-t)$ is the Jacobian $\frac{\partial X(-t,(x)_N)}{\partial (x)_N}$. The operators $J_N(-t)S_N(-t)$ possess the semigroup property. The function $D_N(t,(x)_N)$ defines the state of the system of N point particles at time t if the initial state at $t = 0$ is $D_N((x)_N) = D_N(0,(x)_N)$.

The function $D_N(t,(x)_N)$ satisfies the Itô–Liouville equation

$$\frac{\partial D_N(t,(x)_N)}{\partial t} = -\sum_{i=1}^{N} p_i \cdot \frac{\partial}{\partial q_i} D_N(t,(x)_N) \tag{8.4.5}$$

with the initial condition $D_N(t,(x)_N)|_{t=0} = D_N((x)_N)$ and the boundary condition according to which, for $q_i = q_j$ and $\eta_{ij} \cdot (p_i - p_j) \geq 0$ ($\eta_{ij} \subset S_+^2$), the momenta p_i, p_j in (8.4.5) should be replaced by p_i^*, p_j^* (8.4.1) and

$$\begin{aligned} &D_N(t,x_1,\dots,x_N)|_{q_i=q_j,\,\eta_{ij}\in S_+^2} \\ &\quad = \frac{1}{(1-2\varepsilon)^2} D_N(t,x_1,\dots,x_i^*,\dots,x_j^*,\dots,x_N), \end{aligned} \tag{8.4.6}$$

$$x_i^* = (q_i, p_i^*), \quad x_j^* = (q_j, p_j^*), \quad i,j \in \{1,\dots,N\}.$$

Equation (8.4.5) and the boundary condition are derived in full analogy with those for the system of hard spheres in Chapter 7.

8.4.3 Functional average

Consider a real symmetric test function $\varphi_N(x_1,\dots,x_N)$ as an observable. Then the average of the observable $\varphi_N(x_1,\dots,x_N)$ over the state $D_N(t,x_1,\dots,x_N)$ for infinitesimal time Δt is defined as follows:

$$\begin{aligned} &(D_N(\Delta t),\varphi_N) \\ &\quad = \int D_N(q_1 - p_1\Delta t, p_1,\dots,q_N - p_N\Delta t, p_N)\varphi_N(x_1,\dots,x_N)dx_1\dots dx_N \\ &\qquad + \int \Big\{ \int_0^{\Delta t} d\tau \sum_{i<j=1}^{N} \int_{S_+^2} d\eta_{ij}\,\eta_{ij}\cdot(p_i - p_j)\delta(q_i - p_i\tau - q_j + p_j\tau) \\ &\qquad \times \Big[\frac{1}{(1-2\varepsilon)^2} D_N(q_1 - p_1\Delta t, p_1,\dots,q_i - p_i\tau - p_i^*(\Delta t - \tau), \\ &\qquad\qquad p_i^*,\dots,q_j - p_j\tau - p_j^*(\Delta t - \tau), p_j^*,\dots,q_N - p_N\Delta t, p_N) \end{aligned}$$

$$- D_N(q_1 - p_1\Delta t, p_1, \ldots, q_i - p_i\Delta t, p_i, \ldots, q_j - p_j\Delta t,$$

$$p_j, \ldots, q_N - p_N\Delta t, p_N)\Big]\Big\}\varphi_N(x_1, \ldots, x_N)dx_1 \ldots dx_N$$

$$= \int \bar{D}_N(\Delta t, x_1, \ldots, x_N)\varphi_N(x_1, \ldots, x_N)dx_1 \ldots dx_N, \tag{8.4.7}$$

where

$$\bar{D}_N(\Delta t, x_1, \ldots, x_N) = \bar{S}_N(-\Delta t)D_N(x_1, \ldots, x_N)$$

$$= D_N(q_1 - p_1\Delta t, p_i, \ldots, q_N - p_N\Delta t, p_N)$$

$$+ \int_0^{\Delta t} d\tau \sum_{i<j=1}^{N} \int_{S_+^2} d\eta_{ij}\eta_{ij} \cdot (p_i - p_j)\delta(q_i - p_i\tau - q_j + p_j\tau)$$

$$\times \Big[\frac{1}{(1-2\varepsilon)^2} D_N(q_1 - p_1\Delta t, p_1, \ldots, q_i - p_i\tau - p_i^*(\Delta t - \tau),$$

$$p_i^*, \ldots, q_j - p_j\tau - p_j^*(\Delta t - \tau), p_j^*, \ldots, q_N - p_N\Delta t, p_N)$$

$$- D_N(q_1 - p_1\Delta t, p_1, \ldots, q_i - p_i\Delta t, p_i, \ldots, q_j - p_j\Delta t,$$

$$p_j, \ldots, q_N - p_N\Delta t, p_N)\Big], \tag{8.4.8}$$

The operator $\bar{S}_N(-\Delta t)$ is defined by (8.4.8).

According to the duality principle, the function $\bar{D}_N(\Delta t, x_1, \ldots, x_N)$ is the generalized function corresponding to the usual function $D_N(\Delta t, (x)_N)$ (8.4.4) with discontinuities on the hyperplanes where point particles interact with inelastic collisions.

The generalized function $\bar{D}_N(\Delta t, (x)_N)$ shows how to integrate the usual function $D_N(\Delta t, (x)_N)$ with test functions $\varphi_N((x)_N)$ and to take into account the hypersurfaces of lower dimension where point particles interact.

If the function $\bar{D}_N(t, (x)_N)$ is defined for arbitrary time t, then the function $\bar{D}_N(t + \Delta t, (x)_N)$ is defined by formulas (8.4.8) if one puts the function $\bar{D}_N(t, (x)_N)$ instead of $D_N(\Delta t, (x)_N)$, or

$$\bar{D}_N(t, (x)_N) = \lim_{n\to\infty} \prod_{i=1}^{n} \bar{S}_N(-\Delta t_i)D_N((x)_N), \quad \sum_{i=1}^{n} \Delta t_i = t.$$

The function $\bar{D}_N(t, (x)_N)$ satisfies the Kolmogorov–Itô–Liouville equation

$$
\begin{aligned}
&\frac{\partial \bar{D}_N(t,x_1,\dots,x_N)}{\partial t}\\
&\quad = -\sum_{i=1}^{N} p_i \cdot \frac{\partial}{\partial q_i}\bar{D}_N(t,x_1,\dots,x_N)\\
&\qquad + \sum_{i<j=1}^{N} \int_{S_+^2} d\eta_{ij}\, \eta_{ij}\cdot(p_i-p_j)\delta(q_i-q_j)\\
&\qquad\quad \times \Big[\frac{1}{(1-2\varepsilon)^2}\bar{D}_N(t,x_1,\dots,q_i,p_i^*,\dots,q_j,p_j^*,\dots,x_N)\\
&\qquad\qquad - \bar{D}_N(t,x_1,\dots,q_i,p_i,\dots,q_j,p_j,\dots,x_N)\Big] \qquad (8.4.9)
\end{aligned}
$$

with initial condition $\bar{D}_N(t,(x)_N) = D_N((x)_N)$. Note that the boundary condition has already been taken into account by the term with δ-functions.

Equation (8.4.9) shows how to take into account the boundary condition (8.4.6) in equation (8.4.5) when one integrates equation (8.4.5) with test functions $\varphi_N((x)_N)$.

8.4.4 Hierarchy for correlation functions

By analogy with the stochastic Boltzmann hierarchy for elastic hard spheres, the stochastic Boltzmann hierarchy for inelastic hard spheres can be derived from (8.4.5)–(8.4.6) and (8.4.9). It has the form

$$
\begin{aligned}
&\frac{\partial F_s(t,x_1,\dots,x_s)}{\partial t} = -\sum_{i=1}^{s} p_i\cdot\frac{\partial}{\partial q_i}F_s(t,x_1,\dots,x_s)\\
&\quad + \sum_{i=1}^{s}\int dp_{s+1}\int_{S_+^2} d\eta_{is+1}\,\eta_{is+1}\cdot(p_i-p_{s+1})\\
&\qquad \times\Big[\frac{1}{(1-2\varepsilon)^2}F_{s+1}(t,x_1,\dots,q_i,p_i^*,\dots,x_s,q_i,p_{s+1}^*)\\
&\qquad\quad - F_{s+1}(t,x_1,\dots,q_i,p_i,\dots,x_s,q_i,p_{s+1})\Big], \quad s\ge 1 \qquad (8.4.10)
\end{aligned}
$$

with initial data $F_s(t,(x)_s)|_{t=0} = F_s(0,(x)_s)$ and the boundary condition according to which, for $q_i = q_j$ and $\eta_{ij}\cdot(p_i-p_j)\ge 0$, the momenta p_i, p_j should be replaced by p_i^*, p_j^* in $-\sum_{i=1}^{s} p_i\cdot\frac{\partial}{\partial q_i}F_s(t,(x)_s)$ and $F_s(t,x_1,\dots,x_i^*,\dots,x_j^*,\dots,x_s)$ should be replaced by $\frac{1}{(1-2\varepsilon)^2}F_s(t,x_1,\dots,x_i^*,\dots,x_j^*,\dots,x_s)$. Hierarchy (8.4.10) is the *stochastic Boltzmann hierarchy* for inelastic hard spheres.

8.4.5 Solution of the stochastic Boltzmann hierarchy

The formal solution of the stochastic Boltzmann hierarchy can formally be represented by the series of iterations

$$
\begin{aligned}
&F_s(t,(x)_s)\\
&\quad=\sum_{n=0}^{\infty}\int_0^t\int_0^{t_1}\dots\int_0^{t_{n-1}} dt_1\dots dt_n\Big[\frac{\partial X(-t+t_1,(x)_s)}{\partial(x)_s}\Big]^2\\
&\qquad\times S_s(-t+t_1,(x)_s)\sum_{i=1}^{s} b_{s+1}\delta(q_i-q_{s+1})\int_{S_+^2} d\eta_{is+1}\\
&\qquad\times \eta_{is+1}\cdot(p_i-p_{s+1})\Big\{\frac{1}{(1-2\varepsilon)^2}\Big[\frac{\partial X(-t_1+t_2,(x)^*_{s+1})}{\partial(x)^*_{s+1}}\Big]^2\\
&\qquad\times S_{s+1}(-t_1+t_2,(x)^*_{s+1})-\Big[\frac{\partial X(-t_1+t_2,(x)_{s+1})}{\partial(x)_{s+1}}\Big]^2\\
&\qquad\times S_{s+1}(-t_1+t_2,(x)_{s+1})\Big\}\dots\sum_{i=1}^{s+n-2} b_{s+n-1}\delta(q_i-q_{s+n-1})\\
&\qquad\times\int_{S_+^2} d\eta_{is+n-1}\,\eta_{is+n-1}\cdot(p_i-p_{s+n-1})\Big\{\frac{1}{(1-2\varepsilon)^2}\\
&\qquad\times\Big[\frac{\partial X(-t_{n-1}+t_n,(x)^*_{s+n-1})}{\partial(x)^*_{s+n-1}}\Big]^2 S_{s+n-1}(-t_{n-1}+t_n,(x)^*_{s+n-1})\\
&\qquad-\Big[\frac{\partial X(-t_{n-1}+t_n,(x)_{s+n-1})}{\partial(x)_{s+n-1}}\Big]^2 S_{s+n-1}(-t_{n-1}+t_n,(x)_{s+n-1})\Big\}\\
&\qquad\times\sum_{i=1}^{s+n-1} b_{s+n}\delta(q_i-q_{s+n})\int_{S_+^2} d\eta_{is+n}\,\eta_{is+n}\cdot(p_i-p_{s+n})\\
&\qquad\times\Big\{\frac{1}{(1-2\varepsilon)^2}\Big[\frac{\partial X(-t_n,(x)^*_{s+n})}{\partial(x)^*_{s+n}}\Big]^2 S_{s+n}(-t_n,(x)^*_{s+n})\\
&\qquad-\Big[\frac{\partial X(-t_n,(x)_{s+n})}{\partial(x)_{s+n}}\Big]^2 S_{s+n}(-t_n,(x)_{s+n})\Big\}F_{s+n}(0,(x)_{s+n}). \qquad (8.4.11)
\end{aligned}
$$

Recall that the solution of the Itô–Liouville equation (8.4.5) with the initial and boundary condition (8.4.6) is the function

$$D_N(t,(x)_N) = \Big[\frac{\partial X(-t,(x)_N)}{\partial (x)_N}\Big]^2 S_N(-t,(x)_N) D_N((x)_N).$$

Consider series (8.4.11) for $(x)_s$ outside the hypersurfaces where the stochastic point particles interact, i.e., outside all hypersurfaces

$$q_i - p_i\tau = q_j - p_j\tau, \quad i,j \in \{1,\dots,s\}. \tag{8.4.12}$$

Then all trajectories $X(-t+t_1,(x)_s),\dots,X(-t_n,(x)_{s+n})$ can be replaced by the trajectories of the free system $X^0(-t+t_1,(x)_s),\dots,X^0(-t_n,(x)_{s+n})$, and all operators of shift along the stochastic trajectories can be replaced by the operators of shift along the trajectories of free systems $S_s^0(-t+t_1,(x)_s),\dots,S_{s+n}^0(-t_n,(x)_{s+n})$.

As in the case of stochastic point particles with elastic collisions, the last assertion follows from the fact that, under the integral sign (after integrations with respect to $q_{s+1},\dots,q_{s+n}$ with help of δ-functions), we have the Lebesgue integral with respect to $p_{s+1},\dots,p_{s+n}$, and, thus, the hypersurfaces of lower dimension with respect to $p_{s+1},\dots,p_{s+n}$ with fixed $q_1,\dots,q_s$ where the stochastic particles interact can be neglected. If phase points $x_1,\dots,x_s$ are outside the hypersurfaces where the s-th stochastic point particles interact, then all trajectories, $X(-t_1+t,(x)_s),\dots,X(-t_n,(x)_{s+n})$, and the operators $S_s(-t+t_1,(x)_s),\dots,S_{s+n}(-t_n,(x)_{s+n})$ can be replaced by those of the free systems.

Taking into account that the Jacobians of free trajectories are equal to one, we obtain the following representation for $F_s(t,(x)_s)$ (8.4.11) outside hypersurfaces (8.4.13):

$$\begin{aligned}
F_s(t,(x)_s) &= \sum_{n=0}^{\infty}\int_0^t\int_0^{t_1}\dots\int_0^{t_{n-1}} dt_1\dots dt_n S_s^0(-t+t_1,(x)_s))\\
&\times\sum_{i=1}^{s} b_{s+1}\delta(q_i-q_{s+1})\int_{S_+^2} d\eta_{is+1}\,\eta_{is+1}\cdot(p_i-p_{s+1})\\
&\times\Big[\frac{1}{(1-2\varepsilon)^2}S_{s+1}^0(-t_1+t_2,(x)_{s+1}^*)-S_{s+1}^0(-t_1+t_2,(x)_{s+1})\Big]\dots\\
&\times b_{s+n-1}\sum_{i=1}^{s+n-2}\delta(q_i-q_{s+n-1})\int_{S_+^2} d\eta_{is+n-1}\,\eta_{is+n-1}\cdot(p_i-p_{s+n-1})\\
&\times\Big[\frac{1}{(1-2\varepsilon)^2}S_{s+n-1}^0(-t_{n-1}+t_n,(x)_{s+n-1}^*)\\
&\qquad -S_{s+n-1}^0(-t_{n-1}+t_n,(x)_{s+n-1})\Big]
\end{aligned}$$

$$\times b_{s+n} \sum_{i=1}^{s+n-1} \delta(q_i - q_{s+n}) \int_{S_+^2} d\eta_{is+n}\, \eta_{is+n} \cdot (p_i - p_{s+n})$$

$$\times \Big[\frac{1}{(1-2\varepsilon)^2} S_{s+n}^0(-t_n, (x)_{s+n}^*) - S_{s+n}^0(-t_n, (x)_{s+n})\Big]$$

$$\times F_{s+n}(0, (x)_{s+n}). \tag{8.4.13}$$

8.4.6 Ordinary Boltzmann hierarchy

Consider the hierarchy

$$\frac{\partial F_s(t, x_1, \dots, x_s)}{\partial t}$$

$$= -\sum_{i=1}^{s} p_i \cdot \frac{\partial}{\partial q_i} F_s(t, x_1, \dots, x_s)$$

$$+ \sum_{i=1}^{s} b_{s+1} \delta(q_i - q_{s+1}) \int_{S_+^2} d\eta_{is+1}\, \eta_{is+1} \cdot (p_i - p_{s+1})$$

$$\times \Big[\frac{1}{(1-2\varepsilon)^2} F_{s+1}(t, x_1, \dots, q_i, p_i^*, \dots, x_s, q_i, p_{s+1}^*)$$

$$- F_{s+1}(t, x_1, \dots, q_i, p_i, \dots, x_s, q_i, p_{s+1})\Big], \quad s \geq 1 \tag{8.4.14}$$

with initial data $F_s(t, x_1, \dots, x_s)|_{t=0} = F_s(0, x_1, \dots, x_s)$ but without any boundary condition in the term $-\sum_{i=1}^{s} p_i \cdot \frac{\partial}{\partial q_i} F_s(t, x_1, \dots, x_s)$.

We call this hierarchy the *ordinary Boltzmann hierarchy*. Consider the Boltzmann equation for hard spheres with inelastic collision

$$\frac{\partial F_1(t, x_1)}{\partial t}$$

$$= -p_1 \cdot \frac{\partial}{\partial q_1} F_1(t, x_1) + \int dx_2 \delta(q_1 - q_2) \int_{S_+^2} d\eta_{12}\, \eta_{12} \cdot (p_1 - p_2)$$

$$\times \Big[\frac{1}{(1-2\varepsilon)^2} F_1(t, q_1, p_1^*) F_1(t, q_1, p_2^*) - F_1(t, q_1, p_1) F_1(t, q_1, p_2)\Big] \tag{8.4.15}$$

with initial data $F_1(t, x_1)|_{t=0} = F_1(0, x_1)$. It is easy to verify that the correlation functions

$$F_s(t, x_1, \dots, x_s) = F_1(t, x_1) \dots F_1(t, x_s) \tag{8.4.16}$$

satisfy hierarchy (8.4.14), and, vice versa, if the initial correlation functions possess the chaos property

$$F_s(0, x_1, \dots, x_s) = F_1(0, x_1) \dots F_1(0, x_s), \tag{8.4.17}$$

then hierarchy (8.4.14) possesses the chaos property and its solution is equal to (8.4.16), where $F_1(t, x)$ is a solution of the Boltzmann equation (8.4.15). It is easy to verify that the mild solution of the ordinary Boltzmann hierarchy (8.4.14) can be represented by series (8.4.13) in the entire phase space.

We now prove that series (8.4.13) is convergent.

Theorem 8.4. *Series (8.4.13) are uniformly convergent with respect to $(x)_s$ on compact sets on a finite time interval for initial data $F_s(0, (x)_s)$, $s \geq 1$, from the space $E_{\xi,\beta}$ and globally in time for initial data $F_s(0, (x)_s)$, $s \geq 1$, from the space $\tilde{E}_{\xi,\beta}$.*

The proof of this theorem completely coincides with the proof of Theorems 4.1 and 4.2. The only difference is the new factor

$$A'(\beta, \beta') = \frac{1}{(1 - 2\varepsilon)^2} A(\beta, \beta'),$$

where $A(\beta, \beta')$ is defined according to (4.2.14). Note that Lemma 4.1 remains true for the stochastic dynamics with inelastic collisions.

Remark 8.1. If initial phase points $(x)_s$ are arbitrary and not necessarily lie outside the hypersurfaces $q_i - p_i\tau = q_j - p_j\tau$, $0 \leq \tau \leq t$, $i, j \in \{1, \dots, s\}$, then, in (8.4.11), one can neglect all collisions of n particles that have initial phase points $x_{s+1}, \dots, x_{s+n}$ with one another and with s particles with initial phase points $x_1, \dots, x_s$. Indeed, n particles can collide only on hypersurfaces of lower dimension with respect to $p_{s+1}, \dots, p_{s+n}$ ($x_1, \dots, x_s$ are fixed), and one can neglect these hypersurfaces in (8.4.11). This means that, in (8.4.11), one can consider only collisions of s particles with initial phase points $x_1, \dots, x_s$.

If one of s particles, say, with number i, has the momentum $p_i^* = p_i - \eta_{ij}\, \eta_{ij} \cdot (p_i - p_j)$, where $j > s$, then one can neglect the collisions of this particle with the other of these s particles because collisions can happen only for p_j lying on the hypersurfaces $q_i - p_i^*\tau = q_l - p_l\tau$, $1 \leq l \neq i \leq s$, with respect to momenta p_j, and one can neglect this hypersurface in the integral with respect to p_j. In this terms, the number of particles that can collide is less than s. The collisions of particles with momenta p^* are omitted.

It is obvious that the number of collisions of less than s particles is not greater than the number of collisions of exactly s particles with initial phase points $x_1, \dots, x_s$ on a given time interval t. If, at a given point $(x)_s$, the Jacobian $\frac{\partial X(-t,(x)_s)}{\partial (x)_s}$ is less than

$\frac{1}{(1-2\varepsilon)^{2k}}$, $1 \leq k < \infty$, then the product of all Jacobians with s stochastic point particles

$$\left[\frac{\partial X(-t+t_1,(x)_s)}{\partial (x)_s}\right]^2 \left[\frac{\partial X(-t_1+t_2,(x)_s)}{\partial (x)_s}\right]^2 \cdots \left[\frac{\partial X(-t_n,(x)_s)}{\partial (x)_s}\right]^2$$

is also less than $\frac{1}{(1-2\varepsilon)^{2k}}$. Series (8.4.11) is convergent and equal to the product of series (8.4.13) and the factor $\frac{1}{(1-2\varepsilon)^{2k}}$. Series (8.4.11) represents the mild solution of the stochastic Boltzmann hierarchy (8.4.10).

References

Aizenman M., Goldstein S., and Lebowitz J. L.

[AGL] *Ergodic properties of infinite systems*, in: Lecture Notes in Physics, **38**, Springer, Berlin (1975), 112–144.

Akhiezer A. I. and Peletminsky S. V.

[AkP] *Methods of Statistical Physics*, Nauka, Moscow (1977) (in Russian).

Alexander R.

[Ale1] *Dynamics of ideal hard–core particles*, Commun. Math. Phys., **49**, 81–98 (1976).

[Ale2] *Time evolution for infinitely many hard spheres*, Commun. Math. Phys., **49**, 217–232 (1976).

Arkeryd L., Esposito R., and Pulvirenti M.

[AEP] *The Boltzmann equation for weakly inhomogeneous data*, Commun. Math. Phys., **111**, 393–407 (1987).

Arkeryd L., Lions P. L., Markowich P. A., and Varadhan S. R. S.

[ALMV] *Nonequilibrium Problems in Many-Particle System*, Lecture Notes in Mathematics, **1551**, Springer (1993).

Arlotti L., Bellomo N., De Angelis E., Lachowicz M.

[ABDL] *Generalized Kinetic Models in Applied Sciences*, World Sci., New Jersey (2003).

Arnold V. I.

[Arn1] *Mathematical Methods of Classical Mechanics*, Nauka, Moscow (1974) (in Russian); **English translation:** Springer, New York (1978).

[Arn2] *Ordinary Differential Equations*, Nauka, Moscow (1975) (in Russian).

Arsen'ev A. A.

[Ars1] *Lectures on Kinetic Equations*, Nauka, Moscow (1992) (in Russian).

[Ars2] *On the approximation of the solution of the Boltzmann equation by solution of ito stochastic differential equations*, Zh. Vychisl. Mat. Mat. Fiz, **27**, 400–410 (1987).

[Ars3] *Approximation of the Boltzmann equation by stochastic equations*, Zh. Vychisl. Mat. Mat. Fiz, **28**, 560–567 (1988).

Balescu R.

[Bal] *Equilibrium and Non-Equilibrium Statistical Mechanics*, Wiley, New York (1975).

Belavkin V. P., Maslov V. P., and Tariverdiev S. E.

[BMT] *Asymptotic dynamics of a system of a large number of particles described by Kolmogorov–Feller equations*, Teor. Mat. Fiz., **49**, 298–306 (1982).

Bellomo N., Lachowicz M., Polewczak J., and Toscani G.

[BLPT] *Mathematical Topics in Nonlinear Kinetic Theory II: Enskog Equation*, World Scientific, Singapore (1991).

Bellomo N., Palczewski A., and Toscani G.

[BPT] *Mathematical Topics in Nonlinear Kinetic Theory*, World Scientific, Singapore (1988).

Benedetto D., Caglioti E., and Pulvirenti M.

[BKP] *A kinetic equation for granular media,* Math. Model. Numer. Analysis, **33**, No. 2, 439–441 (1999).

Bezandry P. H., Ferhique X., Giroux G.

[BFG] *A functional central limit theorem for a nonequilibrium model of interacting particles with unbounded intensity*, J. Statist. Phys., **72(1/2)**, 339–353 (1993).

Bogolyubov N. N.

[Bog1] *Problems of Dynamical Theory in Statistical Physics*, Gostekhizdat, Moscow (1946) (*in Russian*); see also Bogolyubov N. N., Selected Works, Vol. 2, Naukova Dumka, Kiev (1970), pp. 99–196.

[Bog2] *On some problems arising in justification of statistical mechanics*, in: Proceedings of the 2nd International Symposium in Selected Problems of Statistical Mechanics, JINR, Dubna (1981), pp. 5–11.

Bogolyubov N. N. and Bogolyubov N. N., Jr.

[BoB] *Introduction to Statistical Mechanics*, Nauka, Moscow (1984) (in Russian).

Bogolyubov N. N. and Khatset B. I.

[BoK] *On some mathematical problems in the theory of statistical equilibrium*, Dokl. Akad. Nauk SSSR, **66**, 321–324 (1949).

Bogolyubov N. N., Petrina D. Ya., and Khatset B. I.

[BPK] *Mathematical description of the equilibrium state of classical systems based on the canonical ensemble formalism*, Teor. Mat. Fiz., **1**, 251–274 (1969).

Bogolyubov N. N., Jr., and Sadovnikov B. I.

[BoS1] *Green functions and distribution functions in statistical mechanics of classical systems*, Zh. Eksp. Teor. Fiz., **4**, 667–675 (1962).

[BoS2] *Some Problems of Statistical Mechanics*, Vysshaya Shkola, Moscow (1975) (*in Russian*).

Boldrighini C., Dobrushin R. L., and Sukhov Yu. M.

[BDS] *One-dimensional hard rod caricature of hydrodynamics*, J. Stat. Phys., **31**, 577–616 (1983).

Boltzmann L.

[Bol1] *Weitere Studien über das Wärmegleichwicht unter Gasmolekülen*, Sitzungber. Acad. Wiss. Wien., **66**, 275–370 (1872).

[Bol2] *Über die Beziehung eines allgemeinen mechanischen Satzes zum zweiten Hauptsatze der Wärmetheorie*, Sitzungsberichte Akad. Wiss, **75**, 67–73 (1877).

[Bol3] *Entgegnung auf die wärmetheoretischen Betrachtungen des Hrn. E. Zermelo*, Annalen der Physik, **57**, 773–784 (1896).

Borgioli G., Gerasimenko V., Lauro G., and Monaco R.

[BGL1] *Many particles dynamics system formulation for the discrete Enskog gas*, Transp. Theor. Stat. Phys., **25**, Nos. 3–5, 581–592 (1996).

[BGL2] *A discrete velocity model of a gas: Global-in-time solutions of the BBGKY hierarchy*, Rep. Math. Phys., **40**, No. 3, 431–442 (1997).

Born M. and Green H. S.

[BoG] *A general kinetic theory of liquids*, Proc. R. Soc. Lond., **A188**, 168–201 (1947); Cambridge University Press (1949).

Braun W. and Hepp K.

[BrH] *The Vlasov dynamics and its fluctuations in the limit of interacting particles*, Commun. Math. Phys., **56**, 101–113 (1977).

Bunimovich L. A., Pesin Ya. B., Sinai Ya. G., and Yacobson M. V.

[BPSY] *Ergodic theory of smooth dynamical systems. II*, in: VINITI Series in Contemporary Problems in Mathematics. Fundamental Directions, Vol. 2, Moscow (1985), pp. 113–231.

Calogero F.

[Cal] *Non-linear evolution equations solvable by the inverse spectral transform*, in: Lecture Notes in Physics, **80**, Springer, Berlin (1978), pp. 235–269.

Caprino S., De Masi A., Presutti E., and Pulvirenti M.

[CDPP] *A stochastic particle system modeling the Carleman equation*, J. Stat. Phys., **55**, 625–638 (1989).

Cercignani C.

[Cer1] *On the Boltzmann equation for rigid spheres*, Transp. Theor. Stat. Phys., **2**, 211–225 (1972).

[Cer2] *Theory and Application of the Boltzmann Equation*, Scottish Academic Press, Edinburgh–London (1975).

[Cer3] *The Grad limit for a system of soft spheres*, Commun. Pure Appl. Math., **36**, 479–494 (1983).

[Cer4] *The Boltzmann Equation and Its Application*, Springer, Berlin–Heidelberg (1988).

[Cer5] *Le radici fisishe e mathematiche dell'irreversibilita temporale*, (accompanied by a translation into English), Alma Mater Studiorum **2**, 37–52 (1989)

Cercignani C., Gerasimenko V., and Petrina D.

[CGP] *Many-Particle Dynamics and Kinetic Equations*, Kluwer, Dordrecht (1997).

Cercignani C., Illner R., and Pulvirenti, M.

[CIP] *The Mathematical Theory of Dilute Gases*, Springer, Berlin–Heidelberg (1994).

Cercignani C. and Lampis M.

[CeL] *On the kinetic theory of a dense gas of rough spheres*, J. Stat. Phys., **53**, 655 (1988).

Chandrasekhar S.

[Cha] *Stochastic Problems in Physics and Astronomy*, Izd. Inostr. Lit., Moscow (1949) (*Russian translation*).

Choh S. T. and Uhlenbeck G. F.

[ChU] *The Kinetic Theory of Dense Gases*, University of Michigan (1958).

Chulaevsky V. A.

[Chu] *The inverse scattering method in statistical mechanics*, Funkts. Anal. Prilozh., **17**, 53–62 (1983).

Cohen E. G. D.

[Coh1] *The Boltzmann equation and its generalization to higher density*, in: Fundamental Problems in Statistical Mechanics, North Holland (1961).

[Coh2] *On the connection between various derivations of the Boltzmann equation*, Physica, **27**, 163 (1961).

[Coh3] *On the generalization of the Boltzmann equation to general order in the density*, Physica, **28**, 1025 (1962).

[Coh4] *Cluster expansions and the hierarchy. I. Nonequilibrium distribution functions*, Physica, **28**, 1045 (1962).

[Coh5] *Cluster expansions and the hierarchy. II. Equilibrium distribution functions*, Physica, **28**, 1060–1073 (1962).

[Coh6] *On the kinetic theory of dense gases*, J. Math. Phys., **4**, 183 (1962).

[Coh7] *The kinetic theory of dense gases*, in: Fundamental Problems in Statistical Mechanics, **11**, North Holland (1968), pp. 228–275.

De Masi A. and Presutti E.

[DMP] *Mathematical Methods for Hydrodynamical Limits*, Lecture Notes in Mathematics, **1501**, Springer (1991).

Dobrushin R. L.

[Dob1] *Investigation of conditions for asymptotic existence of configuration Gibbs integrals*, Teor. Ver. Primen., **9**, 626–643 (1964).

[Dob2] *Gibbs fields for lattice systems with pairwise interaction*, Funkts. Anal. Prilozh., **1**, 31–43 (1968).

[Dob3] *The problem of uniqueness of a random Gibbs field and the problem of phase transitions*, Funkts. Anal. Prilozh., **2**, 44–57 (1968).

[Dob4] *Description of a random field by conditional probabilities and conditions for its regularity*, Teor. Ver. Primen., **13**, 201–229 (1968).

[Dob5] *Gibbs states. General case*, Funkts. Anal. Prilozh., **3**, 27–35 (1969).

[Dob6] *Gibbs random fields for particles without hard cores*, Teor. Mat. Fiz., **4**, 101–118 (1970).

[Dob7] *Representation of a system of random variables in terms of conditional distributions*, Teor. Ver. Primen., **15**, 469–497 (1970).

Dobrushin R. L. and Fritz J.

[DoF] *Nonequilibrium dynamics of one-dimensional infinite particle systems with hardcore interactions*, Commun. Math. Phys., **55**, 275–292 (1977).

Dobrushin R. L., Maslova N. B., Sinai Ya. G., and Sukhov Yu. M.

[DMSS] *Dynamical systems in statistical mechanics and kinetic equations*, in: VINITI Series in Contemporary Problems in Mathematics. Fundamental Directions, Vol. 2, Moscow (1985), pp. 233–307.

Dobrushin R. L. and Minlos R. A.

[DoM] *Existence and continuity of pressure in classical statistical physics*, Teor. Ver. Primen., **12**, 595–618 (1967).

Dobrushin R. L. and Sukhov Yu. M.

[DoS] *Time asymptotics for some degenerate models with infinite number of particles*, in: VINITI Series in Contemporary Problems in Mathematics, Vol. 14, Moscow (1979), pp. 148–255.

Dorfman J. R. and Cohen E. G. D.

[DoC] *Difficulties in the kinetic theory of dense gases*, J. Math. Phys., **8**, 282–297 (1967).

Dorfman J. R. and Ernst M. H.

[DoE] *Hard-sphere binary-collision operators*, J. Stat. Phys., **57**, 581–193 (1989).

Dunford N. and Schwarz G. T.

[DuS] *Linear Operators. General Theory*, Izd. Inostr. Lit., Moscow (1962) (*Russian translation*).

Dyson F. J.

[Dys] *Ground-state energy of a finite system of charged particles*, J. Math. Phys, **8**, 1538–1545 (1967).

Dyson F. J. and Lenard A.

[DyL] *Stability of matter. I*: J. Math. Phys., 8, 423–434 (1967); *II*: J. Math. Phys., **9**, 698–711 (1967).

Ehrenfest P. and Ehrenfest T.

[EE] Phys. ZS.**8**, 311 (1907).

Ernst M., Dorfman J., Hoegy W., and van Leeuwen J. M. J.

[EDHL] *Hard-sphere dynamics and binary collision operators*, Physica, **45**, 127–146 (1969).

Feller W.

[Fel] *An Introduction to Probability Theory and Its Applications*. Vol. 1, Mir, Moscow (1984) (*Russian translation*).

Ferziger J. H. and Koper H. G.

[FeKo] *Mathematical Theory of Transport Processes in Gases*, North Holland (1972).

Fisher M. E. and Ruelle D.

[FiR] *The stability of many-particle systems*, J. Math. Phys., **7**, 260–270 (1966).

Fritz J.

[Fri1] *Some remarks on non-equilibrium dynamics of infinite particle systems*, J. Stat. Phys., **33**, 397–412 (1984).

[Fri2] *Gradient dynamics of infinite point systems*, Ann. Probab., **15**, 478–514 (1987).

Fritz J. and Dobrushin R. L.

[FrD] *Nonequilibrium dynamics of two-dimensional infinite particle systems with singular interactions*, Commun. Math. Phys., **55**, 67–82 (1977).

Gallavotti G.

[Gall] *On the mechanical equilibrium equation for classical systems*, Nuovo Cim., **B57**, 208–211 (1968).

Gallavotti G., Lanford O., and Lebowitz J. L.

[GLL] *Thermodynamic limit of time-dependent correlation functions for one-dimensional systems*, J. Math. Phys., **11**, 2898–2905 (1970).

Gallavotti G. and Miracle-Sole S.

[GaM] *Absence of phase transitions in hard core one-dimensional systems with long-range interactions*, J. Math. Phys., **11**, 147–155 (1970).

Galperin G. A.

[Galp] *Elastic scattering of particles*, Usp. Mat. Nauk, **33**, 199 (1978).

Gelfand I.M., Graev M.I., and Vilenkin N.Ya.

[GGV] *Generalized Functions*, **5**, *Integral Geometry and Problems of Representation Theory Connected with It,* Moscow (1962).

Georgii H.-O.

[Geo1] *Canonical Gibbs states. Their relation to Gibbs states and applications to two-valued Markov chains*, Z. Wahrsch. Verw. Gebiete, **32**, 277–300 (1975).

[Geo2] *Canonical and grand canonical Gibbs states for continuum systems*, Commun. Math. Phys., **48**, 31–51 (1976).

[Geo3] *Canonical Gibbs Measures*, Lecture Notes in Mathematics, **760**, Springer (1979).

Gerasimenko V. I.

[Ger1] *Thermodynamic Limit in Classical Non-Equilibrium Systems. Nearest-Neighbour Interaction*, Preprint No. 58P, Institute for Theoretical Physics, Kiev (1980) (*in Russian*).

[Ger2] *Evolution of an infinite system of particles with nearest-neighbour interaction*, Dokl. Akad. Nauk Ukr. SSR, Ser. A, No. 5, 10–13 (1982).

[Ger3] *On the solution of the BBGKY hierarchy for a one-dimensional system of hard spheres*, Teor. Mat. Fiz., **91**, 120–128 (1992).

Gerasimenko V. I. and Petrina D. Ya.

[GeP1] *Statistical mechanics of quantum-classical systems. Nonequilibrium systems*, Teor. Mat. Fiz., **42**, 88–100 (1980).

[GeP2] *Thermodynamic limit for nonequilibrium states of a three-dimensional system of hard spheres*, Teor. Mat. Fiz., **64**, 130–149 (1985).

[GeP3] *Existence of the Boltzmann–Grad limit for an infinite system of hard spheres*, Teor. Mat. Fiz., **83:1**, 92–114 (1990).

[GeP4] *On the generalized kinetic equation*, Rep. NASU, No. 7, 7–12 (1997).

[GeP5] *The generalized kinetic equation generated by the BBGKY hierarchy*, Ukr. Fiz. Zh., **13**, No. 6, 697–702 (1998).

Gibbs G. W.

[Gib] *Thermodynamics: Statistical Mechanics*, Nauka, Moscow (1982) (*Russian translation*).

[Gss] *Gibbs States in Statistical Mechanics*, Collection of Works, Mir, Moscow (1978) (News of Foreign Science, Mathematics Series, Issue 11).

Ginibre G.

[Gin] *Rigorous lower bound on the compressibility of a classical system*, Phys. Lett., **24A**, 223–224 (1967).

Gonchar N. S.

[Gon] *N. N. Bogolyubov equations for reduced distribution functions and their solutions for arbitrary values of density of particles*, Teor. Mat. Fiz., **57**, 35–96 (1983).

Grad H.

[Gra1] *Principles of the kinetic theory of gases*, in: Handbuch der Physik, Vol. 12, Springer, Berlin, 205–294, (1958).

[Gra2] *On the kinetic theory of rarified gases*, Pure Appl. Math. **2**, 331–407 (1949).

Green H. S.

[Gre] *Statistical mechanics and the Boltzmann equation*, in: Proceedings of the International Symposium on Transport Processes in Statistical Mechanics, Brussels (1956), pp. 52–127; Interscience, New York (1958).

Green M. S.

[Gre1] *The non-equilibrium pair distribution function at low densities*, Physica, **24**, 393 (1958).

[Gre2] *Surface integral form for three-body collisions in the Boltzmann equation*, Phys. Rev., **136**, 905 (1964).

Green M. S. and Piccirelli R. A.

[GrP] *Basis of the functional assumption in the theory of the Boltzmann equation*, Phys. Rev., **132**, 1388–1410 (1963).

Grünboum F.A.

[Grn] *Propagation of chaos for the Boltzmann equation*, Arch. Rational Mech. Anal., **42**, 323–345 (1971).

Gurevich B. M.

[Gur] *Additive integrals of motion of point particles in one dimension*, Math. Forsch., **12**, 59–64 (1982).

Gurevich B. M. and Chulaevsky V. A.

[GuC] *On Gibbs random fields invariant under Hamiltonian dynamics*, in: Proceedings of the 4th USSR–Japan Symposium on Probability Theory and Mathematical Statistics, Tbilisi (1982), pp. 248–249.

Gurevich B. M. and Sukhov Yu. M.

[GuS1] *Invariant measures for Hamiltonian systems with infinitely many degrees of freedom and stationary solutions of the Bogolyubov chain of equations*, Usp. Mat. Nauk, **30**, 199–200 (1975).

[GuS2] *Stationary solutions of the Bogolyubov chain of equations in classical statistical mechanics*, Dokl. Akad. Nauk SSSR, **223**, 276–279 (1975).

[GuS3] *Stationary solutions of the Bogolyubov hierarchy of equations in classical statistical mechanics*, Commun. Math. Phys., **49**, 69–96 (1976); **54**, 81–96 (1977); **56**, 225–236 (1977); **84**, 336–376 (1982).

[GuS4] *Time evolution of Gibbs states in one-dimensional classical statistical mechanics*, Dokl. Akad. Nauk. SSSR, **242**, 276–279 (1978).

[GuS5] *Additive integrals of motion for particles with polar interaction*, Usp. Mat. Nauk, **36**, 218 (1981).

Halmos P.

[Hal] *Measure Theory*, Izd. Inostr. Lit., Moscow (1953) (*Russian translation*).

van Hove L.

[Hov1] *Quelques proprietes generales de l'integrale de configuration d'un systeme de particules avec interaction*, Physica, **15**, 951–961 (1949).

[Hov2] *Sur 1'integrale de configuration pour les systemes de particules a une dimension*, Physica, **18**, 137–143 (1950).

Hurd A. E.

[Hur1] *Existence of time evolutions of non-equilibrium states in statistical mechanics*, Arch. Ration. Mech. Anal., **85**, 201–212 (1984).

[Hur2] *Global existence and validity for the BBGKY hierarchy*, Arch. Ration. Mech. Anal., **98**, 191–209 (1987).

Illner R.

[Ill] *On the number of collisions in a hard sphere particle system in all space*, Transp. Theor. Stat. Phys., **18**, 71–86 (1989).

Illner R. and Neunsert M.

[IlN] *The concept irreversibility in the Kinetic Theory of Gases*, Transp. Theor. Stat. Phys., **18**, 71–86 (1989).

Illner R. and Pulvirenti M.

[IlP1] *Global validity of the Boltzmann equation for a two-dimensional rare gas in vacuum*, Commun. Math. Phys., **105**, 189–203 (1986).

[IlP2] *A derivation of the BBGKY hierarchy for hard sphere particle system*, Transp. Theor. Stat. Phys.,**16:7**, 997–1012 (1987).

Illner R. and Shinbrot M.

[IlS] *The Boltzmann equation. Global existence for a rare gas*, Commun. Math. Phys., **95**, 217–226 (1984).

Kac M.

[Kac1] *Foundation of kinetic theory*, In Third Berkeley Symposium on Mathematical Statistics and Probability Theory, **3**, 171–197 (1959).

[Kac2] *Probability and Related Topics in Physical Sciences*, Interscience, London, (1959).

Kato T.

[Kat] *Perturbation Theory for Linear Operators*, Springer, New York (1966).

Khinchin A. Ya.

[Khi] *Mathematical Fundamentals of Classical Statistical Mechanics*, Gostekhizdat, Moscow–Leningrad (1943) (*in Russian*).

Kipmis C. and Landim C.

[KiL] *Sealing Limits of Interacting Particle Systems*, Springer, Berlin (1999).

Kirillov A. A. and Gvishiani A. D.

[KiG] *Theorems and Problems of Functional Analysis*, Nauka, Moscow (1979) (*in Russian*).

Kirkwood J. G.

[Kir] *The statistical mechanical theory of transport processes*, J. Chem. Phys, **14**, 180–201 (1940); **15**, 72–76 (1947).

Klein D.

[Kle1] *Convergence of grand canonical Gibbs measures*, Commun. Math. Phys., **92**, 295–308 (1984).

[Kle2] *Uniqueness of continuum one-dimensional Gibbs states for slowly decaying interaction*, J. Stat. Phys., **43**, 207–218 (1986).

Klein W.

[Kle] *Spectrum of the Kirkwood-Salsburg operator and phase transitions*, J. Math. Phys., **16**, 1482–1487 (1975).

Klimontovich Yu. L.

[Kli] *Statistical Physics*, Nauka, Moscow (1982) (*in Russian*).

Kochubey A. N.

[Koc] *The Liouville operator*, Ukr. Mat. Zh., **43**, 1664–1671 (1991).

Kornfeld I. P., Sinai Ya. G., and Fomin S. V.

[KSF] *Ergodic Theory*, Nauka, Moscow (1980) (*in Russian*).

Kozlov V. V.

[Koz] *Integrability and nonintegrability in Hamiltonian mechanics*, Usp. Mat. Nauk, **83**, 3–63 (1983).

Krein S. G.

[Kre] *Linear Differential Equations in Banach Spaces*, Nauka, Moscow (1967) (*in Russian*).

Kuni P. M.

[Kun] *Statistical Physics and Thermodynamics*, Nauka, Moscow (1981) (*in Russian*).

Lachowicz M.

[Lac1] *A system of stochastic differential equations modeling the Euler and the Navier–Stokes hydrodynamic equations*, Jpn. J. Industr. Appl. Math., **10**, 109–131 (1993).

[Lac2] *Stochastic differential equations for the kinetic and hydrodynamic equations*, Nonlin. Dynamics, **5**, 393–399 (1994).

[Lac3] *Asymptotic analysis of nonlinear kinetic equations: the thermodynamic limit*, in: Theory of the Boltzmann Equation, Lecture Notes in Mathematics, World Scientific, Singapore (1995), pp. 65–148.

[Lac4] *On the hydrodynamic limit of the Enskog equation*, Publ. RIMS, Kyoto Univ., **34**, 191–210 (1998).

[Lac5] *Stochastic semigroup and coagulation equations*, Ukr. Mat. Zh., **57**, 770–777 (2005).

[Lac6] *From Microscopic to Macroscopic Description for Generalized Kinetic Models*, Math. Models and Methods in Applied Sciences, **12**, 985–1005 (2002).

[La7] *From microscopic to macroscopic descriptions of complex systems*, Comp. Rend. Mecanique (Paris) **331**, 733–738 (2003).

[La8] *On bilinear kinetic equations. Between micro and macro descriptions of biological populations*. Banach Center Publ. **63**, 217–230 (2004).

[La9] *General population systems. Macroscopic limit of a class of stochastic semigroups*, J. Math. Anal. Appl. **307**, 585–605 (2005).

Lachowicz M. and Pulvirenti M.

[LaPu] *A stochastic particle system modeling the Euler equations*, Arch. Ration. Mech. Anal., **109**, 81–93 (1990).

Lampis M. and Petrina D. Ya.

[LaPe1] *The Boltzmann–Grad limit for equilibrium states of systems of hard spheres in the framework of canonical ensemble*, Math. Phys. Analysis Geom. (1996).

[LaPe2] *Spatially hormogeneous Boltzmann hierarchy as averaged spatially in homogeneous stochastic Boltzmann hierarchy*, Ukr. Mat. Zh., **54**, No. 1, 78–94 (2002).

[LaPe3] *Stochastic Dynamics and Hierarchy for the Boltzmann Equation with Arbitrary Differential Scattering Cross Section*, Ukr. Mat. Zh., **54**, No. 12, 1629–1653 (2004).

[LaPe4] *Stochastic dynamics as a limit of Hamiltonian dynamics of hard spheres*, Ukr. Mat. Zh., **51**, No. 5, 614–636 (1999).

Lanford O. E.

[Lan1] *The classical mechanics of one-dimensional systems of infinitely many particles. 1. Existence theorem*, Commun. Math. Phys., **9**, 169–180 (1968).

[Lan2] *The classical mechanics of one-dimensional systems of infinitely many particles. 2. Kinetic theory*, Commun. Math. Phys., **11**, 257–292 (1969).

[Lan3] *Time evolution of large classical systems*, in: Lectures Notes in Physics, **38**, Springer, Berlin (1975), pp. 1–111.

Lanford O. E. and Ruelle D.

[LaR] *Observables at infinity and states with short range correlations in statistical mechanics*, Commun. Math. Phys., **13**, 194–215 (1969).

Lang R.

[Lan] *On the asymptotic behavior of infinite gradient systems*, Commun. Math. Phys., **65**, 129–148 (1979).

Lebowitz J. L. and Montroll E. M. (editors)

[LeM] *Nonequilibrium Phenomena 1: The Boltzmann Equation: Studies in Statistical Mechanics*, Vol. X, North-Holland, Amsterdam (1983).

Leontowich M.A.

[Leo] *Basic equations of the kinetic gas theory from the point of view of the theory of random processes*, Teor. Eksper. Fiziki, **5**, 211–231 (1935).

Lewis R.

[Lew] *Solution of the equations of statistical mechanics*, J. Math. Phys., **2**, No. 2, 222–231 (1961).

Lieb E. H.

[Lie] *The stability of matter: From atoms to stars*, Bull. Amer. Math. Soc, **22**, No. 9, 1–49 (1990).

Lieb E. H., Loss M., and Siedentor H.

[LLS] *Stability of relativistic matter via Thomas–Fermi theory*, Helv. Phys. Acta, **69**, 974–984 (1996).

Lieb E. H. and Thirring W. E.

[LiT] *Bound for the kinetic energy of fermions which proves the stability of matter*, Phys. Rev. Lett., **35**, 687–689 (1975); Erratum, Phys. Rev. Lett, **36**, 1116 (1975).

Ligget T. M.

[Lig] *Interacting Particle Systems*, Springer, New York (1985).

Loschmidt J.

[Los] *Über den Zustand de Wärmegleichgewichtes eines Systems von Körpern mit Rucksicht auf die Schwerkraft*, Wien. Ber. **73**, 128 (1876).

Malyshev P. V.

[Mal] *Mathematical description of the evolution of an infinite classical system*, Teor. Mat. Fiz., **44**, 63–74 (1980).

Malyshev P. V. and Petrina D. Ya.

[MaP] *Thermodynamic limit for nonequilibrium distribution functions for three-dimensional classical systems of interacting particles*, Dokl. Akad. Nauk SSSR, **301**, 585–589 (1988).

Marchioro G., Pellegrinotti A., and Presutti, E.

[MPP1] *Existence of time evolution for ν-dimensional statistical mechanics*, Commun. Math. Phys., **40**, 175–186 (1975).

[MPP2] *Remarks on the existence of nonequilibrium dynamics*, in: Random Fields. Rigorous Results in Statistical Mechanics and Quantum Field Theory, Vol. 3, North-Holland, Amsterdam (1981), pp. 733–746.

Maslov V. P.

[Mas] *Equations of self-consistent field*, in: VINITI Series in Contemporary Problems in Mathematics, Vol. 11, Moscow (1982), pp. 85–125.

Maslov V. P. and Shvedov O. Yu.

[MaS] *An asymptotic formula for the N-particle density function as $N \to \infty$ and the violation of the chaos hypothesis*, Rus. Zh. Mat. Fiz., **2**, No. 2, 217–234 (1994).

Maslov V. P. and Tariverdiev S. E.

[MaT] *Asymptotics of the Kolmogorov–Feller equation for a system of a large number of particles*, in: VINITI Series in Probability Theory, Mathematical Statistics, and Theoretical Cybernetics, **19**, Moscow 85–126 (1982).

McKean

[MK] *A class of Markov processes associated with nonlinear parabolic equations*, Proc. Nat. Acad. Sci., **56**, 1907–1911 (1966).

Minlos R. A.

[Min1] *Limit Gibbs distribution*, Funkts. Anal. Prilozh., **1**, 60–73 (1967).

[Min2] *Regularity of limit Gibbs distribution*, Funkts. Anal. Prilozh., **3**, 40–53 (1967).

Moraal H.

[Mor1] *Derivation and spectrum of a generalized Kirkwood–Salsburg operator*, Physica, **A81**, 469–474 (1975).

[Mor2] *Spectral properties of Kirkwood–Salsburg operators*, Commun. Math. Phys., **87**, 331–343 (1977).

Murphy T. J. and Cohen E.G.D.

[MuC] *Maximum number of collisions among identical hard spheres*, J. Stat. Phys., **71**, 1063 (1993).

Mynbaev K. T. and Otebaev

[MyO] *Weighted Functional Spaces and Spectra of Differential Operators*, Nauka, Moscow (1988) (*in Russian*).

Olla S., Varadhan S. R. S., and Yan H. T.

[OVY] *Hydrodynamical limit for Hamiltonian systems with weak noise*, Commun. Math. Phys, **155**, 523–560 (1993).

Onsager L.

[Ons] *Electrostatic interactions of molecules*, J. Phys. Chem., **43**, 189–196 (1939).

Pastur L. A.

[Pas] *Spectral theory for the Kirkwood–Salsburg equations in a finite volume*, Teor. Mat. Fiz., **18**, 233–242 (1974).

Petrina D. Ya.

[Pet1] *On solutions of the kinetic Bogolyubov equations: quantum statistics*, Teor. Mat. Fiz., **13**, 391–405 (1972).

[Pet2] *Mathematical Description of the Evolution of Infinite Systems of Classical Statistical Physics*, Preprint No. 37E, Institute for Theoretical Physics, Kiev (1978) (in Russian).

[Pet3] *Mathematical description of the evolution of infinite systems of classical statistical physics: locally perturbed one-dimensional systems*, Teor. Mat. Fiz., **38**, 230–250 (1979).

[Pet4] *Mathematical Foundations of Quantum Statistical Mechanics. Continuous Systems*, Kluwer, Amsterdam (1995).

[Pet5] *Methods of derivation of the stochastic Boltzmann hierarchy,* Ukr. Mat. Zh., **52**, No. 4, 474–492 (2000).

Petrina D. Ya. and Caraffini G.

[PeC1] *Analogue of Liouville Equation and BBGKY hierarchy for a System of Hard Spheres with Inelastic Collisions*, Ukr. Mat. Zh., **57**, No. 6, 818–840 (2005).

[PeC2] *Solution of the BBGKY hierarchy for a system of Hard Spheres with inelastic Collisions*, Ukr. Mat. Zh. **58**, No. 3, 371–381 (2006).

Petrina D. Ya. and Gerasimenko V. I.

[PeG1] *Mathematical description of the evolution of states of infinite systems of classical statistical mechanics*, Usp. Mat. Nauk., **38**, 3–58 (1983).

[PeG2] *Evolution of states of infinite systems in classical statistical mechanics*. Sov. Sci. Rev., Ser. C, Math. Phys., **5**, 1–60 (1984).

[PeG3] *The Boltzmann–Grad limit for an infinite system of hard spheres*, Dokl. Akad. Nauk SSSR., **297**, 336–340 (1987); Dokl. Akad. Nauk Ukr. SSR, Ser. A, No. 12, 17–19 (1987).

[PeG4] *Mathematical problems of statistical mechanics of a system of hard spheres*, Usp. Mat. Nauk., **45:3**, 159–211 (1990).

Petrina D. Ya., Gerasimenko V. I., and Malyshev P. V.

[PGM1] *Thermodynamic limit for solutions of the Bogolyubov equations*, Sov. Sci. Rev., Ser. C, Math. Phys., **7**, 281–337 (1988).

[PGM2] *Solutions of Bogolyubov equations for infinite three-dimensional systems of particles*, in: Proceedings of the 4th International Conference on Probability Theory and Mathematical Statistics, Vol. 2, Vilnius (1987), pp. 451–461; Proceedings of the 2nd Soviet–Italian Symposium on Mathematical Problems in Statistical Physics, Lvov (1985).

[PGM3] *Mathematical Foundations of Classical Statistical Mechanics. Continuous Systems*, Gordon and Breach, New York (1989). *Mathematical Foundations of Classical Statistical Mechanics. Continuous Systems*, Second Edition, Taylor and Francis, London and New York (2002).

Petrina D. Ya. and Petrina K. D.

[PeP1] *On existence of equilibrium states of systems of hard spheres in the Boltzmann–Enskog limit*, Math. Phys. Anal. Geom. (1996).

[PeP2] *Stochastic dynamics and Boltzmann hierarchy, I*, Ukr. Mat. Zh., **50**, No. 2, 195–210 (1998).

[PeP3] *Stochastic dynamics and Boltzmann hierarchy, II*, Ukr. Mat. Zh., **50**, No. 3,372–388 (1998).

[PeP4] *Stochastic dynamics and Boltzmann hierarchy, III*, Ukr. Mat. Zh., **50**, No.4, 552–570 (1998).

Petrina D. Ya. and Skrypnik V. I.

[PeS] *Kirkwood–Salsburg equations for the coefficient functions of the scattering matrix*, Teor. Mat. Fiz., **8**, 369–380 (1971).

Petrina D. Ya. and Vidybida A. K.

[PeV] *Cauchy problem for the Bogolyubov chain of equations*, Trudy MIAN SSSR, **136**, 370–378 (1975).

Pogorelov Yu. G.

[Pog] *Convergence of virial expansions for a classical canonical ensemble*, Teor. Mat. Fiz. **24**, 248–254 (1975).

Poincaré H.

[Poi1] *Sur le probleme des trois corps et les équation de la dynamique*, Acta Mathematica, **13**, 1–270 (1890).

[Poi2] *Le méhanisme et l'experience*, Revue de Metaphysique et de Morale, **1**, 534–537 (1893).

Povzner A.Ya.

[Pov] *The Boltzmann equation in the kinetic theory of gases*, Amer. Math. Soc. Transl., **47**, 193 (1962).

Presutti E., Pulvirenti M., and Tirrozzi E.

[PPT] *Time evolution of infinite classical systems with singular long range two-body interactions*, Commun. Math. Phys., **47**, 81–95 (1976).

Pulvirenti M.

[Pul] *On the time evolution of infinitely extended particle systems*, J. Stat. Phys., **27**, 693–709 (1982).

Reed M. and Simon B.

[ReS1] *Methods of Modern Mathematical Physics*, Vol. 1: Functional Analysis, Academic Press, New York–London (1972).

[ReS2] *Methods of Modern Mathematical Physics*, Vol. 2: Fourier Analysis, Self-Adjointness, Academic Press, New York–London (1975).

[ReS3] *Methods of Modern Mathematical Physics*, Vol. 3: Scattering Theory, Academic Press, New York–London (1978).

[ReS4] *Methods of Modern Mathematical Physics*, Vol. 4: Analysis of Operators, Academic Press, New York–London (1979).

Resibois P. and de Leener H.

[ReL] *Classical Kinetic Theory of Fluids*, Wiley, London (1977).

Ruelle D.

[Rue1] *Classical statistical mechanics of a system of particles*, Helv. Phys. Acta, **36**, 183–197 (1963).

[Rue2] *Correlation functions of classical gases*, Ann. Phys., **25**, 109–120 (1963).

[Rue3] *Rigorous results in statistical mechanics*, in: Lectures in Theoretical Physics, Vol. 6, University of Colorado Press, Boulder (1964), p. 208.

[Rue4] *States of classical statistical mechanics*, J. Math. Phys., **8**, 1657–1668 (1967).

[Rue5] *Statistical Mechanics. Rigorous Results*, Benjamin, New York–Amsterdam (1969).

[Rue6] *Superstable interactions in classical statistical mechanics*, Commun. Math. Phys., **18**, 127–159 (1970).

Sengers J. V. and Cohen E. G. D.

[SeC] *Statistical mechanical derivation of the generalized Boltzmann equation for a fluid consisting of rigid spherical molecules*, Physica, **27**, 230 (1961).

Shinbrot M.

[Shi] *Local validity of the Boltzmann equation*, Math. Meth. Appl. Sci, **6**, 539–549 (1984).

Simyatitsky I. L.

[Sim] *A remark on the work "Mathematical description of equilibrium states of classical systems" by N. N. Bogolyubov, D. Ya. Petrina, and B. I. Khatset*, Teor. Mat. Fiz.,**6**, 230–237 (1971).

Sinai Ya. G.

[Sin1] *Construction of dynamics in one-dimensional systems of statistical mechanics*, Teor. Mat. Fiz., **11**, 248–258 (1972).

[Sin2] *Construction of cluster dynamics for dynamical systems of statistical mechanics*, Vestn. Mosk. Univ., Ser. I, Mathematics, Mechanics, No. 1, 152–158 (1974).

[Sin3] *Theory of Phase Transitions*, Nauka, Moscow (1980) (*in Russian*).

Sinai Ya. G. and Sukhov Yu. M.

[SiS] *On the theorem on existence of solutions of the Bogolyubov chain of equations*, Teor. Mat. Fiz., **19**, 344–363 (1974).

Skorokhod A. V.

[Sko] *Stochastic Equations for Complex Systems*, Nauka, Moscow (1983) (*in Russian*).

Spohn H.

[Spo1] *Kinetic equations from Hamiltonian dynamics: Markovian limits*, Rev. Mod. Phys.,**55**, 569–615 (1980).

[Spo2] *Large Scale Dynamics of Interacting Particles*, Springer (1991).

Streltsova E. A.

[Str] *Non-stationary processes in electrolyte theory*, Ukr. Mat. Zh., **11**, 83–92 (1959).

Sukhov Yu. M.

[Suk1] *Matrix method for continuous systems of classical statistical mechanics*, Trudy Mosk. Mat. Obshch., **24**, 175–200 (1971).

[Suk2] *Strong solution of the Bogolyubov chain of equations in one-dimensional classical statistical mechanics*, Dokl. Akad. Nauk SSSR, **224**, 1081–1084 (1979).

[Suk3] *Stationary solutions of the Bogolyubov chain of equations and first integrals of motion for classical systems of particles. One-dimensional case*, Teor. Mat. Fiz, **55**, 78–87 (1983).

Sznitman, A. S.

[Szn1] *Equations de type de Boltzmann, spatialment homogenes*, Z. Wahrsch. Verw. Gebiete, **66**, 559–592 (1984).

[Szn2] *Topics in propagation of chaos*, in Lecture notes in Mathematics, Springer, Berlin, **1964**, 165–251 (1991).

Takahashi Y.

[Tak1] *A class of solutions of the Bogolyubov system of equations for classical statistical mechanics of hard core particles*, Sci. Pap. Coll. Gen. Educ. Univ. Tokyo, **26**, 15–26 (1976).

[Tak2] *A class of Bogolyubov equations and time evolution in classical statistical mechanics*, in: Random Fields. Rigorous Results in Statistical Mechanics and Quantum Field Theory, Vol. 2, North-Holland, Amsterdam (1981), pp. 1031–1056.

Tanoka H.

[Tan] *Probabilistic treatment of the Boltzmann equation of Maxwellian molecules*, Z. Wahrsch. Verw. Geb., **46(1)**, 67–105 (1978)

Thirring W.

[Thi] *A Course in Mathematical Physics*, Vols. 1–4, Springer, Vienna–New York (1978–1981).

Truesdell C. and Muncaster R.

[TrM] *Fundamentals of Maxwell's Kinetic Theory of a Simple Monatomic Gas*, Academic Press (1980).

Uchiyama K.

[Uch1] *Derivation of the Boltzmann equation from particle dynamics*, Hiroshima Math. J.,**18**, 245–297 (1988).

[Uch2] *Atagged particle process in the Boltzmann-Grad limit for the Broadwell model*, Probab. Th. Rel. Fields, **82**, 419–433 (1989).

Uhlenbeck G. E. and Ford G. W.

[UhF] *Lectures in Statistical Mechanics*, Lectures in Applied Mathematics, 1, AMS, Providence (1963).

Valton M., Guillaume J. L., Hoffmann P., and Roger C.

[VGHR1] *Algebres de Lie sous-jacentes a la hierarchie BBGKY (Bogolyubov, Born, Green, Kirkwood, Yvon)*, C.R. Acad. Sci. Paris, **291**, 105–108 (1980).

[VGHR2] *Uhe solution formelle de la hierarchie BBGKY (Bogolyubov, Born, Green, Kirkwood, Yvon)*, C.R. Acad. Sci. Pris, **291**, 137–140 (1980).

Varadhan S. R. S.

[Var] *Large Deviation and Applications*, SIAM (1994).

Vlasov A. A.

[Vla1] *Many-Particle Theory*, Gostekhizdat, Moscow–Leningrad (1950) (*in Russian*).

[Vla2] *Statistical Distribution Functions*, Nauka, Moscow (1966) (*in Russian*).

Wagner W.

[Wag] *A functional law of large numbers for Boltzmann type stochastic particle systems*, Preprint, (1994).

Wild E.

[W] *On Boltzmann's equation in the kinetic theory of gases*, Proc. Cambridge Philos. Soc. **47**, 602–609 (1951).

Yosida K.

[Yos] *Functional Analysis*, Springer, Berlin (1995).

Yukhnovsky I. R. and Golovko M. F.

[YuG] *Statistical Theory of Classical Equilibrium Systems*, Naukova Dumka, Kiev (1980) (*in Russian*).

Yvon S.

[Yvo] *La Theorie Statistique des Fluids et l'Equation d'Etat, Hermann*, Paris (1935).

Zagrebnov V. A.

[Zag1] *On the solutions of correlation equations for classical continuous systems*, Physica, **A109**, 403–424 (1981).

[Zag2] *Spectral properties of Kirkwood–Salsburg and Kirkwood–Ruelle operators*, J. Stat. Phys., **27**, 477–491 (1982).

[Zag3] *Bogolyubov–Ruelle theorem: A new proof and generalizations*, Teor. Mat. Fiz., **51**, 389–402 (1982).

Zagrebnov V. A. and Pastur L. A.

[ZaP] *On singular interaction potentials in classical statistical mechanics*, Teor. Mat. Fiz., **36**, 352–372 (1978).

Zermelo E.

[Zer] *Über einen Satz der Dynamik und die mechanische Wärmetheorie*, Annalen der Physik, **54**, 485–494 (1896).

Zessin H.

[Zes] *Equilibrium time evolution and BBGKY hierarchy equations of large classical systems*, Z. Wahrsch., **68**, 461–472 (1985).

Zubarev D. N.

[Zub] *Nonequilibrium Statistical Thermodynamics*, Nauka, Moscow (1971) (*in Russian*).

Index